Chinese Literature and Culture in the World

Series Editor
Ban Wang
Stanford University
Stanford, CA, USA

As China is becoming an important player on the world stage, Chinese literature is poised to change and reshape the overlapping, shared cultural landscapes in the world. This series publishes books that reconsider Chinese literature, culture, criticism, and aesthetics in national and international contexts. While seeking studies that place China in geopolitical tensions and historical barriers among nations, we encourage projects that engage in empathetic and learning dialogue with other national traditions. Imbued with a desire for mutual relevance and sympathy, this dialogue aspires to a modest prospect of world culture. We seek theoretically informed studies of Chinese literature, classical and modern—works capable of rendering China's classical heritage and modern accomplishments into a significant part of world culture. We promote works that cut across the modern and tradition divide and challenge the inequality and unevenness of the modern world by critiquing modernity. We look for projects that bring classical aesthetic notions to new interpretations of modern critical theory and its practice. We welcome works that register and analyze the vibrant contemporary scenes in the online forum, public sphere, and media. We encourage comparative studies that account for mutual parallels, contacts, influences, and inspirations.

More information about this series at
http://www.palgrave.com/gp/series/14891

Chia-ju Chang
Editor

Chinese Environmental Humanities

Practices of Environing at the Margins

palgrave
macmillan

Editor
Chia-ju Chang
Brooklyn College
Brooklyn, NY, USA

Chinese Literature and Culture in the World
ISBN 978-3-030-18633-3 ISBN 978-3-030-18634-0 (eBook)
https://doi.org/10.1007/978-3-030-18634-0

Cover illustration: Photographer is my life. / Getty

This Palgrave Macmillan imprint is published by the registered company Springer Nature Switzerland AG
The registered company address is: Gewerbestrasse 11, 6330 Cham, Switzerland

Foreword: The Middle Place, *Ziran*, and *Huanjing*

Over the course of the last decade and a half, my friend and colleague, Chen Chi-szu (1962–2017), took me to the National Palace Museum in Taiwan many times to learn about one of China's most important arts—*shanshui* or landscape painting. Chen was a prolific painter, trained in all the classical forms of Chinese art. He saw himself primarily as an artist and secondarily as a literary critic interested in ecocriticism. He taught at Tamkang University, one of the universities in Asia that had been networking with other environmental literary critics from around the world since the 1990s. Professor Chen and his colleagues, Department Chair and Professor Yang Ming-tu and Professor Robin Tsai, were on the cutting edge of ecocriticism and well networked with scholars across the Taiwan Straits in China. As detailed in the introduction to *Chinese Environmental Humanities*, from the early 1990s forward, there was great interest in eco-aesthetics, nature-oriented philosophy, and ecocriticism (*shengtai piping*) in China and Taiwan. By 2001, China had convened the first conference in Asia on "Globalization and Ecocriticism" at Tsinghua University in Beijing. During my first 2002 visit to Tamkang University, it was already clear that strong foundations for what would later become recognized as the environmental humanities were being laid.

Chinese Environmental Humanities will immediately be seen as a landmark book because it maps the ideological influences, disciplinarian differences, and methodological/theoretical orientations of the environmental humanities as they have been developing in China, Taiwan, and Hong Kong since the early 1990s. Over the last decade and a half, I have had the pleasure of witnessing this growth and expansion as I have met with and worked with scholars from

China, Taiwan, Korea, Japan, and India who have been laboring, with much love and foresight, to build this new field. For my 2002 visit to Tamkang University, I had been asked to lecture on ecocriticism, and more specifically on the concept of the "middle place" that I had developed in my book *American Indian Literature, Environmental Justice and Ecocriticism* (2001).[1] I recount this experience here because, after reading Chai-ju Chang's groundbreaking collection, I believe that what I learned at the National Palace Museum from Chen Chi-szu offers some insight into *Chinese Environmental Humanities.*

The "middle place," as I define it, is an ancient concept based on indigenous North American peoples' philosophy that the earth is not some separate "natural" realm; it is "home." Indeed, among indigenous North America peoples, there was no concept of a "pristine wilderness" where humans do not live. Rather, they conceived of themselves as living in a middle place, or home, in which nature is sacred and respected but also the source of all life and well-being. Thus, human well-being is dependent upon understanding the animals, plants, and the patterns of the natural world.

In early ecocritical work in the United States and Britain, there was a notable focus on the concepts of "pristine nature" or "wilderness." However, the trouble with the concept of "wilderness," I explained in my Tamkang lecture, is that it assumes there was a time before human history in which ecosystems functioned in balanced, harmonious ways. But, in the twenty-first century, or the Anthropocene, we are coming to understand that it is nearly impossible to find a place on the Earth untouched by humans in either beneficial or detrimental ways. Most ecosystems have developed in relationship to human activities. Also, the political and economic forces that draw lines of protection around some areas of wilderness are the same forces which designate other places for mining, logging, or dam development. After the mines are dug, the forests logged, and the dams built, these places often become perceived as "fallen" or "corrupt," and no longer worthy of protection. In the United States, during the presidential administration of Richard Nixon in the 1970s, lands that were being mined or logged were even legally designated as "sacrifice zones" where resources were extracted for the good of the nation. People who inhabited those regions, despite long claims to ownership, were designated "sacrifice peoples" and removed.

As I explained in my lecture, wilderness lands and sacrifice zones, as antithetical as they might at first appear, are really mirror images of each other. Both are enclosed by economic and political processes that draw lines of protection around one while designating the other as disposable. Narratives are then created about how some enclosures benefit the market or the nation while others benefit tourism and protect endangered species. The real power of narratives about overly simplified concepts, including "wilderness" and "sacrifice zone," is their very abstraction. They give humans permission to behave as if they have no relationship to the processes that protect one area and sacrifice the other. People living in cities, or in developed nations, come to think of themselves as separated from wilderness and often fail to acknowledge their well-being as connected to the activities occurring in "sacrifice zones." The concept of the "middle place," therefore, calls upon ecocritics to become as aware of the connections between social and environmental injustices in sacrifice zones as they are of the need to protect and sustain ecological processes in wilderness areas. The concept of the "middle place" calls upon ecocritics to teach in ways that encourage students to gain a more ethical and just understanding of the human relationship to other humans and to the natural world (Adamson 2001, 183–184).

After my lecture, Chen Chi-szu explained to me that the "middle place," as a concept, was an idea he believed resonated with Asian religious and philosophical concepts. He took me to the National Palace Museum to introduce me to the same "mountain-water" (*shan shui*) genre of paintings that Chia-ju Chang references in the introduction to *Chinese Environmental Humanities.* These paintings were created by multiple artists from different dynasties, many with the title *Xishan wujin tu*, or "Streams and Mountains without End."[2] In many of these paintings, humans are represented as small figures against the grand backdrop of nature, which represents the cosmos. They pass along routes that take them through majestic mountains, beautiful valleys, and dense forests, or along winding rivers or over placid lakes, but also through villages with houses, towers, pavilions, and bridges where people are building, making, and doing.

As Professor Chen explained to me, and as Chai-ju Chang also discusses in her "Introduction," these paintings introduce the viewer to the concepts of "environment," or surroundings, and *ziran*, one of the most important keywords in Chinese eco-aesthetics, nature-oriented philosophy, and ecocriticism. *Ziran* is understood as not a mere external phenomenon, but as an interfused conglomerate of the natural world, the body, and the mind, and as Chang explains, involves a fusion or *ganwu* (a form of knowing through sensorial engagement). It is a celebration of the harmonious fusion of human and nature. As the traveler walks through the mountains and along the streams, s/he experiences *ziran*.

In comparing the concepts conveyed in "Mountains and Streams without End" to the concept of the middle place, Professor Chen and I agreed there seems to be no binary demarcation between "wilderness" and "sacrifice zones" in "mountain-water" (*shanshui*) paintings. They depict humans coming down from the mountains and entering villages, engaging in mundane daily tasks and activities, crossing over bridges and entering houses that have been built from materials acquired in the natural world.

I believe my ensuing conversations with Professor Chen about "mountain-water" painting over the last decade resonate with Chai-ju Chang's introduction to *Chinese Environmental Humanities* and her provocative call for more consideration not only of *ziran*, but also of another concept from Chinese culture that has been less considered in ecocriticism but which might come to the fore in a robust, interdisciplinary environmental humanities lexicon. *Huanjing*, Chang explains, is an "earthier term" evoking the activities of human beings and is a concept that connotes managerial and practical engagement with the material world (*shiran*). This term would spur deeper study and understanding of the "practice of environing" or "encircling" or "marking territory" especially, as argued earlier, when it comes to the practices of designating sacrifice lands or peoples. In the Anthropocene, humanists must lead in the study of anthropogenic processes and, as Chang observes, China's geopolitical standing as one of the world's economic powerhouses, and increasingly, as a leader in addressing climate change, cannot be understated. It is critical for humanists to understand not only literary, historical, and philosophical texts, but documents such as those outlining the Belt and Road Initiative (BRI) which propose a "massive infrastructure and trade project that will link countries across Asia, Europe, and Africa."[3] If the BRI is juxtaposed with images from "Mountains and Streams without End," it becomes more clear why humanists must be at the table with politicians and economists in discussions about what the road ahead, for China and for the world, promises. Such a juxtaposition calls for both humanists and their colleagues in other disciplines and in government to work for a future that ensures the well-being of humans, nonhumans, and natural and social systems.

Chinese Environmental Humanities makes the reasons why environmental humanists pay attention to both scholarly and political developments more visible. China is one of the world's oldest living "civilizations" or *wenming* (literally, "the light of words), with continuous records dating back more than 35,000 years. This is a resilient nation that has weathered many challenges and thus has many lessons—cultural and political—to teach. Emerging from this history is the concept of "ecological civilization" (*shengtai wenming*), which took a prominent place in China's political sphere in 2007 at the 17th National Congress of the Communist Party of China (CPC). The notion of an ecological civilization encourages and promotes more creative and fast-tracked response to unprecedented ecological pressures arising from high-speed economic growth. In 2012, it was written officially into the CPC constitution.[4] Thus, "ecological civilization" becomes one of the cultural, political, and economic forces at work in the "middle place," or this planet we call home that Professor Chen and I began discussing at the turn of the twenty-first century. At both the local and global scales, as Chang observes, we must be able to more ethically engage with the processes of "environing" or *huanjing* that will be at work as we create our future.

Most of the chapters in *Chinese Environmental Humanities: Practices of Environing at the Margins* engage in some way with the environing processes at work in both natural areas and sacrifice zones. Contributors invite readers to delve into knowledge from the periphery that might address these processes, such as the values and teachings of Daoism, Buddhism, and minority cultures that advocate for a deeper appreciation of the concept of *ziran* originating in ancient Chinese culture. Several chapters also invite readers to engage with the concept of *huanjing* in order to better understand "large-scale economic, political, cultural, historical, ecological, and spiritual forces affecting both the places where people live and where they do not" (Adamson 2001, 184). Chia-ju Chang's brilliant second chapter on ecocriticism, "Building a Post-Industrial *Shangri-la*: Lu Shuyuan, Ecocriticism, and Tao Yuanming's 'Peach Blossom Spring,'" explores a type of "Chinese syncretism" that insists on incorporating foreign cultures into the Chinese indigenous ones in ways that creatively forge new ideas while at the same time recuperating traditional ideas. This is a must-read chapter for every environmental humanist who wishes to understand how diverse regions around the world are expanding the field in important, and distinct, new directions. Other chapters are equally fascinating, as they examine topics such as the transformation of China from an agrarian society to an industrial one or explore the writings of ethnic minority women on the human relationship to nonhuman species. Contributors also investigate the impact of dams, advocate for eco-communities in hyper-urbanized centers, and trace the history of silkworms as both biological entities and cultural symbols. In short, each of the chapters calls upon readers to think carefully about how humans might quickly recalibrate their behavior, motivations, and desires in the Anthropocene so that all life in the middle place can adapt and flourish well into the future.

Since my visit with Professor Chen at the National Palace Museum in 2002, the field of the environmental humanities has taken shape with astonishing speed. The reasons for this can be attributed to the increasing recognition that global efforts to monitor, measure, and reverse the drivers of climate change will require science, but that the analytical power of science stops short of investigating the main driver of planetary change—the human (Holm et al. 2015, 978). What humans believe and value, how they organize themselves and behave, and what they are willing to invest to achieve their goals are factors that lie largely outside scientific calculation. Thus, growing numbers of scientists, policymakers, and business and education leaders are declaring the environmental humanities crucial to addressing the anthropogenic factors contributing to dramatic environmental changes.

In response to these calls, humanists around the world are building networks such as the Humanities for the Environment consortium of international observatories which are engaging in collaborative research and community projects.[5] These activities are supporting fast-accelerating international institutionalization and program building in universities around the world. In July 2018, evidence of the field's growing influence could be seen at the Rachel Carson Center in Munich, Germany, where the first international meeting of programs in the environmental humanities brought together leaders from 5 continents and 32 countries to explore new ways of cooperating and sharing expertise. At the conclusion of this meeting, delegates agreed that environmental humanists around the world needed to do as much as possible to break out of the Euro-American comfort zone, to learn new languages, and to become much more globally diverse and inclusive.

To meet this goal, environmental humanists will need to become much more knowledgeable about both the original and syncretic ideas, concepts, and contributions to the environmental humanities of China and its Asian sister countries. This makes *Chinese Environmental Humanities* an incredibly timely publication. As Chang observes, the concept of *shengtai wenming* not only is advocating for the design of a sustainable and harmonious society based on China's premodern cultural and religious traditions but is also guiding the creation and institutionalization of eco-philosophical units and programs supporting the growth of the environmental humanities in the Chinese Academy. This vision of the future requires a modification of the anthropocentric worldview which is the root cause of ecological destruction, and this makes the role of the environmental humanities in this discussion incredibly productive and requisite.

The imagery of roads winding through both majestic nature and bustling cities that Chen Chi-szu opened up to me over a decade ago and which Chang places centrally before readers in *Chinese Environmental Humanities* suggests productive ways of thinking about how humanists in Asia are leading interdisciplinary discussions of traditional cultural values and philosophy in the contexts of specific histories, politics, and material realities. These mountains and streams without end call us to think carefully about how we might all prepare for a more harmonious future in the middle place.

Arizona State University
Tempe, AZ, USA

Joni Adamson 琼妮·亚当逊

Notes

1. For an explanation of the ancient Zuni concept of the "middle place," see Adamson 2001, 47–48, 190 N. 13.
2. See, for example, Ma Yuan's (1190–1279) version of "Streams and Mountains without End," http://blog.sina.com.cn/s/blog_82aa88a501010scn.html. Accessed 20 Dec. 2018.
3. Joshi, "The Belt and Road Initiative aka One Belt One Road Scheme."
4. "Ecological Civilization is Meaningful to China," *ChinaDaily.com*, http://www.chinadaily.com.cn/opinion/2012-11/19/content_15942603.htm. Accessed 20 Dec. 2018.
5. See Adamson and Davis, eds. *Humanities for the Environment: Integrating Knowledge, Constellating New Practices.*

Bibliography

Adamson, Joni. 2001. *American Indian Literature, Environmental Justice and Ecocriticism*. Tucson: The University of Arizona Press.

Adamson, Joni and Davis Michael, eds. 2017. *Humanities for the Environment: Integrating Knowledge, Constellating New Practices.* New York: Routledge.

Holm, Poul, Joni Adamson, Hsinya Huang, et al. 2015. Humanities for the Environment—A Manifesto for Research and Action. *Humanities* 4 (4): 977–992.

Joshi, Manoj. The Belt and Road Initiative aka One Belt One Road Scheme. https://www.orfonline.org/wp-content/uploads/2018/01/The-Belt-Road-Initiative-.pdf. Accessed 26 Dec 2018.

Acknowledgments

Many people contributed to the production and refinement of this volume. First, I would like to sincerely thank Patrick D. Murphy for his interest in seeing this project through to completion. I am deeply grateful to Joni Adamson and Scott Slovic for their support and invaluable comments. The generosity of Prof. Lu Shuyuan's support is deeply appreciated. I was moved after learning about his personal involvement with local communities in helping restore traditional Chinese eco-cultures. One example is his frequent visits to the Wanshan ("Ten Thousand Cedar") Buddhist monastery in Mt. Lu to educate the Buddhist nuns on various topics of environmental humanities. Lu showcases Chinese environmental humanities scholars bridging academia and society; they are not only ecocritical theorists but also practitioners who integrate their knowledge into the communities about which they deeply care. I would like to thank my mentor, Prof. Kurt Spellmeyer, and my husband, Steven Peterson, for their endless patience and encouragement. Finally, I want to thank the editor-in-chief Ban Wang, who invited me to submit to the Chinese series of Palgrave Macmillan. I appreciate the professional guidance of the editors, Allie (Bochicchio) Troyanos, Rachel Jacobe, and other publishing crew. This project would not have come to fruition without all of you.

Praise for *Chinese Environmental Humanities*

"This book makes a tremendous contribution to global environmental humanities by zeroing in on a vast body of ideas and practices about the environment, nature, earth, ecology, humanity, and animals in China's long tradition. The reader will discover a series of illuminating, wide-ranging case studies on literature, philosophy, religion, art, film, and media as they pertain to the environment. The contributors together make a timely intervention in the central themes and topics of contemporary humanities and cultural studies."

—Sheldon Lu, *Professor of Comparative Literature, University of California, Davis, USA*

"Chia-ju Chang's timely collection *Chinese Environmental Humanities: Practices of Environing at the Margins* is comprised of 14 fascinating chapters that examine silkworms, migrant workers, ecomedia, animal rights, Buddhism, and sustainability. The chapters also explore traditional Chinese concepts, such as *ziran* (Nature), *huanjing* (environment) and *shengtai wenming* (ecological civilization), to forge new environmental perspectives in the Anthropocene. Although "environing at the margins", as the subtitle playfully indicates, this collection powerfully illustrates how "environing" in China and Asia is never marginal and how concepts and symbols from the Chinese culture can help enhance the field in new directions."

—Serpil Oppermann, *Professor of Environmental Humanities, Cappadocia University, Turkey*

"*Chinese Environmental* Humanities is *not* categorized within Sinology. It demonstrates an innovatively comparative and refreshingly transnational approach to Environmental Humanities. The book's focus on "the discourse of environing at the margins" offers a new angle from which to see the politics and aesthetics of environmental inclusion and exclusion. A must-read book for those who are interested in ecocriticism, environmental studies, multispecies studies, Asian studies, and other related fields."

—Masami Yuki, *President of ASLE-Japan, Routledge Environmental Humanities Series Co-editor*

Contents

Notes on Contributors

Chia-ju Chang is Professor of Chinese Languages and Literature at Brooklyn College, the City College of New York, USA. Her research interests include Chinese and comparative green literary and cultural studies, animal studies, and eco-Buddhist studies. Her publications include *Global Imagination of Ecological Communities* (2013) and many articles in both English and Chinese, which have appeared in many peer-reviewed journals and scholarly collections in the U.S., China, and Taiwan. She co-edited the volume *Ecocriticism in Taiwan: Identity, Environment, and the Arts* (2016) and guest-edited the special issue on animal writing in the journal *Taiwan Literature Translation Series* (2018). She was the Kiriyama Professor in the Asia Pacific Center at the University of San Francisco in 2016.

Kiu-wai Chu is Assistant Professor of Green Humanities at Nanyang Technological University. He got his PhD in Comparative Literature from the University of Hong Kong, and his previous degrees from SOAS University of London and the University of Cambridge. He was a visiting Fulbright Scholar at the University of Idaho, and completed his postdoctoral fellowships at the University of Zurich and Western Sydney University. His research focuses on ecocriticism, and contemporary cinema and visual art in Chinese and Southeast Asian contexts. His work has appeared in *Transnational Ecocinema*, *Oxford Bibliographies*, *Journal of Chinese Cinemas*, *The Palgrave Handbook of Asian Cinema* and others.

Haomin Gong is Associate Professor of Chinese at Lingnan University, Hong Kong, China. He used to teach at the College of William and Mary, St. Mary's College of Maryland, and Case Western Reserve University in the USA. His most recent publications include *Essays on Chinese Ecocinema* (coedited with Sheldon Lu, Wuhan University Press, 2017, in Chinese) and *Reconfiguring Class, Gender, Ethnicity and Ethics in Chinese Internet Culture* (coauthored with Xin Yang, 2017). He is also the author of *Uneven Modernity: Literature, Film, and Intellectual Discourse in Postsocialist China* (2012).

Dong Isbister is Associate Professor of Women's and Gender Studies at University of Wisconsin-Platteville, USA. Her research and teaching interests include collective memory and immigration, transnational feminism, environmental humanities, US multiethnic studies, women's literature, and translation and interpreting studies. She is the coauthor, along with Xiumei Pu and Stephen Rachman, of "(Re)connecting People and the Land: Ecomemory in Environmental Writings by Ethnic Minority Women Writers in China." She is also the recipient, along with Xiumei Pu and Stephen Rachman, of a 2016 Association for the Study of Literature and Environment translation grant.

Ralph Litzinger is Associate Professor of Cultural Anthropology at Duke University, USA. He is the author of *Other Chinas: The Yao and the Politics of National Belonging* (2000) and the coeditor of *Ghost Protocols: Development and Displacement in Global China* (2016). His research engages questions of border ecologies, bio-politics and ethics, and migrant labor, education, and health. He is completing two book projects, *Migrant Futures: Education and Labor in Global China* and *Black Lung: An Ethnography of Dust.*

Xinmin Liu is Associate Professor of Chinese and Comparative Cultures at Washington State University, USA. His research has been chiefly cross-cultural and interdisciplinary, and since 2005, he has focused on cultural geography, nature writing, and ecocriticism in China and the West. He is completing two book projects: a single-author book entitled *Agential Landscapes, Embedded Imaginary: Aspects of China's Ecocriticism* which focuses on the aesthetic, ontological, and ethical issues underlying the cluster of land-landscape-living habitats in China and a coedited volume of critical essays named *Embodied Memories, Embedded Healing: New Ecological Perspectives from East Asia* forthcoming from Lexington Books in 2019.

Jeffrey Nicolaisen is a doctoral candidate in Asian Religions at Duke University, USA. He has a master's degree in Civil Engineering from Nagoya University and a master's degree in Asian Studies from University of Illinois at Urbana-Champaign. He worked as an environmental consultant with Environmental Resources Management before returning to graduate school to study Asian religions and ecology. Nicolaisen has returned from one year of research as a Fulbright-Hayes Fellow in Taiwan, where he conducted a multi-species ethnography centered on the Buddhist-inspired organization Life Conservation Association. He is completing his dissertation as a Charlotte Newcombe Doctoral Dissertation Fellow.

Xiumei Pu is Assistant Professor of Environmental Studies at Westminster College, USA, where she teaches indigenous environmental thoughts, multiethnic environmental literature, ecowomanist and ecofeminist theories, and other related areas of inquiries. She is the author of "Turning Weapons into Flowers: Ecospiritual Poetics and Politics of Bön and Ecowomanism" (2016) and other publications about transcultural understanding of gender and the environment. She is the coauthor, along with Dong Isbister and Stephen Rachman, of "(Re)connecting People and the Land: Ecomemory in Environmental Writings by Ethnic Minority Women Writers in China" (2017).

Stephen Rachman is associate professor in the Department of English, and Co-Director of the Digital Humanities Literary Cognition Laboratory at Michigan State University, USA. He is the editor of *The Hasheesh Eater* by Fitz-Hugh Ludlow (2006). He is a coauthor of the award-winning *Cholera, Chloroform, and the Science of Medicine: A Life of John Snow* (2003) and the coeditor of *The American Face of Edgar Allan Poe* (1995). In 2012–2013, he began a project on Pearl S. Buck as a transnational Sino-American figure exploring her concepts of comparative democracy, feminism, and literature.

Stephen Roddy is Professor of Modern Languages and Director of Asian Studies at University of San Francisco, USA. His research focuses on the intersections between institutional change and literary expression in late-traditional China and Japan, the cultural history of tea connoisseurship throughout East Asia, and ethnographic poetry of borderland regions of the Qing Empire. He has held research fellowships at Sogang University, the National Library in Taipei, Nanjing University, and Peking University.

Chao-hwei Shih is a Buddhist nun, activist, and scholar of Buddhism, serving as Professor of Religious Studies and Dean of the College of Social Sciences at Hsuan Chuang University, Taiwan. She has written 28 books and numerous scholarly articles in Buddhist studies and ethics. Her activism includes advocacy for a variety of movements including gender equity in Buddhist communities, same-sex marriage, animal protection, anti-gambling, and anti-nuclear power. Chao-hwei has received numerous awards, including the Chinese Literature and Arts Medal for Cultural Discourse (2007), International Outstanding Woman in Buddhism (2009), and Social Movement Person of the Year (Taiwan, 2012).

Scott Slovic is Professor of Literature and Environment, Professor of Natural Resources and Society, and a faculty fellow in the Office of Research and Economic Development at the University of Idaho, USA. He served as founding president of the Association for the Study of Literature and Environment in the early 1990s, and since 1995 he has edited the journal *ISLE: Interdisciplinary Studies in Literature and Environment*. The author, editor, or coeditor of 27 books, his most recent publication is *The Routledge Handbook of Ecocriticism and Environmental Communication*. He is currently coeditor of the book series Routledge Studies in World Literatures and the Environment and Routledge Environmental Humanities.

Darryl Sterk is Assistant Professor of Translation at Lingnan University, Hong Kong, China. He is hoping to find a sub-discipline with an upcoming monograph entitled *Indigenous Cultural Translation Studies*, on the translation of the screenplay of the feature film *Seediq Bale* (dir. Wei Te-sheng, 2011) into Seediq. He also has an interest in the grammatical analysis of Chinese and English; see "The grammatical artistry of Chinese English translation" in *The Routledge Handbook of Chinese Translation*. He is a translator of ecoliterature from Taiwan, most notably of Wu Ming-Yi's *The Stolen Bicycle* (2017), which was long-listed for the Booker International.

Christopher K. Tong is an assistant professor at the University of Maryland, Baltimore County, in the United States. Prior to joining the University of Maryland, he was a postdoctoral fellow at Washington University in St. Louis. He is a recipient of grants from the National Endowment for the Humanities, American Philosophical Society,

Association for Asian Studies, and the Fulbright U.S. Scholar Program (Nanjing, China, 2018–2019). He is working on a book manuscript on ecological thinking in early twentieth-century China.

Fan Yang is an associate professor in the Department of Media and Communication Studies at University of Maryland, Baltimore County, USA. She is the author of *Faked in China: Nation Branding, Counterfeit Culture, and Globalization* (2016). An interdisciplinary scholar, Yang works at the intersection of cultural studies, transnational media studies, globalization, postcolonialism/postsocialism, and contemporary China. Yang's research has appeared in *positions: asia critique*; *Critical Studies in Media Communication*; *Theory, Culture & Society*; *Journal of Asian American Studies*; *Quarterly Review of Film and Video*; and *Verge: Studies in Global Asias*, among others.

Winnie L. M. Yee is Assistant Professor of Comparative Literature and at the University of Hong Kong, China. Her research interests are contemporary Chinese literature and film, Hong Kong culture, environmental humanities, and postcolonial theories. She has published on Hong Kong independent cinema, Chinese filmmaker Jia Zhangke, and Chinese writers Natalia Chan and Dung Kai-cheung. Her works have appeared in *Communication and the Public, Environment, Space, Place*; *Jump Cut*, *Journal of Urban Cultural Studies*, among other places. She is working on a book project exploring the relationship between ecopoetics and Chinese independent film scene and an edited volume on Asian ecocinema.

Runlei Zhai is a postdoctoral teaching scholar in the Department of Foreign Languages and Literatures at North Carolina State University, USA. She received her first Ph.D. in American literature from Beijing Foreign Studies University and her second Ph.D. in comparative literature from Purdue University. Her research interests include environmental humanities and Chinese language cinema. Her publications cover both Chinese and American environmental studies. They include books and journal articles. Her recent project is on Chinese eco-heroes. She hopes to discuss whether there are eco-heroes on Chinese screen and what eco-ethics and aesthetics these eco-heroes convey.

List of Figures

List of Tables

CHAPTER 1

Environing at the Margins: *Huanjing* as a Critical Practice

Chia-ju Chang

The twelfth-century handscroll painting, titled *Xishan wujin tu*, or "Streams and Mountains without End,"[1] rendered a Chinese vision of an ideal landscape: it depicts a spiritual and ecological cosmology predicated on the notion of "the fusion of heaven and humanity" (*tianren heyi*). The harmonization of energy (*qi*) is actualized by a balanced interplay between the invisible force *ying* and its material manifestation *yang*. This painting visualizes the Northern Song Dynasty painter Guo Xi's (1020–1090) aesthetic theory of "moving step by step, observing from side to side" (*bubuyi, mianmiankan*).[2] For the ancient Chinese, the natural world was not a mere external and material phenomenon, but rather an interfused conglomerate of the environment, the body, and the mind.

This artistic vision has both sustained China's aesthetic tradition and inspired Gary Snyder's poem "Endless Streams and Mountains" in the

All the Chinese names in this chapter, except the author of this chapter, follow the Chinese order, namely, surname first, followed by first name. This applies to the rest of the chapters.

C. Chang (✉)
Brooklyn College, Brooklyn, NY, USA

C. Chang (ed.), *Chinese Environmental Humanities*,
Chinese Literature and Culture in the World,
https://doi.org/10.1007/978-3-030-18634-0_1

collection, *Mountains and Rivers without End*. However, it would seem that such a vision has come to an end in the current age of climate change. Behind the façade of the world's glitzy modernization and development in the late twentieth century are toxic landscapes, displaced communities, and rising numbers of endangered species. Take air pollution as an example. As the world's No. 1 carbon dioxide emitter, outdoor air pollution contributes to the deaths of an estimated 1.6 million people every year in China.

The endgame story of nature and its manifestations in various eco-disasters are not exclusively Chinese, however. Global industrialization has also contributed plenty of such anthropogenic eco-disasters, for example, the Bhopal Gas Tragedy in India. They compel humanists to theorize critical or aesthetic strategies vis-à-vis global eco-disasters. In effect, the emergence of Anglo-European or global environmental humanities[3] is a collective, humanist response to this realization.

The term "environmental humanities" covers a multidisciplinary movement that includes history, philosophy, literature, film and media studies, and religious studies in order to "address anthropogenic factors contributing to the understanding and reassessment of dramatic environmental change."[4] Joni Adamson notes the gradual recognition of the importance of the humanities in environmental studies and writes, "scientists, policy makers, and business and educational leaders have begun to recognize the indispensability of humanities."[5] In the 2012 report of the Responses to Environmental and Societal Challenges for Our Unstable Earth (RESCUE) initiative,[6] commissioned by the European Science Foundation and Europe's intergovernmental Cooperation in Science and Technology program, the humanities are placed in the spotlight. The science on global climate change has been settled for decades, at least among the scientific community. Not only do the facts fail to motivate action (the U.S. pulled out of the Paris Climate Agreement, despite its already watered-down requirements), but the facts themselves are subject to constant attack by pseudo-scientists and reactionary political and media figures. However, a photo of a lonely polar bear on a chunk of ice or a documentary about penguins struggling to keep their children alive in Antarctica can inspire action because they inspire emotion. Philosopher Donald Davidson[7] argued that actions are caused by an agent having a belief combined with a desire. Science provides the belief (ideally), but it is art that provokes the desire. The environmental historian Sverker Sörlin even went as far to say that "hopes are tied to the humanities"[8] by pointing out that the humanist thinkers participated not only in the first phase of the environmentalist movement,[9] the concern for the place of humanity in the more-than-human world far preceded the onset of the industrial revolution where the environmental crisis began to be identified. In fact, the roots of the environmental humanities, argued by the scholars of indigenous studies and postcolonial environmental justice, "may be traced back to some of the earliest cosmological narratives, stories and symbols among the world's oldest cultures."[10] Humanities scholars in our current time continue to contribute to the conversation and discuss solutions to environmental problems, which tend to be viewed as belonging to the compartment of natural sciences and engineering. In contrast to the scientific approaches, humanist thinkers understand that environmental problems are not simply technological or policy issues. Rather, they are implicated in a world where cultural symbols, political systems, and religious values still rule the way people produce, consume, and govern their lives. Hence, the current environmental crisis is a crisis of human habitus, narrative, and imagination.

In examining the root crises of humanity that are leading to a planetary system collapse, humanist methodologies tend to be hermeneutic and phenomenological. Humanities afford interventionist channels to help forge "new environmental imaginaries, formulat[e] new discursive practices, and mak[e] changes in economic and political structures."[11] In addition, the humanist approach also encourages interdisciplinary integration, which often involves global, transnational, comparative, or deep-time perspectives. For example, the digital Humanities for the Environment (H*f*E) projects are on the forefront of experimenting with the collaborative production of environmentally relevant knowledge on a global scale.[12]

Chinese Environmental Humanities: Practices of Environing at the Margins is a collaborative humanities project which brings together 16 scholars from anthropology, literary criticism, aesthetics and cultural criticism, ecocinema, ecomedia, religion, and minority and translation studies. The chapters directly or indirectly engage what I call "the discourse of environing at the margins," which is defined as consideration of modes, aesthetics, ethics, and politics of environmental inclusion and exclusion, however the margins are conceptualized. Such a discourse evokes the Chinese term for the English word "environment": ***huanjing***. As will be elucidated in the section on keywords found in the introductory chapter, ***huanjing*** can be understood as a type of environmental/material practice (tending to one's own surroundings or territory, especially at the borders) that cannot be detached from the premodern politics of boundary making in historical China. As such, the practice of *huanjing* as "environing" or "place-ma(r)king" inevitably becomes intertwined with questions of onto-epistemology, aesthetics, identity, politics, and ethics concerning the self in relation to the other. Instead of viewing *huanjing* as a noun ("environment"), I repurpose this term as a verb phrase with the "verb-object" structure to interrogate contemporary literary and artistic forms, concepts, or practices of environing at the margins where the self and other are contested. Such a practice of environing at the margins (however that center and margin are figured) as a Chinese Environmental Humanities (CEH) mode of critical inquiry conceives a correlating "ethics at the margins" that addresses issues of marginality as the result of environing practice. "The discourse of environing at the margins," recuperated from the premodern word *huanjing*, remains relevant in both China and beyond, and becomes even more poignant in the current era of globalization where we begin to see nation-state or transnational corporations encircling/marking territories and fencing off marginalized groups. "The discourse of environing at the margins" contributes to CEH and global EH that aims to question the past and present "normative dimensions of current environmental practices."[13] By questioning normative practices, EH "open[s] up wider possibilities for redefining environmental scholarship beyond, in between, or outside of disciplinary conventions."[14]

1.1 Why Does China Matter? From Industrial Modernity to Ecological Civilization

The national humiliation experienced since China's First Opium War in 1842 ignited an enthusiastic espousal of modernist ideas about science and technology and the ideology of progress for nation building, escaping from poverty, and transitioning from an agricultural to an urban state. Industrialization, which lasted more than 300 years in the West, took China just 30 years. After joining the World Trade Organization in 2001, China surpassed the U.S. to become the world's largest economy in terms of purchasing power parity, according to the International Monetary Fund.[15] But the open market economy and the urbanization process were accompanied by countless devastating ecocatastrophes. The People's Republic of China's (PRC) choice of the slogan "an ecological civilization" (*shengtai wenming*) is a response to unprecedented ecological pressures arising from a prolonged period of high-speed economic growth. Employing the term *wenming* or "civilization" to remind the people of China's long civilizational history and heritage, the state has mobilized it as a guiding principle that unifies material, spiritual, political, and ecological considerations. The idea of "the construction of ecological civilization" came into the political domain in 2007 at the 17th National Congress of the Communist Party of China (CPC). In 2012, it was written into the CPC constitution for the first time at the 18th National Congress.

"Ecological civilization" is more than political rhetoric or "an attempt to counteract the growing and widespread criticism of environmental degradation."[16] The rhetoric of reshaping the society along ecological lines taps into a collective anxiety about toxins and a desire for a pollution-free environment, from the top to the bottom. Both the government and people are striving to transform the lived environment into a sustainable one, as seen in various commitments and engagements such as cutting carbon emissions, producing electricity in wind farms and solar-powered water heaters, innovating green urban designs, inventing eco-fertilizers, building eco-villages, and protesting to halt industrial projects (also see Chap. 2). Given the size and population of the country, China's geopolitical and economic importance as the world's economic powerhouse cannot be understated, especially now with the Belt and Road Initiative (BRI). On the environmental front, after the withdrawal from *the Paris Agreement* under the Trump administration, it has become increasingly clear that China has taken over the role of climate leadership. In this light, global environmental humanities will prove to be under-realized without the contribution and participation of China and scholars of Chinese studies in both China and beyond.

China certainly responds to the civilizational wake-up provoked by Naomi Klein. What is worth noting is that the notion of ecological civilization was first broached by the agricultural economist Ye Qianji (1909–2017) for building a sustainable agriculture in China.[17] As shown in Chap. 2, Chinese humanists realized that environmental issues are issues of production mode and lifestyle, which accounts for the inefficacy of environmental law enforcement. To address the way the Chinese people live and relate to the environment, they realize an ethical void in society at large and turn to traditional philosophy as "an antidote to China's worship of growthmania" and preparing for a harmonious post-GDP society. This is an example of the way hopes are tied to humanities. Here, humanists contribute to an examination of cultural patterns or certain traditional ways of life or practices that are resources for redressing the current destructive environmental behavior. With regard to the idea of reconnecting to premodern cultural traditions for conceiving China's ecological civilization, Scott Slovic writes, "When we speak today of the *emergence* of an ecological civilization in China, we are, in a sense, referring to a reassertion of traditional Chinese values rather than the creation of entirely new concepts, vocabularies, or attitudes."[18] Unlike Western civilization, China's standing as the oldest living "civilization" or *wenming* (literally, "the light of words"), with continuous historical records of more than 35,000 years of history, means China does not share an anxiety about reinventing itself. There is something to learn from China's resilience and the ability to continue to survive and thrive after so many dynastic changes, the Cultural Revolution, and even Western imperialism.[19]

History teaches us to ask questions about how a given idea, cultural pattern, or "collective unconscious" was formed and became the governing moral principles for later generations. It also reveals to us what values, practices, and customs in the past are worth revisiting and renewing for building a more-than-human civilization and society. Here, China and CEH have a lot to offer. The reassertion or preservation of traditional Chinese views vis-à-vis Western industrial modernity is a poignantly recurring theme throughout modern Chinese history. Placing the current climate change crisis into the perspective of a *longue durée*, Stephen Roddy, for example, shows that late-nineteenth-century Chinese intellectuals had already been very wary of Western modernity and warned against China's "uncritical acceptance of the materialistic worldview [...] which was evident in the technological and military aggressiveness of foreign business and governments" (see Chap. 3).

1.2 Sketching a Field: Chinese Environmental Humanities

Though there has not yet emerged a field called "Chinese environmental humanities," "environmental humanities in Greater China," "environmental humanities in the Sinosphere,"[20] or something along these lines, it does not mean that green humanities scholarship in China is nonexistent. On the contrary, in places like China, Hong Kong, Taiwan, the U.S., and Europe, literary and cultural scholars have been actively pursuing such a green scholarship, using ecocriticism, eco-philosophy, and so on, as critical and aesthetic approaches. Given the fact that these "green scholarships" in different geopolitical centers are practiced independently, an umbrella term to encompass such a broad spectrum of scholarship risks sacrificing their respective developmental trajectories, ideological influences, disciplinarian differences, and methodological/theoretical orientations. For the sake of simplification, I tactically use the term "Chinese environmental humanities" to represent this emerging field of scholarly inquiry and practice in the world.

However, there is a potential problem with CEH. To qualify this field as "Chinese," it intuitively necessitates perspectives, approaches, or methodologies related to something that can be recognized as "the Chinese" or "the Chinese people." But does that mean scholarship "about China," undertaken only by the Chinese themselves and employing only "Chinese approaches"? What about scholarship on China "by the non-Chinese others" and/or employing "non-Chinese theories"? Unless we are careful to avoid doing so, "Chinese" can be reduced to one of the local examples of concern to a universalizing or Westernizing theory. One might argue that such a project does not qualify as CEH at all. According to this view, scholarship that employs Western theory should at most be considered as "environmental humanities perspectives on Chinese culture." Indeed, one could draw a sharp line between these two ways of engaging environmental humanities and Chinese studies.

Having said that, the essentializing of Chinese identity, however, can be just as reductive as a colonizing imposition of Western theory. It might be more fruitful to consider these two approaches as representing strong and weak forms of a single CEH. Incorporating "environmental humanities perspectives on Chinese culture" under the banner of CEH allows one to explore the full spectrum of engagement, ranging from the strong-form CEH in Chap. 2, which demonstrates a conscious effort to construct Chinese eco-theories, to Chap. 10, where the authors introduce Western media materialism studies as an interpretive method to broaden the existing field of inquiry. Here, the practice of "holding on[to] Chinese philosophy and learning the methodology from the West" (*zhongti xiyong*) proposed by Feng Guifen (1809–1874) when China was at the juncture of modernity is a historical model for the weaker form of CEH.

The tension between Chinese ontology and Western practical application, or vice versa, will remain a core theoretical concern for the field of Chinese or comparative EH. A contemporary exploration of this tension can be found in Sheldon H. Lu's articulation of "Chinese ecocinema." Lu envisions "Chinese ecocinema" as a new critical paradigm and characterizes "the Chinese" as a context "that may be different from Euro-American settings in significant ways."[21] Here, Lu broadens Chinese characteristics beyond premodern cultures to include the dimension of modern and postmodern socio-historical Chinese contexts. This formulation further blurs the boundary between CEH and "environmental humanities perspectives on Chinese culture." Like Lu's project, this volume as a whole will respond to the binaries in a rather "Chinese" way, with a synthesis that preserves the distinctness of its constitutive oppositions.

1.3 Institutionalization: Conferences, Workshops, Institutions, and Anthologies

Scholarly activities on CEH (specifically referring to ecocriticism, ecocinema studies, and ecoaesthetics[22]) can be found in China, Hong Kong, and the West.[23] I now provide an outline for a preliminary sketch of the field to introduce key conferences, workshops, and major figures in the field of CEH.

1.3.1 China and Hong Kong

The CPC's promotion of the construction of an ecological civilization ripples down to lower sectors of governance, including the proliferation of eco-related terms such as eco-cities and the establishment of eco-civilization research centers in several major universities. It also helps sponsor and legitimize disciplines such as ecological literary studies and ecoaesthetics. For example, the 2001 symposium on "Globalization and Ecocriticism" held by Tsinghua University marks the formal appearance of the phrase "ecocriticism" in Chinese literary and cultural studies, thus further legitimizing the field by the authorities in China. Other landmark conferences include the 2008 first cross-strait eco-conference, "Beyond Thoreau: American and International Responses to Nature" at Tsinghua University, Beijing, organized by Song Lili. In 2009, the inaugural meeting of the World Ecoculture Organization was held at Peking University, organized by Zhao Baisheng, a distinguished professor and chair of the Institute of World Literature at Beijing University. The institutional home for Chinese ecoaesthetics—Shandong University Research Center for Ecoaesthetics and Ecoliterature, directed by Zeng Fanren—organized its first international ecoaesthetics conference in 2005, "Human and Nature: The International Symposium on Aesthetics and Literature from the Perspective of Contemporary Eco-Civilization." The center has hosted three more ecoaesthetics conferences, and it also launched a *Newsletter on Ecoaesthetics and Ecocriticism* under the editorship of Cheng Xiangzhan. One scholar who is worth mentioning here is Cheng Hong at the Capital University of Economics and Business. She has written and translated several scholarly books and articles and also hosted several high-level ecocriticism conferences in recent years. In addition to those mentioned, other important scholars include Hu Zhihong, Liu Bei, Wang Nuo, and Wei Qingqi. All of these scholars contribute to the development of ecocriticism and ecofeminism in China by introducing Western eco-theories and eco-literature; many of them also turn to their own cultural resources to develop Confucian, Daoist, and Buddhist ecocriticism.

Hong Kong in recent years has also begun to develop environmental humanities approaches and topics. In 2016, Winnie L. M. Yee and Kiu-wai Chu at the University of Hong Kong hosted "Contextualizing Asian Ecocinema: Past and Future." Two years later, Ka-ming Wu at the Chinese University of Hong Kong organized an environmental humanities workshop, "Environmental Humanities in Asia: Ecological Crisis and Cultural Responses."

Currently, teaching in the area of EH has mostly focused on "ecological" approaches or environmental issues through the study of literature, culture, and philosophy. Courses such as ecocriticism, ecoaesthetics, environmental history, and ecological and environmental philosophy are offered at the graduate level at many major institutions in China, such as Beijing University, Beijing Normal University, Hunan University, Renmin University, Shandong Normal University, Shanghai Normal University, and Tsinghua University. In recent years, Hong Kong has also begun actively developing ecological humanities curricula under various labels. As will apply to the remaining part of this section, the list for the conferences, major scholars, and publications are not complete, but at least the reader will get a sense of the robust progress of this academic movement in China and beyond.

Curriculum-wise, many universities and colleges have set up EH courses. For example, Lu Shuyuan, one of the most important ecocritics in China, began offering units on ecological literature and arts in the mid-1990s. The History and Philosophy Department of Peking University has provided environmental history and environmental philosophy courses for more than ten years. Zhao Baisheng has been operating his ecocritical program under the auspices of the Center for World Literature at Peking University for a number of years. Beijing Normal University and Renmin University all have courses in environmental history and environmental philosophy. Ecocriticism and ecoaesthetics courses have also been regularly offered at Shandong University and Shanghai Normal University, to name a few.

Western ecocriticism and ecocritics have contributed significantly to the development of Chinese ecocriticism and ecoaesthetics. The shift from the early native terms such as "the studies of ecological literature and the arts" (see Chap. 2) to the single-word "ecocriticism" has marked a more pronounced cross-cultural communication between Chinese and Western academia and helped to officially recognize it as a legitimate field of literary study. Western and Asian humanist scholars such as Joni Adamson, Arnold Berleant, Allen Carlson, Simon Estok, Greta Gaard, Greg Garrard, Serenella Iovino, Patrick D. Murphy, Serpil Oppermann, Yrjö Sepänmaa, and Scott Slovic have been invited to China since the early 2000s to offer keynote lectures, teach courses, and so on. Many of the scholars' works have been translated into Chinese. These cross-cultural communications have catalyzed comparative studies and eventually formed Chinese developed humanist theories.

1.3.2 Beyond China and Hong Kong

In areas outside of China, conference, workshops, online courses, and now programs have also arisen above the horizon in the U.S. and Europe. First, in 2014, *ISLE: Interdisciplinary Studies in Literature and Environment* published a special cluster of articles devoted to ecocriticism in China, coordinated by distinguished literary theorist Wang Ning at Tsinghua University, which showcases China's accomplished and up-and-coming ecocritics. In Switzerland, a workshop, "Humanistic Scholarship in the Anthropocene: Approaching China from a Sustainability Paradigm," was organized at the University of Zurich in 2015 by Andrea Riemenschnitter. This workshop resulted in the publication of a special issue of *International Communication of Chinese Culture*, "Anthropocene matters: envisioning sustainability in the Sinosphere," edited by Andrea Riemenschnitter and Jessica Imbach. Another workshop, "Local Climate Change Governance & Resource Management in China," took place at University of Vienna in 2016. Finally, Prasenjit Duara, the director of the Global Asia Initiative at Duke University organized the first Duke Kunshan conference on environmental humanities in Asia: Environmental Justice and Sustainable Citizenship at Duke Kunshan University, China in 2017.

Thus far, two volumes have been devoted to CEH: Sheldon H. Lu and Jiayan Mi's *Chinese Ecocinema: In the Age of Environmental Challenge* (2009) and Chia-ju Chang and Scott Slovic's *Ecocriticism in Taiwan: Identity, Environment, and the Arts* (2016). There are also numerous essays, published in various peer-reviewed journals, either scattered or grouped under the section of CEH in various anthologies. One anthology that comes to mind is Simon C. Estok and Won-Chung Kim's *East Asian Ecocriticism: A Critical Reader* (2013). The contributors in that collection and this collection are not heavily concerned with traditional culture or language as methodologies, instead drawing critical methodologies from Europe and the U.S. *Chinese Environmental Humanities: Practices of Environing at the Margins* distinguishes itself from these volumes on the following grounds: first, it includes a diverse range of disciplines; second, it draws from both Western and Chinese methodologies; finally, it is loosely woven together through the CEH keywords (particularly *huanjing*) developed below.

1.4 Two Keywords in CEH

In this section, I introduce two important keywords that concern humans' relation, perception, and interaction with the environment: *huanjing* (environment) and *ziran* (自然; "Nature" with an emphasis on the capital "N"). As will become clear, the dynamic between these two keywords is akin to that of deep ecology and social ecology in the West. While *ziran* probes into the question of humans' eco-cosmological identity, the "earthier term"—*huanjing*—evokes the activities of human beings. *Huanjing* connotes the notion of managerial and practical engagement with the material world (*shiran*), which inevitably involves the issue of social and environmental injustice. Here, I recuperate the premodern meaning of *huanjing* to develop an environmental discourse—"the discourse of environing at the margins"—to shed light on a certain environmental mindset, engagements or practice, both modern and premodern times, that are in alliance with the exclusionist ideology, such as ethnocentrism or nationalism that makes manifest at the margins as one of the intended consequences of "the multiple processes through which human beings…modify their surroundings."[24]

1.4.1 Huanjing *and the Practice of Environing*

Huanjing in modern Chinese is used to represent the English word "environment." Currently, usage of *huanjing* in China (as well as how it is conceptualized in the field of Chinese environmental history in both China and beyond) syncs with a modern and contemporary understanding of the term "environment," which refers to the natural *environment* with which living organisms are said to interact in the twentieth century.[25] This use of *huanjing* tracks the Czech and Canadian environmental historian Vaclav Smil's definition of the term "environment" as including both the physical setting and "the essential extractive activities such as farming and the harnessing of energy."[26] From here, we get a glimpse into how environment and *huanjing* have become effectively interchangeable in modern times, indicating a natural or artificial site of production, complicated material conditions, and surrounding things with which humans (and living organisms) interact.

If we return, however, to premodern times, *huanjing* immediately reveals something more than just a natural or artificial site of production or surrounding things with which humans (and living organisms) interact. As a compound word, *huanjing* consists of *huan* (环; a verb that means to "encircle") and *jing* (境; a noun that means "land" or "territory"). The verb-noun phrase *huanjing* hence by and large denotes "to surround or to mark the territory" or "to encircle the jurisdictional area."[27] One of the early instances of *huanjing* as "encircling territory" can be traced back to the Song Dynasty (960–1279). "The Biography of Wang Ning" in *The New History of Tang Dynasty* (*Xin Tang Shu*) mentions *huanjing* as "the surrounding areas (*huanjing*) of Jiangnan (aka the south side of the lower reaches of the Yangtze River) [were then] bandit territory." In the next example, a sense of marking and tending to one's own jurisdictional territory is evident. In "The Biography of Yu Que" of *The History of Yuan* (*Yuan Shi*), an excerpt is translated as: "No sooner had Yu Que arrived at the municipality for ten days than the anti-Yuan rebels began to attack. After repelling the rebels, Yu ordered soldiers to construct camps and *cultivate the land* [my emphasis]. Yu also decided to *build a wall around the surrounding jurisdictional area*" (my emphasis). In this specific context, *huanjing* or "the practice of environing" during the Yuan Dynasty (1271–1368) correlates with the establishment of an imperial or ethnic boundary and with defending or expanding one's own territory by securing the border via the practice called *tuntian* (i.e. using a military settlement to keep out invaders and helping with the conversion of the area into cultivated farmlands). Such a practice (e.g. enclosing of space or marking off territories)[28] led the Chinese historian Robert B. Marks to believe that the Chinese conception of *huanjing* is essentially an ethnocentric, or more specifically, Hanocentric one.[29] The ethnocentric practice of "encircling" or "marking the territory," as observed by Marks, conjures up another English term, "colony,"[30] which "invoke[s] a sense of closure, of space cordoned off by human activity."[31] And these activities are of environmental concern, as they are related to interacting and transforming the surrounding place, space, and things for the sake of production, consumption, and waste generation.

These lines of thinking about *huanjing* as the "practice of environing" from the Chinese historical-linguistic context informs us of a certain type of environmental practice, which is still prevalent in and detrimental to the construction of ecological civilization, locally, globally, and transnationally. It sheds light on practices such as the privatization of commons, public space, or lands that originally belong to other human communities or species before the rise of imperial powers, nation-states, and transnational capitalism. It then becomes clear that the practice of environing is not merely a matter of environmental science or of economics, but, more importantly, also of politics of what is included and excluded. A consummate contemporary example of such a practice of environing (in regard to exclusion) is the construction of an anti-immigrant wall on the border between the U.S. and Mexico. This type of "enclosing of space or marking off territories" is none other than "blood and soil" nationalism.

The recuperated meaning of *huanjing* as the "practice of environing" can, first of all, serve as a critical tool for CEH and global EH to think about environmental exclusion: to investigate multiple forms of exclusive socio-environmental practices at the margins, however that marginality is defined and constituted and regardless of whether that marginality is conscious or unconscious, historical or contemporary. It can be a marginalized identity (e.g. class/caste, gender, race/ethnicity, and nation); it can be disputed, remote or frontier places, colonies, settlements, ghettos in urban centers, or other contact zones,[32] occupied by indigenous or disenfranchised people, and communities, where their resources are taken away for other uses; it can be that biodiversity is threatened with expulsion or extinction in the name of industrial agriculturalism; it can be a form of premodern or indigenous knowledge that is appropriated for commercial or bio-capitalist gains (here, Vandana Shiva's notion of bio-piracy is a case in point); it can be a discipline or field at the margins of environmental humanities whose contributions to ecological and global discourses are crucial and yet often dismissed (e.g. ecotranslation studies).

Secondly, *huanjing*, conceptualized as "practice of environing," also helps theorize environmental ethics of inclusion or ethics of care at the margins: to examine various viewpoints, voices, and methods of "taking-in." This involves thinking in different scales in places, space and time, as well as in different languages, modes of rhetoric, knowing, and nomenclature (see Chaps. 4, 6, and 13). Such an environmental practice of inclusion involves risk and painstaking negotiations in its insistence on deconstructing all detrimental centric thinking, such as ethnocentrism, nationalism, anthropocentrism/androcentrism, and so on, and therefore, an inclusive practice of environing, which takes the excluded subjects into ethical considerations, marks the beginning of the ethics of care at the margins. In this light, the margins are the site of vulnerability and compassion. A practice of environing that transcends the "skin-bound" ethics and embraces the "more-than-human" dimension will be articulated in the next keyword section.

1.4.2 Ziran *and Nature*

What can the concept of *ziran* contribute to contemporary global EH or environmentalism? Through the work of deep ecologist Arne Naess, *ziran* has already influenced Western thinking,[33] but *ziran* also provides an alternative, nonanthropocentric approach to the idea of sustainability. In demonstrating the relevance of the deep ecological Daoist notion of *ziran* in contemporary conversation about sustainability, the CEH scholar James Miller writes, "a Daoist vision of sustainability [is] not simply as the maintenance or preservation of a status quo, as though nature were no more than a resource to be conserved for future generations. Rather, it considers what global environmentalism might look like if we consider that it is not the responsibility of humans to 'save nature' but rather to enable the natural world to continue its work of bringing forth the flourishing of things."[34] Here, Miller articulates a non-mainstream view or an "ethics at the margins" that puts trust in nature's ability to recuperate, thrive, and flourish.

Huanjing and *ziran* denote two contrasting modes of environmental-oriented discourse. While *huanjing* (recuperated as a verb) suggests the political dimension of a material process or "what things are like" (*shiran*), *ziran* focuses on the metaphysical aspect of reality, which is an "all-inclusive whole, the spontaneously self-generating life process."[35] Such a harmonious "great unity" (*datong*) is regarded as how things "should be" (*yingran*), not how they are. Tu Weiming observes that when premodern Chinese thinkers envision "great unity" (*datong*), they are also aware that the world "is laden with disruptive forces including humanly caused calamities and natural catastrophes" and that "history is littered with internecine warfare, oppression, injustice, and numerous other forms of cruelty."[36]

Ziran, as the deep structure of the Dao (or "the Way"), is hidden beneath the strife of this world like "ocean."[37] If one confuses or conflates these two aspects of reality—aspiration (what things aspire to be) and practicality (what realities are)—one falls prey to unnecessary disputes, if not misunderstanding or unmerited criticism.[38] The conflation further risks engendering a puzzling, disingenuous, if not schizophrenic, and particularly reductive view of China, one of the oldest civilizations in the world, that splits it into two polarized images: as a nature-loving culture of unison and harmony between heaven and humanity, on the one hand, and a nature-hating empire, now decried as the "Yellow Eco-Peril," on the other (see Chap. 10). To return to the metaphor of the "ocean," *Chinese Environmental Humanities: Practices of Environing at the Margins* occasionally taps into the serene depths (the "Dao") but stays mostly engaged with "the surface of the ocean" that is in constant motion and filled with strife, injustice, and violence.

Like *ziran*, the Chinese *huanjing* can make an important contribution. It is an important keyword not only in this volume but also in global EH in the era of globalization and climate change. It inquiries into humans' relationship with the land or the planet, addresses prevalent malign forms of exclusion, elimination, appropriation, or negligence at the frontiers and borderlands, and bears witness to a whole planet now entangled in a global economy and environmental migration. It brings attention to our tendency to enclose, rather than open up to, vulnerability. *Huanjing* demands that we think and act on the margins away from the centers of self-identity. While the margins may be the place of injustice, uncertainties, and vulnerability, they are also sites of plurality, transformation, wonder, creativity, and care. As Haomin Gong rightly claims in this volume, the practice of environing is not simply about the environment or economics; it is also about social and political processes (see Chap. 8). Hence, an EH scholarship should keep such processes in mind and, at the same time, move beyond them to make room for new thinking about sustainability, diversity, openness, resistance, and alliance for generations to come.

1.5 Chapter Outline

Chinese Environmental Humanities showcases scholarship that engages the theme of environmental inclusion and exclusion at the margins. While some authors reflect on the normative or mainstream "practice of environing," which undermines ecological and human diversities, others discuss alternative theories and imagine the praxis of care at the margins to create a more just and sustainable society. Here, the idea of the margins, though conceptualized differently by each author, serves as a critical site for thinking about environmental practices of inclusion and exclusion. It is not limited to the conception of places, but rather enlarged to include marginalized (visible or invisible) voices, disciplines, viewpoints, and phenomena.

There are in total 14 chapters, divided into three sections: Section I: Chinese Ecocriticism and Ecotranslation Studies; Section II: Chinese Ecocinema and Ecomedia Studies; Section III: Sustainability, Organic Community, and Buddhist Multispecies Ethics.

In Chap. 2, Chia-ju Chang focuses on the post-Mao CEH scholarship (mainly ecocriticism) in China and considers the way in which CEH scholars respond to China's runaway urbanization. The chapter first traces the emergence of ecological consciousness and discourses in China in the 1980s and early 1990s as a reaction to worsening environmental conditions. Special attention is paid to Lu Shuyuan and his homegrown ecocritical thinking. Chang first situates Lu's Chinese-oriented "spiritual ecology" methodology in the nexus of China's industrial modernity, historical/postcolonial consciousness, and current political discourse on ecological civilization. In search of an alternative postmodern eco-paradigm, Lu looks to the politically marginalized premodern "peasant-poet," Tao Yuanming (365–427 CE). Lu's Tao Yuanming studies exemplify the conceptualization of "the discourse of environing at the margins" that speaks for/with China's largest marginalized group, rural migrant workers (*nongmingong*). Chang examines the way in which Lu addresses issues of disenfranchisement (e.g. social injustice and inequality) in a poetic genre called "moonlight poetry" (*dagong shi*) composed by rural migrant workers through the prism of his Tao Yuanming studies. Here, Lu lays bare how the contemporary marginalization of migrant workers is in actuality the demise of the Daoist pastoral/organic way of life, and therefore, a crisis of China's agricultural civilization. Refusing to subscribe to the Western linear conception of progress, Lu argues that China's pre-industrial past and the organic life process of *ziran* continue to linger in a spectral or ghostly form. Lu proposes a post-industrial, de-growth ethics of care based on Tao's "Peach Blossom Spring," a pastoral vision of a small-scaled harmonious village that has evolved into a Sinospheric archetypal image of "arcadia." Chapter 2 showcases the modernist practice of environing at the margins in which marginalized subjects are located not in remote, underdeveloped areas but in urban centers. It contributes to the current discussion of global sustainability by illustrating the importance of the role premodern cultural tradition plays in the contemporary postmodern environmental imagination and ecological construction that embraces equality and inclusion, as opposed to the capitalist model of exploitation and exclusion.

In Chap. 3, Stephen Roddy considers a critique of the Western industrial practice of environing (e.g. cordoning territory for coal mining) that transpired in the late-Qing period from the mid-nineteenth century to the early twentieth century. The late-Qing period is the time where China was beginning its transformation from an agrarian society to an industrial one, following the West's modernization trajectory. Unsurprisingly, this is also the period where early critical voices of modernity and the burgeoning environmentalist discourse began to surface. Roddy examines four Chinese intellectuals, members of the so-called National Essence school (*guocui pai*) of thinkers: Yu Yue (1821–1907), Yuan Zuzhi (1827–1898), Zhang Binglin (1868–1936), and Liang Shuming (1893–1988). Harshly criticizing the developmental mode of industrial-scale extractivism and mass consumption, they were wary of the environmentally destructive modernization that had already become an inescapable trend. Against this grain, they reaffirmed premodern Chinese values, visions, practices, and ways of living, such as a simple life and "rural construction" based on the Mencius model and Buddhist teachings. Chapter 3 contributes to the theme of the practice of environing and ethics of care at the margins by showing how the EH questions are asked and answered. Like the ecocritic Lu Shuyuan discussed in Chap. 2, these early CEH thinkers identify Western-style of consumption and industrial modernity as the root problem that jeopardizes the well-being of both humans and the planet. Roddy's chapter showcases the relevancy of history and the questions asked in the past for reevaluating the present. He demonstrates that while non-Western humanists have diagnosed the causes and identify solutions of environmental problems based on philosophical and epistemological premises fundamentally different from those of Western scientists, they have arrived at very similar conclusions to those of contemporary scientists. Here Roddy is in agreement with Sinologist Benjamin Schwartz that although these thinkers who questioned the project of total modernization in China are called conservative, their responses are in fact very prescient.[39] In a word, these Chinese classicists recognized both the environmental and the societal unsustainability of many of the practices of modern industrial civilization, a century before the western scientific community finally caught up with them.

Chapter 4 on Chinese ethnic minority literature is written by Dong Isbister, Xiumei Pu, and Stephen Rachman. In analyzing two ethnic minority women's writings—the *Tujia* writer Chen Danling's "Herbs Living in the Body" (2012) and the *Hui* writer Mao Mei's "Snow Lotus" (2012)—the authors demonstrate a limitation of the widely accepted geopolitical binary conception of margins and centers in post-socialist China (e.g. the Han vs. other ethnic minorities). Such a construct is oftentimes articulated through the prism of Han ethnocentrism, if not Euro-American ethnocentrism, and risks oversimplifying ethnic minority experience as well as the complex interplay of ethnic histories in China. Moreover, it puts a constraint in imagining a more-than-human ethnic experience in relation to the practice of environing and ethics. Through the interpretive analysis of these two ethnic minority women's works, the authors illustrate the ways in which these women writers refuse to be situated in a default view of ethnic minorities as marginal. On the contrary, these writers reconfigure the center and margin, where they create a nondual genderspace and situate "the feminine" within a nexus of gendered human-human and human-nature interactions. Here, the chapter contributes to the reconceptualization on the practice of environing at the margins where the concept of the margins, or even periphery ("the outside of boundary"), becomes problematic. From the perspective of ethnic minority, what is often considered the peripheral by the dominant ethnic group or the central government tends to be poly-ethnic space, where the natural habitats are home and the central locus of healing. Here, plants such as the Tian Shan Snow Lotus bring together "multiple ethnic groups across differences in language, ethnicity, religion and ideology." Such a reconceptualization of center and margin from "the margins" helps transcend modernist notions of individualism and anthropocentrism. It also sheds light on how such a borderline area can be a productive breeding ground for forging an ecospiritual identity. Relevant to the literary practice (or ecocriticism) is that the natural world (here the vegetal world) becomes the center of literary imagination and crucial part of the project of "imaginative act of recovery." The "ethnobotanical healing," as the authors call it, imagines culture and health in broader indigenous (their land, history, and cosmology) and multispecies contexts, and therefore, it provides a model to think about indigenous practice of environing and "ethics of care."

Chapter 5 by Runlei Zhai also concerns the issue of translation and ecocriticism. The chapter is a study of a Western cinematic adaptation of Jiang Rong's *Wolf Totem* (*Lang tuteng*). This 2004 semi-autobiographical novel narrates the experiences of a young student from Beijing who is sent to Inner Mongolia at the beginning of the Cultural Revolution. Zhai argues that when the French director Jean-Jacques Annaud makes *Wolf Totem* into a film, it ceases to be a simple rendition, but instead a conscious ecological revision. There are pros and cons to Annaud's translation. While the film, in Zhai's view, pioneers translating Chinese ecological and cultural ethics and aesthetics for the global audience, it "neglects the social and cultural backgrounds." In other words, this highlighted ecological theme can function as a bridge in transnational and transcultural communication, but cultural specificity, including the writer's ethnocentrism for a political purpose, is obscured or lost in the process of transnational and trans-media translation. In thinking about the chapter's contribution to the less well developed CEH, Zhai shows the necessity for comparative and multicultural eco-communication strategies to help negotiate between both source and target cultures, which are needed for the current transnational world.

Chapter 6, written by Darryl Sterk, engages with the discourse of environing at the margins from the perspectives of translation studies and ecocriticism. The chapter addresses the issue of the mass extinction of bionym ("names of plant and animal") in modern literature. Viewing language and translation through the lens of ecology, Sterk notes that there are just as many species now threatened with extinction as there are bionyms. He argues that, as we seek to protect biological diversity, we should protect bionym diversity, because it is a part of humanity's cultural diversity that can help solve the problems of today and tomorrow. Sterk presents a "bionym translation manifesto" as a way of protecting bionym diversity and, hence, embracing the polyglot. Here, he poses a question, "Is there an ecology of language or translation?" Beginning with Taiwanese nativist writer Huang Chun-min's essay "Waiting for the Name of a Flower" to consider one bionym in particular, Sterk moves to the polyglot translation of bionyms in fellow Taiwanese nature writer Wu Ming-yi's works to demonstrate that a nature writer, as a spokesperson of the natural world, pays much closer attention to nature than a nativist. Through the process of translating Wu's works, Sterk begins to wonder if there is a bionymic ecology analogous to natural ecology, wherein it would behoove us to maintain nomenclatural diversity, just as diversity in nature is the source of the resilience of nature. Sterk concludes the answer to his aforementioned question in the positive: to the extent that language is a system, and that it often models living systems in the world, it is for sure ecological. Viewing language and translation as contact zones where nature-humanity or different human cultures meet, Sterk's "bionym translation manifesto" can be considered as a statement of "the ethics of care at the margins" in that it calls attention to the important aspect of cultural diversity in the current era of global homogenization. The manifesto also contributes to the field of ecotranslation, a much-marginalized field within translation and EH that aims to "establish criteria for assessing translations from an ecocritical perspective."[40]

The section concludes in Chap. 7 with Kiu-wai Chu. Chu's chapter expands the current CEH scholarship by including contemporary avant-garde art. This chapter on visual arts is included in the first section as an acknowledgment of the traditional Chinese holistic view of literature and the arts (see Chap. 2), a view which does not treat them as a separate entity. In this chapter, Chu examines Xu Bing's installation art series *Silkworm Books* (1994–2014), and several artworks that make use of living insects, to call for a "material turn" in contemporary Chinese art that explores the agential capacity of material entities through installation art assemblages. Silkworms, together with the silk they produce, appear both as material biological entities and as a cultural symbol that represent China's history and civilization. The art assemblage thus depicts a "material-semiotic reality" (Haraway) that "emerges as an intertwined flux of material and discursive forces."[41] Chu highlights the temporal aspect of eco-materialism inspired by the tiny, slow-moving silkworms and urges us to perceive the material world differently through the reconfiguration of temporality and artifacts. Cultivating an appreciation of slowness through an extended duration of observation, "silkworm ecology" offers a new, ecocritical way to perceive life in a world that is increasingly defined by speed, efficiency, and developmentalism. The artistic attention to insects, particularly silkworms (*Bombyx mori*), Chu asserts, attests to the fact that art has always been a multispecies concept. In the case of Xu's eco-art, Chu contends that human artmaking consists of interspecies interactions, tensions, and intervention. Chu views art as a metaphor for the Anthropocene, a concept that has emerged in several chapters (e.g. Chap. 9); it is a term that refers to a proposed new geological epoch dating from *the* inception of significant human impact on *the* Earth's ecosystems and geology. Silkworm installations can be viewed as "performances of care," where viewers are enabled to see "the miniature of a world co-habited and shared by human and nonhuman beings." Evoking van Dooren's theorization of care as a practice of worlding, Chu contributes to thinking of arts as a creative "contact zone" that carry out or "perform" multispecies worlding where care is the major mode of engagement.

In Section II, "Chinese Ecocinema and Ecomedia Studies" syncs with the critique of Western industrial modernity established in Chap. 3 but extends to examine the human-place and human-matter relationships in China's reform era (1980s–present) on both local and transnational contexts. As a thematic frame, "place" and "matter" are not simply catalogued as a green addition to the current list of "red-critical categories" such as gender, class/caste, race, and ethnicity. As we will see in the following chapters, the discussion of two camps are inseparable. Scholars in this section scrutinize ill forms of *huanjing*—"the practice of environing at the margins" in which marginalized places (e.g. the mined, contaminated, and wasted land) and matter (e.g. e-waste) become ineluctably intertwined with what Zygmunt Bauman calls the "wasted human" or "environmentalized bodies" (to recast Sean Cubbit's phrase; see Chap. 10).[42]

Chapter 8 by Haomin Gong analyzes Wang Jiuliang's *Beijing Besieged by Waste* (*Laji Weicheng*, 2012). The term "besieged by trash" evokes the idea of *fortress assiégée* in Qian Zhongshu's (1910–1998) novel *Besieged by City* (*Weicheng*) but is now recast with an environmental overtone. Gong considers waste from a socio-geographical standpoint, observing a "process of social structuration" (John Agnew) that turns humans, animals, and matter into waste, and peripheral places into wasteland. Gong illustrates that the production of waste is not merely an ecological issue, but a social one; it is a product (or I shall add "a metabolic process") of modern urbanization, commercialization, and consumerism. In the vein of Mary Douglass's formulation of dirt/dirtiness, Gong further notes the relevancy of the notion of "waste as displaced or misplaced objects" to the discourse of environing: that is, waste as a misplaced object is always already implicated in the politics of environmental inclusion and exclusion. Here, Gong conceptualizes the practice of environing in contemporary China quite differently from the premodern expansionist, military-agricultural mode (which is *tuntian*) mentioned above. In the age of the garbage-locene, waste that is relocated to the margins eventually returns to confine and define urban space, like a wall or fortress. Gong demonstrates this by showing Wang Jiuliang's visualization of the city's besiegement by waste literally via a google map image of Beijing surrounded by a yellow ring which is composed of hundreds of yellow dots signifying the waste dump sites Wang visited. The idea of a city being hemmed in by its own waste (or misplaced objects) not only highlights the uneven center-peripheral spatial organization between urban centers and their surroundings, but also problematizes cities as "metabolic sites" in which urbanites greedily covet and digest natural "resources" at the margins without reciprocity. Here, Gong's posthumanist take zooms in on "urban waste," much like Sterk in Chap. 6, Litzinger and Yang in Chap. 10, and Nicolaisen in Chap. 13, articulates how nonhuman matter, species, an event, or even a language can be a productive site for EH scholars to address issues of environmental/multispecies justice and care. Once viewed as social actors enmeshed in cultural and social networks and chains of signification, waste can be conceptualized as a broken social chain, or what Marx and Engels described as a "metabolic rift."

In Chap. 9, Xinmin Liu examines a contemporary ethics-free practice of environing that cordons off the remote places (regions distant from the political center)—natural habitats with vibrant ecosystems but home to "poly-ethnic groups" (also see Chap. 4)—for various lucrative mega-projects. Usually, these places are chosen for the "cash changeover" (or, a practice of capitalizing on nature). In the eyes of the technocrats, Liu uses the decade-long debate (2003–2013) over "Damming the Nu River" in southwest China as a concrete example of the practice over *bianxian* for fast-track economic gains. The Nu River, also known as the Salween in Thailand and the Thanlwin in parts of Myanmar, is home to one of China's most biodiverse regions. A hydraulic-dam-building plan for energy production was proposed in 2003, but suspended in 2005.[43] The construction of the dam would have not only destroyed local ecosystems but caused widespread displacement of minorities and changed their traditional indigenous culture, lifestyle and livelihood. In this chapter, Liu looks at two documentaries, *Chenmo de Nu jiang* and *Nu Jiang: Cong kaifa dao tingjian*, to demonstrate the filmic practice as a form of activism and expression of care for both indigenous people and nature, expressed through the testimonial capacity of documentary films. Liu evokes the idea of re-enchantment, reminiscent of Lu Shuyuan's invocation of Tao Yuanming's "Peach Blossom Spring," not as a naïve pastoral, but as a critique of market-driven economic practices that fail to recognize land-human interdependence and symbiotic relationship with nature.

Chapter 10 by Ralph Litzinger and Fan Yang shifts ground from film studies to ecomedia or eco-materialism to reassess the oft-racialized "Yellow Eco-Peril" discourse, rhetoric, and representation (i.e. viewing China as "a polluting and polluted Other" and now a threat to the planet's sustainability) in the era of global digital capitalism. The "Yellow Eco-Peril" or Beijing's "airmageddon" rhetoric exemplifies the persistent Western perception of China as "a threat to the ecological health of the planet," instead of the other way around: that is, seeing the addicted digital consumption of the "First World" as posing danger for the well-being of people living in the "global south." In countering such a biased view, the authors propose the concept of the "eco-event"—a moment in which "the material processes of digital production link the 'old' forms of resource extraction with our electronic gadgetry" to illuminate multilayered mediacy and complicity. It is mobilized as a critique of "global digital capitalism and consumption"; the material processes of digital-capitalist production-consumption are linked to resource extraction, chemical plant construction, and to an obsession with a media-saturated lifestyle. Here, they not only debunk the myth of the "Yellow Eco-Peril" but also further blur the clear-cut line between activism and complicity. Their analysis helps in thinking about CEH from a transnational perspective. It highlights the global dimension to account for local eco-disasters (here, China) as "the material production of global capitalism" and "China's key role in the process" in its capacity as world's factory. It also invites the reader to reflect on nonhuman agency and vitality, which "offer a glimpse into the workings of digital global capitalism" and should be "part of any debate about digital media in the environmental humanities."

The last part, Section III: Sustainability, Organic Community, and Buddhist Multispecies Ethics, reflects on alternative and constructive modes of environing. It extends a human-centered tunnel vision to embrace multispecies approaches that tend to "ten thousand things" (*wanwu*). It examines the question of China's sustainability, organic farming in urban centers, and Buddhist animal ethics as well as the monastic praxis of human-vegetal multispecies holism.

Chapter 11 by Christopher K. Tong investigates the paradoxical nature of China's longevity as a civilization by asking whether it makes sense to call a civilization "sustainable," if it has thrived for several millennia by modifying and exploiting its natural environment. Some scholars claim that China's millennia-long transformation of the environment was indeed unsustainable over the long run, while others argue that China managed to provide for its population for a long period of time, largely through labor-intensive agriculture within its borders. Still others point to the discrepancy between the appreciation of Nature in Chinese texts and actual behaviors of Chinese society. Surveying a spectrum of positions in Chinese environmental history, literature, and thought, Tong analyzes the rhetoric that various scholars use to discuss what they perceive to be the paradox of China's sustainability. In the process, Tong draws on a wide array of disciplines ranging from ethnic studies to philosophy to deconstruct familiar narratives in CEH scholarship, reinvigorating the debate on China's sustainability with renewed attention to detail and argumentation.

Chapter 12 by Winnie L. M. Yee examines alternative ways of living and thinking in post-urban Hong Kong. Yee first reflects on the city's colonial history and urbanization and views rurality not as a metaphoric or social construct, but instead as a form of the soil's material agency. She observes a post-urban, "rural" turn in Hong Kong through the emergence of an ecological movement that is characterized by the construction of small-scale communities. This chapter contributes to the reconceptualization of the "practice of environing" by examining the constant struggle between the developmental model of the use of land in Hong Kong and the practices of organic communities at the margins. Here, Yee uses the organic farm *Sangwoodgoon* as a case study for thinking about the possibility of creating Hong Kong's sustainable future. Yee sees the rural area not as a metaphoric or social construct, but instead focuses on soil's material agency. She shows how land is perceived as a physical, experimental locale for organic methods and permaculture design principles, as well as a site of local creativity and ecocosmopolitan imagination to cultivate ecocritical thinking that addresses environmental issues at various scales (local, national, transnational, and global). Here, Yee contributes to CEH and EH by going beyond ecocriticism and into a discussion of recent organic culture and community-based activism around the world. By reviewing the shared concerns and struggles of various community-based activist movements depicted in documentaries featured at Sangwoodgoon's Food and Farming Film Festival, Yee's discussion highlights the importance of including the voices at the margins and the necessity to transcend national boundaries to forge alliances for future generations.

The next two chapters center around Taiwanese Buddhism and multispecies studies (including animal studies and plant studies). They focus on the question regarding the ethical status of animals and plants, as well as human-nonhuman relationships; they also consider issues such as vegetal agency and multispecies community from the perspectives of religious studies, which is a marginalized field within Chinese and global EH. Religion, particularly Buddhism, is a strong shaping voice for Taiwan's environmental ethics and multispecies sensibility.

Chapter 13 by Jeffrey Nicolaisen examines a Buddhist conservation practice of environing from the perspective of multispecies ethnography. Special attention is given to the way in which the ethics of Taiwan's

Buddhism are applied in actual environmental practice. Nicolaisen zooms in on the Taiwanese Buddhist organization Dharma Drum Mountain (DDM) and its environmental practices such as tree preservation and animal release. Underneath these practices tending to the environment and nonhuman animals, he reveals that there is a complex relationship among human and nonhuman actors who all actively produce ecology at DDM's World Center for Education. In this ecology, humans as well as trees, spirits, dharma protectors, earthquakes, and forests all sculpt the landscape of the Center. There are multiple views of animals, plants, fungi, and sentient beings even within a single organization, and just like multiple-species that interact, so do multiple ecologies. Dharma Drum Mountain, its clerics, and its scientists integrate modern ecology with traditional Buddhist notions. Nicolaisen's multispecies ethnography is reminiscent of both Bruno Latour's network theory and the Buddhist Huayan philosophy, which articulate multispecies relationships, ethics, and agency.[44] While the chapter contributes to the Buddhist system theory, it also sheds light on the discourse of environing at the margins where the margins are sites where multispecies negotiations take place.

Chapter 14, translated by Jeffrey Nicolaisen, is written by Chao-hwei Shih. As a scholar and activist, she is one of the most outspoken Taiwanese Buddhist nuns, who has made it her mission to defend all those at the margins. By using the word "life," she does not separate sentient beings into human and nonhuman groups. She advocates not only for human concerns such as gender equality and LGBQT+ rights but also for the protection of nonhuman animals. Chao-hwei makes a philosophical case for a version of bioethics based on Buddhist teachings. Critical of Western theory's imperialism because of her adherence to the basic Buddhist principles of the equality of all life and non-violence (Skt. ahimsa), Chao-hwei employs three core Buddhist principles—dependent origination (*yuanqi*), protecting life (*husheng*), and middle way (*zhongdao*)—to formulate a Buddhist ethics of equality. She argues that though all things are equal, sentient beings are the primary ethical subjects, because they all equally possess sentience. Hence, she rejects a human-nature binary that marginalizes nonhuman animals and rechart the ethical territories to include all sentient beings, not just humans. In Chao-hwei's Buddhist environing, nonhuman animals are not part of the environment because they are co-equal with humans. While in this ultimate reality, all things are equal, Chao-hwei also recognizes the suffering of those sentient beings that are pushed to the margins of an anthropocentric Western ethics. Chao-hwei is a voice from the margins that reformulates traditional Chinese Buddhist notions of equality for a contemporary audience and offers EH a Chinese Buddhist ethics that challenges hegemonic forms of equality that protect only humans.

To conclude, scholars in this volume document marginal(ized) human and nonhuman groups or communities in various circumstances. Beyond raising awareness about existing unsustainable practices, many of these scholars presage a posthumanist shift that addresses the agency of nonhumans or material objects, injecting more-than-human subjectivity into their respective frameworks in order to imagine a new or renewed earth-based, ecohumanist, or multispecies identity. Like global EH scholars, many of these scholars also uphold visions that transcend an uncritical, normalized sense of boundaries. They offer intersectional visions of a sense of place and temporality to enable a shifting of scales, locales, and times, as evidenced in the methodologies of comparative, transnational, indigenous, and multispecies studies that employ trans-historic, deep-time approaches integrating the earth into their articulation of marginality, injustice, and exploitation. The authors also articulate ethics and pro-active praxis at the margins, advocating eco-communities in hyper-urbanized centers, proposing comparative and multicultural eco-communication as well as multilinguistic translation strategies to preserve bio-cultural diversity, reconceptualizing art practice as a form of multispecies care, and finally, elucidating Buddhist more-than-human theory and praxis. Scholars in this volume turn to the marginalized knowledge, value, and teachings such as Daoism, Buddhism, and minority cultures to restore ethics and values originating in Chinese culture.

1.6 Conclusion

In the bigger scheme of things, the story of "mountains-and-rivers-coming-to-an-end" in the current age of climate change is de facto a crisis of human civilization and it cannot be solved by science and technology alone. As mentioned in the section "Why Does China Matter? From Industrial Modernity to Ecological Civilization," China's decision to veer toward a green society and an ecological civilization shows that China understands that this crisis challenges the industrial-era ideology of human exceptionality. Now, China is gradually taking up the world's climate leadership to tackle one of the most important events in history. Just as China has a lot at stake in the survival of the planet if China succeeds, so too would the rest of the world also benefit, not just its own people and land.

China's state promotion of ecological civilization, quite contrary to the policies of the current U.S. administration, helps CEH, or EH in general, get wider recognition and legitimation outside of academy.

Asian cultures have contributed to the development of Anglo-European EH, as noted by Christopher K. Tong in this volume: "environmental philosophers such as J. Baird Callicott and Donald Swearer have long cited the philosophical and religious traditions of East Asia, especially Buddhism, as sources of comparison and inspiration." Anglo-European EH can reciprocally enrich and strengthen CEH. One of the greatest strengths in Anglo-European EH, ecocriticism for example, is critical thinking. As demonstrated in this volume, the majority of the CEH scholars are well versed in a wide range of Western theories. On the critical front, Anglo-European EH contributes significantly to discourses of anti-oppression and biodiversity to help make the multispecies world a more just and sustainable place to live. What current CEH contributes to Anglo-European or global EH is, first of all, to provide non-Western critical theories, vocabularies, and viewpoints. From the keywords I introduce in this chapter, the first keyword *ziran—the co-evolving nature of all things*—can be employed to develop a nondualistic eco-ethics and ecological thought. The second keyword, *huanjing* as "environing" or "place-ma(r)king" scrutinizes issues of marginality, exclusion, privatization, and extinction at the hands of a dominant colonial or imperial power—now, transnational neoliberal capitalism. It also sheds a positive light on how marginality as a conceptual frame can become the center of multicultural and artistic expressions and care in the advent of an ecological civilization and a global ecological movement found in different corners of the world.

All in all, CEH contributes to deconstructing Western enlightenment ideology and offers a postmodern, holistic mode of constructive thinking and developing solutions. The holistic mode brings us back to our early discussion of the *ying* and *yang* dynamics as a governing cosmic-psycho-cultural principle. As embodied in these two keywords, *ziran* and *huanjing* complement each other. The aesthetic-metaphysical realm of *ziran* should be integrated into the current discourse of Chinese environmental imagination and environmental humanities. This does not mean replacing *huanjing* with *ziran* to the extent of dismissing the ecological discourse within the social and science communities. Rather, what the term *ziran*, by way of its messenger *qi*, can teach us ecological piousness, a sense of awe and interconnectedness, and even compassion, in understanding our deep connection to the smallest particles of the grand universe. Though it may seem useless to materialists, however, it is the source of compassion and transcendent wisdom. As in classical Chinese painting, it reminds us of the place of *Homo sapiens* in the unfathomable depth of this universe, here visualized in the form of a sublime landscape. In order to survive as a species as well as to chart a planetary future, Chinese and global EH scholars need to tap into the energetic vibrancy and liveliness of "ten thousand things," both animated and inanimate. And that work of integration could be one of the future directions of CEH and global EH.

Notes

1. "Streams and Mountains without End" is also the title for a two-year-long exhibition featuring Chinese landscape paintings at the Metropolitan Museum of Art in New York from August 26, 2017 to August 4, 2019.
2. See Cheng, *Zhongguo huanjing meixue sixiang yanjiu*, 62–66.
3. Adamson, "Introductory chapter," 5.
4. Ibid., 4.
5. Ibid.
6. This initiative is commissioned by the European Science Foundation and Europe's intergovernmental Cooperation in Science and Technology program.
7. Davidson, "Actions, Reasons, and Causes".
8. Sörlin, "Environmental Humanities," 788.
9. Ibid.
10. Adamson, *Humanities for the Environment*, 5.
11. Oppermann and Iovino, "The Environmental Humanities and the Challenges of the Anthropocene," 3.
12. HfE, https://hfe-observatories.org/
13. Bergthaller et al. "Mapping Common Ground," 268. In their essay, the authors' notion of "environing" is a historicist one, which aims to view the environment as a historical process, instead of "taking the existence of 'the environment' as a given," 272. Hence, environment is "the multiple processes through which human beings (and other species) modify their surroundings as they make their living from and in the natural world, and of the symbolic transformations which configure 'the environment' as a space for human action," 267. The discourse of environing developed in this chapter syncs with this view but provides a more focused critical vantage point to enable a discourse centering around politics and ethics from the margins.
14. Heise, "Comparative Literature and the Environmental Humanities."
15. Bird, "China Just Overtook the US as The World's Largest Economy."
16. Marinelli, "How to Build a 'Beautiful China' in the Anthropocene," 365–386.
17. Marinelli, "How to Build a 'Beautiful China' in the Anthropocene," 373, 380.

18. See Slovic, "Landmarks."
19. Jacques, *When China Rules the World*, 196 *et passim*.
20. Originally, this term "Sinosphere" refers to countries that are historically influenced by Chinese culture. Here, I use it to refer to environmental humanities scholarship practiced by Chinese studies scholars in regions not only limited to Greater China (e.g. the mainland, Hong Kong, Taiwan) and the rest of the world, as a politically more correct term.
21. Lu and Mi, *Chinese Ecocinema*, 2.
22. The term "ecoaesthetics" appeared in China in the early 1990s, the field was developed by Zeng Fanren and Cheng Xiangzhan. See Cheng, "Ecoaesthetics and ecocriticism," 786.
23. For those who are interested in learning more about the development of ecocriticism in Taiwan, see Chang and Slovic, *Ecocriticism in Taiwan*, x–xi.
24. Bergthaller et al. "Mapping," 267.
25. The present-day sense of environment, said to be popularized in the mid-nineteenth century and referring to the "circumstances of an organism," alludes to all the forces that affect and transform an organism and the reciprocal relationship between the organism and the environment. Bonneuil and Fressoz note that the term "environment" is closely associated with *circumfusa* (Latin, "surrounding things") and point out its radical shift from a conception of environment (as external or remote surroundings) to one that emphasizes reciprocity, transcorporeality, and co-evolution. See *The Shock of the Anthropocene*, 173.
26. Smil, *The Bad Earth*, 4.
27. The early use of *huanjing* here finds a striking resemblance with the original meaning of "environment." "Environ" can be both a noun ("the surrounding area") and a verb ("to surround, to form a ring around, to envelop or enclose, to hem in"). Etymologically, "environment" is derived from the French word "environ." It first appeared in the fourteenth-century Old French word *environner* with a combination of "en" (in) and "*viron*" (a circle) to form the meaning of "to surround." The word "environment" first appeared in English in the early seventeenth century, also referring to "the act or state of being encircled." See Online Etymology Dictionary: https://www.etymonline.com/word/environ. Accessed 29 Jan 2018.
28. Marks, *China*, 4.
29. The ethnocentric practice of "encircling" or "marking the territory," as observed by Marks, conjures the other English term "colony," connoting "a sense of closure, of space cordoned off by human activity."
30. Vermonja Alston notes that "colony," or "colōnia" in Latin, renders the meaning of a "public settlement of a newly conquered country, or 'the planting of settlements.'" See "Environment," 93.
31. Ibid.

32. Mary Louise Pratt borrows the term "contact" from linguistics in the context of "contact language" to refer to hybrid languages such as pidgin in the "contact zone" where different groups meet, Pratt uses the term "contact zone" to refer to the spaces of imperial encounter. See *Imperial Eyes*, 8. Synonymous with "colonial frontier," "the space in which peoples geographically and historically separated come into contact with each other and establish ongoing relations, usually involving conditions of coercion, radical inequality, intractable conflict" (2007:8). Different from the other synonymous term, "colonial frontier," contact zone for Pratt aims to invoke "the space and time where the subjects previously separated by geography and history are co-present, the point at which their trajectories now intersect."
33. Arne Naess, the Norwegian philosopher who coins the term "deep ecology," credits Daoism as one of deep ecology's philosophical foundations. See "The Deep Ecological Movement," 79.
34. *China's Green Religion*, xxiii.
35. Tu, "The Continuity of Being," 71.
36. Ibid., 72.
37. Ibid., 71.
38. Lo, *Taiwan xiandaishi ziran meixue*, 19.
39. Benjamin Schwartz (1974, 4) quoted in Prasenjit Duara's "The Critique of Modernity in India and China."
40. Kato and Allen, "Toward an Ecocritical Approach to Translation: A Conceptual Framework," *State of the Discipline Report.*
41. Iovino and Oppermann, "Introduction," 3.
42. Wasted humans refer to "population as 'out of place,' 'unfit,' or 'undesirable.'" See Bauman, *Wasted Lives*, 5.
43. Thanks to the collective efforts of environmental NGOs' relentless campaigning and the overall awakening of environmental consciousness, in 2005, Premier Wen Jiabao suspended the plans due to ecological concerns. While this project is suspended, it is worth noting that the halted project faces pressure to be reopened.
44. Huayan, or Flower Garland school of Mahayana Buddhism, is a school of Buddhist thought found throughout East Asia that first developed during China's Tang dynasty (618–907 CE).

Bibliography

Adamson, Joni. 2016. "Introduction: Integrating Knowledge, Forging New Constellations Practice in the Environmental Humanities." In *Humanities for the Environment: Integrating Knowledge, Forging New Constellations of Practice*, ed. Joni Adamson and Michael Davis, 3–19. New York: Routledge.

Alston, Vermonja R. 2016. "Environment." In *Keywords for Environmental Studies*, ed. Joni Adamson and William A. Gleason, 93–96. New York: New York University.

Bauman, Zygmunt. 2003. *Wasted Lives: Modernity and Its Outcasts*. Cambridge: Polity.
Bennett, Jane. 2010. *Vibrant Matter: A Political Ecology of Things*. Durham: Duke University Press.
Bergthaller, Hannes, et al. 2014. "Mapping Common Ground: Ecocriticism, Environmental History, and the Environmental Humanities." *Environmental Humanities* 5 (1): 261–276.
Bird, Mike. 2014. "China Just Overtook the US as the World's Largest Economy." *Business Insider*, October 8. http://www.businessinsider.com/china-over-takes-us-as-worlds-largest-economy-2014-10. Accessed 30 Nov 2018.
Bonneuil, Christophe, and Jean-Baptiste Fressoz. 2016. *The Shock of the Anthropocene: The Earth, History and US*. Trans. David Fernbach. London: Verso Publisher.
Chang, Chia-ju, and Scott Slovic, eds. 2016. *Ecocriticism in Taiwan: Identity, Environment, and the Arts*. New York: Lexington.
Cheng, Xiangzhan. Autumn 2010. "Ecoaesthetics and Ecocriticism." *ISLE: Interdisciplinary Studies in Literature and Environment* 17 (4): 785–789.
Davidson, Donald. 1963. "Actions, Reasons, and Causes." *Journal of Philosophy*, 60. (Reprinted in Davidson, 2001a.)
Duara, Prasenjit. 2015. *The Crisis of Global Modernity: Asian Traditions and a Sustainable Future*. Cambridge: Cambridge University Press.
———. 1998. "The Critique of Modernity in India and China." In *Across the Himalaya Gap: An India Quest for Understanding China*, ed. Tan Chung, 77–89. New Delhi: Indira Gandhi National Centre for the Arts/Gyan Publishing.
Elvin, Elvin. 1988. *Sediment of Time: Environment and Society in Chinese History*. Cambridge: Cambridge University Press.
———. 2006. *The Retreat of the Elephants: An Environmental History*. New Haven: Yale University Press.
Estok, Simon C., and Won-Chung Kim, eds. 2013. *East Asian Ecocriticism: A Critical Reader*. New York: Palgrave Macmillan.
Heise, Ursula K. 2014. "Comparative Literature and the Environmental Humanities." *State of the Discipline Support*, March 9. https://stateofthedis-cipline.acla.org/entry/comparative-literature-and-environmental-humani-ties. Accessed 30 Nov 2018.
Jacques, Martin. 2009. *When China Rules the World: The End of the Western World and the Birth of a New Global Order*. New York: Penguin Group.
Kato, Daniela, and Allen, Bruce. 2014. "Toward an Ecocritical Approach to Translation: A Conceptual Framework." *State of the Discipline Report*, March. https://stateofthediscipline.acla.org/entry/toward-ecocritical-approach-translation-conceptual-framework. Accessed 30 Aug 2018.
Liu, Jing. 2016. "What Is Nature?—*ziran* in Early Daoist Thinking." *Asian Philosophy* 26 (3): 265–279.
Lo, Ren-ling 羅任玲. 2005. *Taiwan xiandaishi ziran meixue* 台灣現代詩自然美學 (*Natural Aesthetics in Modern Taiwan Poetry*). Taipei: Erya.

Lu, Sheldon H., and Jiayan Mi, eds. 2009. *Chinese Ecocinema: In the Age of Environmental Challenge*. Hong Kong: Hong Kong University Press.

Marks, Robert B. 2017 [2011]. *China: Its Environment and History*. New York: Rowman & Littlefield.

Marinelli, Maurizio. 2018. "How to Build a 'Beautiful China' in the Anthropocene. The Political Discourse and the Intellectual Debate on Ecological Civilization." *Journal of Chinese Political Science* 23 (3): 365–386.

Miller, James. 2017. *China's Green Religion: Daoism and the Quest for a Sustainable Future*. New York: Columbia University Press.

Naess, Arne. 1995. "The Deep Ecological Movement: Some Philosophical Aspects." In *Deep Ecology for the 21st Century*, ed. George Sessions. Boston: Shambhala Publications.

Oppermann, Serpil, and Serenella Iovino. 2016. "Introduction: The Environmental Humanities and the Challenges of the Anthropocene." In *Environmental Humanities: Voices from the Anthropocene*, ed. Serpil Oppermann and Serenella Iovino, 1–21. New York: Rowman and Littlefield.

Pratt, Mary Louise. 2007. *Imperial Eyes: Travel Writing and Transculturation*. New York: Routledge.

Slovic, Scott. 2013. "Landmarks in Chinese Ecocriticism and Environmental Literature: The Emergence of a New Ecological Civilization." *Chinese Social Sciences Today*, July 31. http://www.csstoday.com/Item/268.aspx. Accessed 30 Aug 2018.

Smil, Vaclav. 2017 [1984]. *Bad Earth: Environmental Degradation in China*. New York: Routledge.

Sörlin, Sverker. 2012. "Environmental Humanities: Why Should Biologists Interested in the Environment Take the Humanities Seriously?" *Bioscience* 62 (9/1): 788–789.

Sudilovsky, Judith. 2017. "Faith Leaders Discuss Duty to Protect Rainforests At Interfaith Rainforest Initiative Religious Voices Reinforce the Need to Protect these Vital Resources." *OSVNewsweekly*, June 28. https://www.osv.com/OSVNewsweekly/Story/TabId/2672/ArtMID/13567/ArticleID/22609/Faith-leaders-discuss-duty-to-protect-rainforests.aspx. Accessed 6 Oct 2018.

Tu, Weiming. 1989. "The Continuity of Being." In *Nature in Asian Traditions of Thought: Essays in Environmental Philosophy*, ed. J. Baird Callicott and Roger T. Ames, 67–78. Albany: State University of New York Press.

Williams, Raymond. 2015 [1976]. *Keywords: A Vocabulary of Culture and Society*. New York: Oxford University Press.

SECTION I

Chinese Ecocriticism and Ecotranslation Studies

CHAPTER 2

Building a Post-Industrial *Shangri-La*: Lu Shuyuan, Ecocriticism, and Tao Yuanming's "Peach Blossom Spring"

Chia-ju Chang

Ever since Deng Xiaoping's "economic reform and opening up" (*gaige kaifang*) commencing in 1978, and which integrated China into the global capitalist economy, China has undergone a transformation from an agrarian society to an industrial and economic powerhouse. At present, China has the second-highest number of millionaires after the USA and has a bigger middle class than the USA. Yet China's ascendency to economic juggernaut and the unprecedented high speed of urbanization are a precarious cautionary tale, one that uncritically follows the Western industrial model of "pollute first and clean up later" in the age of global warming. As the world's largest carbon emitter (with the USA falling to second and India in third) accounting for 23% of global CO_2 emissions, China achieved economic prosperity at the expense of people and ecological well-being.[1] In addition to choking air that is responsible for 1.6 million deaths every year, other forms of industrial pollution have also taken a heavy toll on the environment, resulting in environmentally induced illnesses and the so-called

C. Chang (✉)
Brooklyn College, Brooklyn, NY, USA

C. Chang (ed.), *Chinese Environmental Humanities*,
Chinese Literature and Culture in the World,
https://doi.org/10.1007/978-3-030-18634-0_2

cancer villages. In other words, the present-day economic prosperity comes at the expense of its citizens' fundamental relationship with the environment and its primal elements, such as water, air, soil, and minerals—all necessary for people's physical well-being, survival, and sense of belonging.

While "sympathetic" critics of colonialism may argue that developing countries should not be denied the same access and aspiration to material wealth that developed countries have been enjoying, it does not justify following in the same unsustainable and exploitative modernist footsteps. In fact, Western modernity should not be taken as the universal path to begin with, as the success story of Western modernity is not separable from its colonial expansion and militarism.[2] The entire planet's current precariousness cannot be framed only in terms of colonial-postcolonial temporal order. Instead one should ask: How will late-comer industrial countries wisely respond to the existing environmental conditions as a crisis of global modernity and planetary destruction? In the case of China, Jiang Yi, Vice Dean of the School of Architecture at China's Tsinghua University, in a PBS documentary film, *Design China: From Red to Green* (2006), soberly warns that if China does not begin to "turn green" soon, it will suffer severe consequences together with the world at large.

The Chinese Communist Party (CCP) started tackling environmental issues as early as the 1970s. In an article titled "The Ecological Civilization Debate in China: The Role of Ecological Marxism and Constructive Postmodernism—Beyond the Predicament of Legislation," the authors (Comprised of three professors in China from social science and philosophy, two of them directors of the Institute for Postmodern Development of China, while one has served as a former assistant mayor of Kaifeng, Henan Province) note that China has enacted many strict laws against pollution since 1970 and some of them were even stricter than in the USA.[3] Yet, environmental legislation has been ineffective in curbing environmental degradation.[4] Here the authors ask, "Why have these environmental laws and regulations been so ineffective?" They conclude that environmental issues are not only legal but also "political, social, and philosophical issues" that involve factors such as "the role of capital, interest groups, the development model, and an anthropocentric worldview and values."[5] Concurring with Pan Yue, the executive vice president of the Central Academy of Socialism in Beijing, the authors submit that in order to make these laws effective, the most crucial thing to do is to "change the mode of production and the way of life that are driven by capital."[6] Here the authors argue that the current turn of ecological civilization (*shengtai wenming*) of the CCP requires a modification of the anthropocentric worldview, which is the root cause of ecological destruction. Identifying the origin of the contemporary anthropocentric and utilitarian worldview that frames China's economic discourse, as most famously exemplified in Deng Xiaoping's "development-is-the-absolute-principle" (*fazhan shi yingdaoli*), they write, "Deeply influenced by Western modernity, China has predominantly accepted an anthropocentric worldview and values, which regard human beings as totally different from the world of natural things, and accordingly treats the world of nature as a world of objects."[7] According to the authors, Chinese traditional cultural values and philosophy not only serve as an antidote to China's "worship of growthmania," they are also a useful resource in preparing for a harmonious post-GDP society.

Here the authors provide a glimpse into how Chinese environmental humanists identify China's environmental problems and solutions. This chapter turns to literary and cultural studies to show how Chinese literary scholars contribute to the larger conversation regarding China's practice of environing at the margins and the construction of China's eco-civilization. This chapter zooms in on ecocriticism and turns to Lu Shuyuan, one of the most prominent ecocritics, as a case study to illustrate a Chinese-oriented ecocriticism.

I first place Lu's ecocritical theory and praxis in the context of China's emergent eco-consciousness (e.g. toxic discourse, environmental protection, and ecological civilization) in the 1980s and early 1990s, especially in the context of both the popular cultural domain and academia. Against the grain of the dominant ideology of the time, Lu is a relentless defender of spirituality, subjectivity, and traditional cultures. His ecocritical position should be understood in the context of post-Mao intellectuals defying viewing "literature-art" as serving as a mouthpiece for states' political propaganda as well as resisting the uncritical acceptance of Western theories, a blindfolded pursuit of Western modernity that can be traced back to the turn of the twentieth century. Lu's contribution to ecocriticism or, more broadly, Chinese environmental humanities lies in his insistence on the inclusion of "spiritual ecology" and traditional wisdom and values. As a civilization or a state cannot be sustained without its traditional cultural foundation, it is more so for an agriculturally based "civilization-state"[8] with such rich native resources on human-land kinship. Lu, in search of an alternative postmodern eco-paradigm, looks to the much celebrated literary yet politically marginalized premodern "peasant-poet," Tao Yuanming 陶渊明 (365–427 CE).

Lu's Tao Yuanming study deserves attention as a model for critical and poetic thinking about how to create a postmodern eco-community, both locally and globally. It provides a critical frame to conceptualize "the discourse of environing at the margins" that speaks for/with China's largest marginalized group, rural migrant workers (*nongmingong*), vis-à-vis current systematic erasure of the pastoral mode of life. I examine the way in which Lu addresses issues of marginality and disenfranchisement (e.g. social injustice and inequality) in a poetic genre called "moonlight poetry" (*dagong shi*) composed by rural migrant workers through the prism of his Tao Yuanming study. Here Lu resurrected Tao's image of the harmonious village "Peach Blossom Spring" (*taohua yuan*)—a *Shangri-la* that "has endured through the ages as an ideal world in the East Asian imagination and beyond."[9] The evocation of a premodern utopian vision provides an alternative to the Western linear conception of temporality and modernity. In this sense, Lu's Tao Yuanming study showcases a post-industrial, degrowth ethics of care based on Tao's "Peach Blossom Spring," a pastoral vision of a small-scale harmonious village that has evolved into a Sinospheric archetypal image of "arcadia."

This chapter contributes to the current discussion of global environmentalism and sustainability by illustrating the importance of the role that premodern cultural tradition plays in the contemporary postmodern environmental imagination and ecological construction that embraces equality and inclusion, as opposed to the capitalist model of exploitation and exclusion.

2.1 Why Does China Matter? From Industrial Modernity to Ecological Civilization

The term "ecocriticism" (*shengtai piping*) was first introduced to China in early 1990.[10] The 2001 symposium on "Globalization and Ecocriticism" held by Tsinghua University in Beijing marked the formal appearance of the term "ecocriticism" in Chinese literature. It was not until veteran ecocritic Wang Nuo's 2002 *Ecocriticism: Development and Origin* that the term was popularized. The entry of this Western term, however, does not mean there had not emerged any eco-awareness or ecology-related concepts in the cultural and academic domains previously. In fact, the term "ecological civilization" was broached by the agricultural economist Ye Qianji (1909–2017) in 1982 to advocate the practice of sustainable agriculture.[11] Writers and journalists who were concerned about social and cultural issues had already been keenly aware of the ecological degradation. Such awareness led to a mushrooming establishment of "green magazines." In fact, as early as 1981, there had already existed a popular science magazine, *Forest & Humankind*. From the 1980s through the 1990s, several environmental magazines and newspapers were launched. For example, *China Green Times* was launched in 1984, the same year as the coinage of "environmental literature." *Chinese Man and the Biosphere* magazine was established by the Chinese National Committee for Man and Biosphere (MAB) in 1999. The MAB Programme is an intergovernmental and interdisciplinary research program. It was launched by UNESCO in 1971 to tackle the population, resources, and environment problems facing the whole world. The committee was established in 1978, two years after the ending of the Cultural Revolution.[12]

Many writers were involved in promoting environmental protection awareness; they also served as editors of these green literary magazines.[13] In 1984, a female writer, Gao Hua, put forward the concept of "environmental literature" (*huanjing wenxue*) to promote environmental reportage, a dominant literary form during that time. The most salient example of the environmental reportage of the time is Xu Gang's *Wake up, Woodcutters!* Xu's reportage is often praised as the forerunner of Chinese eco-literature, as it shows the serious effects of the chopping down of jungles and forests in some parts of China. The publication subsequently effected the policy of forestry. In addition to these green magazines and examples of environmental literature, many writers' works that deal with the predicament of human existence in modern times reflect what Lawrence Buell calls "environmental unconscious." All these are evidence of the emergence of a modern ecological consciousness during the 1980s.

For literary studies, two names were proposed: "the studies of ecological literature-art" (*shengtai wenyi xue*)[14] and "the studies of literary-artistic ecology" (*wenyi shengtai xue*).[15] Here I shall note that there is a tendency to think of literature and art as a non-separate entity, "literature-art" (*wenyi*), which broadly includes literature, theater, music, visual arts, dance, film, and calligraphy. Thanks to the rapid growth of environmental reportage and green magazines, literary scholars in academia became aware of the scientific field called "ecology" and were exposed to the field of "cultural ecology."[16] In a 1987 essay, "The Studies of Literary and Artistic Ecology," Zhang Songkui noted how he applies the concept of literary/artistic ecology to study the production, existence, and development of art.[17] Here Zhang elaborated on what he calls "the studies of literary and artistic ecology" (*wenyi shengtai xue*) and suggested at least two possible "ecocritical" approaches.[18] First, the interrelationship between literature-art and natural ecology (i.e. the impact of natural environment on the production of the arts or vice versa), and, second, the application of ecological method to the studies of literature-art. On the second one he writes, "'the studies of literary and artistic ecology treat literature-art as an open and mobile system, no different from that of biological ecology" (my translation). While the first approach immediately evokes Cheyll Glotfelty's 1996 definition of ecocriticism as "the study of the relationship between literature and the physical environment,"[19] the second pertains to the studies and understanding of literature-art through the lens of ecosystem or ecological concepts such as symbiosis (namely, the interdependent relationship between the artists, the audience, and the critic). Here the study of literature-art does not simply pertain to a work of literature-art per se, but also its production network, including author, creative activities, dissemination, and consumption. Here Zhang envisions literature-art as an ecosystem where different parts within the organism interact. The idea of harmony or balance is accentuated to form a positive feedback loop among these actors. At one point, the author turns to the direction of what would be later conceptualized as material ecocriticism in the West. In ruminating on the creative chain of literature-art, he asserts, "there must exist a serious ecological issue in such a complicated and grand literary-art system."[20] Zhang, among other scholars, represents early ecocritical thinking of this time.

The emergence of the post-Mao ecological consciousness, the proliferation of green magazines, and eco-related discourses such as ecological civilization and cultural ecology within academia provide a broader framework to situate the coming-of-age ecoscholars such as Lu Shuyuan in the 1990s. The second framework has to do with relaxation of ideological control over the production and studies of literature-art in the wake of the Cultural Revolution (1966–1976) and China's economic Open Door Policy. What China ushered in during this time was not only the massive inundation of foreign investment, but also cultural exchange and academic freedom. The influx of different schools of thought into the domain of literature-art, ranging from Roland Barthes' semiotics, Freud and later Jungian and Lacanian psychoanalytical schools, to Edward Said's postcolonial studies, to name a few, was aimed to exclusively serve the socialist agenda. During this time, China's intellectuals again found themselves at the crossroads of tradition and Western modernity, just like their predecessors in the May Fourth Movement (1915–1921) at the turn of the twentieth century. The freedom allowed them to examine fundamental humanist themes such as love, creativity, and freedom. This explains Lu's early interest in humanist psychology and inquiry into questions of subjectivity, creativity, and agency, particularly the issue of spirituality, one that was once deemed taboo.

The last framework that situates Lu's other methodological orientation is a postcolonial one. As a professor of Chinese literature, Lu was keenly aware of his own positionality as a non-Western, postcolonial scholar. Throughout his career, he has been insisting on "equal footing with Western counterparts" and this distinguishes him from other Chinese ecocritics. As mentioned in the introductory chapter, the humiliation China experienced starting in the second half of the nineteenth century resulted in total rejection of Chinese culture by the Chinese themselves. The majority of intellectuals espoused a modernist ideology of progress for nation building and jettisoned traditional Chinese values, which were viewed as backward, without realizing that China in many respects has been "modern" by Western standards prior to Western modernity (e.g. secular state, a highly self-governed civil society, a written language accessible to both literati and laypeople, and a cohesive culture open to multiculturalism).[21] Lu points out how the May Fourth or New Cultural Movement thinkers such as Lu Xun, Hu Shi, and Liang Qichao at the turn of the twentieth century completely lost confidence in traditional Chinese culture and philosophy vis-à-vis aggressive Western colonial powers.[22] He particularly notes that Qian Xuantong (1887–1939) was so ashamed of Chinese culture that he even proposed to abolish Chinese characters, to burn all Chinese books, and to shoot any Chinese over 40![23] Lu states that, in his nearly four-decade career, he has witnessed Chinese intellectuals either trash Chinese tradition, just like their May Fourth predecessors, or suffer from cultural aphasia, completely submitting themselves to the Western way of thinking. Younger-generation nature writers such as Hai Zi (1964–1989) and Wei An (1960–1999),[24] passionately admired and imitated Western poets such Shelley, Whitman and Emerson but never mention any Chinese poets. Wei An, while considering himself the child of Henry David Thoreau (who ironically was a great admirer of premodern Chinese poetry and philosophy), criticized Tao Yuanming and announced that "The long-standing literature of the motherland has never been able to enter my field of vision."[25] Lu shows another irony by stating that, while the Western thinkers after WWII began to lose faith in their own culture and turn to the East for answers, their Chinese counterparts were doing the opposite.[26] In this regard, Lu can be viewed as a contemporary successor of figures such as Zhang Binglin and Liang Shuming discussed in Chap. 3.

A personal observation attests to Lu's emphasis on equal dialogue with the West. When I first met him at the 2009 ecoaesthetic conference at Shandong University, in a panel discussion, he stressed the importance of engaging in a conversation with the West on an equal footing without losing his own culture and history. Together with the ecoaesthetician Zeng Fanren, Lu employs the terms "to connect" (*jiegui*) and "to dialog" (*duihua*) with the West, instead of following a one-way uncritical reception of Western theories. Commenting on the genesis of ecocriticism and its uniqueness, Lu points out that, uniquely, ecocriticism in the early 1990s arose in China even before the arrival of Western ecocriticism, as opposed to other disciplines or schools of theory, which were imported from the West. The dialogical impulse behind the idea of an equal footing results in a syncretic integration, a trait that is quite "Chinese." This type of "Chinese syncretism" of incorporating foreign culture into the Chinese indigenous one to create something new has been seen throughout Chinese history. It invites new ideas to create something that is new and traditional at the same time. Lu's ecocritical theory can be seen as a contemporary demonstration of traditional Chinese characteristics such as a syncretic way of thinking and cultural tolerance.

2.2 Sketching a Field: Chinese Environmental Humanities

Lu Shuyuan was born in Kaifeng, Henan Province, in 1946, into a family that belonged to the lower stratum of society. He studied Chinese literature at Hunan University at the onset of the Cultural Revolution. In the 1980s, Lu was interested in the question of human subjectivity and creativity, and soon established himself as a scholar of "the studies of psychology of literature-art" (*wenyi xinli xue*). In his system of the psychology of literature-art, he bridges human psychology with traditional Daoist cosmology to articulate a cosmogenic human spiritual world, or what he calls "inner universe." Highly critical of modernist rational, utilitarian ideology, Lu's research centers on how to use premodern philosophy as a foundation for postmodern thinking. Currently he is the director of the Center for Constructive Postmodern and Ecological Studies at Huanghe Science and Technology College, the vice president of the Chinese Association for Theory and Literature and Art, and a member of the literary criticism committee of the Chinese Writers Association. Lu has taught at several institutions, such as Hunan University, Hainan University, Zhengzhou University, and Suzhou University, where he offered courses on "eco-literature and the arts" (*Shengtai wenyi xue*) in the early 1990s. He has written several books relating to the subject of ecology and literature in Chinese, including *The Studies of Ecological Literature and the Arts* (2000), *Space for Ecocriticism* (2006), and *The Ecological Era and Classical Chinese Naturalism: A Case Study of Tao Yuanming* (2012). The last book won him the Lu Xun Literary Prize in 2000.

As a fervent learner of Russian and Anglo-European literature, philosophy, and cultural theories, Lu took a comparative and syncretic approach while firmly grounding his theoretical orientation in Chinese culture. In the early 1990s, Lu expanded his interest in humanist psychology and spirituality to explore the intersection between spirituality and ecology. Such a shift is not a mere intellectual penchant to explore new interdisciplinary territory, but a visceral reaction to China's social change. He described his initiation into ecocriticism as follows. One day, he was walking on the streets of Shenzhen, one of the fastest-growing cities in Southern China and, describing the secular, urban street scene in early 1990s post-Mao China, he wrote:

> The street was full of companies, banks, hotels, dance halls. Music was so busy and noisy, the crowds were grotesque, and there was a mixture of cosmetic powder and barbecue smells. I felt a huge sense of loss. People worked so hard to make money so as to squander it away. What was the value and significance of life? (my translation)

This post-Mao urban flaneur experience triggered a sense of existential crisis. While this passage may seem all too familiar for urbanites in hyper-capitalist societies, such as a New Yorker in Times Square, the context of Lu's sense of bewilderment and crisis should be understood in the compressed time and scale of China's unprecedented urban transformation. In 1950, only 13% of people in China lived in cities. By 2010, the urban share of the population had grown to 45%; it's projected to reach 60% by 2030.[27] The frantic transformation to urban landscape is beyond words. According to the economic voice *World Finance*, there are more than 200 skyscrapers under construction in China today. After five years, the total number of skyscrapers in China will exceed 800, which is four times the total number in the USA today.[28] The urbanization process widens the gap between the poor and nouveau riches in urban and rural households, on the coast and in the interior. China has the second-highest number of millionaires after the USA and about 101 billionaires in Beijing alone,[29] but it has 30.46 million rural people living below the national poverty line as at the end of 2017.[30] The nation's obsession with getting rich and the espousal of the motto "getting-rich-is-the-absolute-truth" (a twist of Deng's "development-is-the-absolute-principle") resulted in prevalent practices of top-down corruption and all kinds of fake commodities/businesses, such as the "poisoned milk powder," "recycled restaurant cooking oil," "fake wines," "fake publications," and "medical scam."

Seeing the connection between the getting rich principle, spiritual nihilism, moral decay, and environmental degradation, Lu ascribes the social abnormalities described above to consumerism. He considers it a form of "spiritual malaise," resulting from the absence of moral authority, or any "transcendent" (to borrow Prasenjit Duara's use of the term) order and guidance, and directly results in all kinds of crises in Chinese society.[31] For someone who grew up in a time prior to the advent of capitalism and urbanization where the line between the urban and rural and the poor and rich was blurred, Lu could not help but question the impact that the economic miracle and fetishization of individual wealth had on people's impoverishment of emotional well-being and spiritual lives, which impart moral values and a genuine sense of happiness. Finding a way to avert the course of Western industrial modernity and global neoliberalism became his lifelong quest. For him, environmental crisis *is* fundamentally a spiritual crisis, and therefore a humanist issue. Environmental degradation and social injustice reflect a distorted collective view of a culture as a whole; they are not something science, environmental regulations, or policy enforcement can fix. Current hyper-consumer culture, spiritual impoverishment, and moral decadence lead him to question whether a "high quality of life with low material consumption" (*di wuzhi haosun de gao pinwei shenghuo*) is possible. This question arises naturally out of his critique of this very stage of China's economic development, where capitalist, utilitarian values prioritize the accumulation of capital more than life itself.

Lu argues that literature-art plays a crucial role as catalyst or as a liminal space for conscious transformation. According to Lu, the mission of literature-art does not lie exclusively in its ability to raise awareness of environmental destruction and social injustice as seen in the genre of the environmental reportage mentioned above, but also to "mend the broken spiritual sphere."[32] In other words, the function of literature-art is something more fundamental, that is, to "conjure one's own lost spirit" (*zhaohun*), in other words, to reconnect with an authentic self that has gone astray in the labyrinth of material pursuit. Here Lu turns to Tao, a premodern peasant-poet, as a model for building a postmodern eco-community.

Lu's *Space for Ecocriticism* is devoted to a systematic study of what he calls "spiritual ecology," examining the "poetic and spiritual dimensions of human nature and human culture in relation to the natural world."[33] According to Lu, ecology or ecosystem should not be limited to a scientific discipline as we see in the study of ecosystems that excludes humans or humanities (e.g. wetland ecology).[34] For him, ecology is life, or the web of life; it is the organic process in which the humanities partakes. "Spiritual ecology" concerns people's inner emotional and spiritual life, and its interconnection with the natural and cosmic realms. In Lu's view, spiritual ecology finds its best formal expression in literature-art, examining questions concerning the place of humanities in the world. Here his spiritual ecological view resonates with Timothy Morton's expansive definition of ecology. Morton elevates ecology to a philosophical level. Thinking about ecology from the perspective of (eco-)existentialism, he writes, "[e]cology includes all the ways we imagine how we live together. Ecology is profoundly about coexistence... thinking ecologically isn't simply about nonhuman things."[35] Here Morton touches on the psychological dimension of ecological thinking such that ecology does not only have to do with the "definitions" of environmental protection and multispecies relationship, but also with "love, loss, despair, compassion and other affects. It has to do with depression and psychosis."[36] Similar to Morton, Lu's eco-existential and eco-psychological approach goes against the grain of dominant environmentalist or materialist approaches to ecocriticism. Situating his methodology in the context of China's modernity experience, Lu "opens up wider possibilities for redefining environmental scholarship beyond, in between, or outside of disciplinary conventions."[37]

How then does Lu's spiritual ecology address issues of marginality such as social injustice and inequality as the result of China's uneven urbanization and agricultural industrialization? To answer that question, I turn to Lu's approach to one of the most marginalized populations in China—migrant workers—and their "moonlight poetry" as an extension of his praxis of spiritual ecocriticism.

2.3 Tao Yuanming's Lost Generation: Migrant Workers' "Moonlight Poetry" and Tao Yuanming's "Specters"

In his reception remarks for the 11th John Cobb Common Good Award, Lu Shuyuan summarizes his lifelong interest in ecology, spirituality, and culture as follows:

> As an ecocritic, what I am trying to do is, on the one hand, to integrate the vibrant variables of human spiritual activities into the ecosystem of the earth. On the other hand, I bring ecology into the domain of culture and the arts as a crucial aesthetic and poetic category.[38] (my translation)

Here he elaborates upon his dual mission as an ecocritic: to humanize the ecological system and ecologize humanity. This inter-permeation and mutual illumination of the earth's ecosystem and human cultural system is a reminder of the harmonizing dynamic *yin* and *yang* diagram, where each side contains the opposing element, rather than a static dualistic compartmentalization. Such a dynamic and holistic ecocritical position does not exclude a critique of environmental injustice, as seen in his sustained critical attention to the violence of urbanization and large-scale agricultural modernization against the peasants and the erasure of premodern agricultural civilization on which the primordial visions of *ziran* (nature) and the *yin* and *yang* are based. His writings on China's migrant workers and the so-called "moonlight poetry" are cases in point.

China's urbanization has produced the largest migration movement in history. "Migrant workers" are people who migrate either within or outside their home country to pursue work. In the case of China, the term refers to peasants who have left the countryside in search of jobs in cities that are mostly on the coast. Usually, they work long hours and are exposed to poor working conditions and occupational hazards. There are 281 million "migrant workers" as of 2017, and among them about 20 million are writing poetry.[39] The so-called "moonlighting poetry" is a genre of poetry written by this most marginalized group of people in China. Their marginality consists of a multilayered exclusion from the current civil-economic system, including class, education, and household registration (*hukou*). The poetry they produce is a collective testimony to their living and working conditions and the existing inequality in post-Mao society. Zheng Xiaoqion, a woman migrant worker from Sichuan, describes her working condition in the poem "Village" (*Cunzhuang*): "Five years of symphony of the machines, the fate of an ant/their turbulent migration/from inland to the coast/from countryside to city."[40] Here she compares her fate to that of "an ant" who works endlessly for five years in a factory enveloped by the "symphony" of machines. Industrialization not only destroys the migrants, but also turns them into machines. In "Life" (*Shenghuo*), Zheng writes:

> you don't know, my name is buried in a worker's ID card/my hands become part of the assembly line/my body surrenders to the contract/through the silent white light/I see my tired shadow projected onto the machine, it moves slowly/turning, bending, silent as a cast iron. (my translation)

Here she is describing her experience of alienation as a worker. Her identity is reduced to nothing but an ID card; her body fuses with the machine, repeating the same motion endlessly on the assembly line.

Lu calls these poets "the fallen offspring of Tao Yuanming," lost in the turbulence of industrial modernity where they are trapped in an urban world with no home to return to. Often compared to American writer Henry David Thoreau, Tao Yuanming (365–427 AD), born to a noble family, lived during hard times when Northern China was under non-Chinese rule during the latter part of the Eastern Jin Dynasty (317–420). Tao served in several official posts during his lifetime, but never quite fit into the political environment. After resigning from his last civil post at the age of 40, Tao returned to his hometown to lead a reclusive lifestyle. He spent the rest of his life at his home estate and kept up his friendships with local and court officials as well as contemporary literary celebrities, drinking and exchanging poetry with them. As he worked in the fields, he read and wrote poetry until the end of his life. His poetry is hailed as the epitome of "pastoral poetry" (*tianyuan shi*), his life practice is regarded as a model of farming culture, and his spirit is regarded as the collective unconscious of Chinese agricultural civilization.

Lu highly praises Tao's preference for the *yin* aspect of life. As shown in his other name, "Tao Qian" ("*qian*" means "to dive or to hide"), which derives from the First Nine of the hexagram *Qian* (the Creative) in *The Book of Change*, "the hidden dragon is not to be employed" (*qianlong wuyong*). The notion of hiddenness not only suggests a retreat from civil engagement as worldly advice, but also has a Daoist implication. Lu notes that "to know the bright but cleave to the dark" and "knowing glory but cleaving to ignominy" (Laozi's *Daodejing*, Chapter 28) suggest a return to the Dao (The Way), or the *yin* aspect, without being ignorant of "the bright" or the *yang/masculine* aspect of life (e.g. pursuit of wealth, fame, power, etc.).

Lu's comparison of migrant-worker poets to Tao Yuanming's descendants might seem a mismatch of class and anachronism, as these poets were neither educated farmer-poets of noble descent nor part of an intellectual circle that embraced an "idling lifestyle" and relished poetry writing, celebrating the pastoral mode of existence and labor. However, the references to Tao and his poetry serve more as points of contrast to shed light on Lu's discontent with the onslaught of industrial modernity against China's peasant class and the capitalization of their lives. He writes:

> In the early memories of these young and sensitive migrant-worker poets, the native land and the pastoral are far from being beautiful and leisurely. Even if there is an ancestral poet Tao Yuanming in the memory, it has also become a bygone and vague dream, like a thread of light swallowed by the darkness. Those profoundly delicate pastoral and faint poems have even become incredible nonsense in their minds.[41]

Evoking Tao's poetic dwelling as an exemplum of the ideal pastoral lifestyle for farmers, Lu shows how peasants nowadays have departed from their agricultural home, roots, and spirit. In contrast with Tao's famous homecoming poetic declaration, "Homeward ho, why not return now that my fields will go into weeds!" ("*guiqu lai si, tianyuan jiangwu, hubugui?*"), which the poet calls out as he returns home to tend to the deserted fields, these migrant-worker poets instead express "home-leaving" *pathos*. Lu points out that, from a thematic standpoint, this runs directly counter to the impulse of traditional pastoral poetry, though Tao's poems and those of the migrant-worker poets are both about homeland. Noting the similarities between Tao Yuanming's homecoming and these poets' home-leaving, Lu states that both migrant workers' home-leaving and Tao Yuanming's "homecoming" grow out of a yearning for the pursuit of an ideal lifestyle and happiness.[42] However, he writes that, because he was content with "honorable poverty" (*qingpin*),

> Tao was able to savor the joy of 'returning to nature after being in the cage for a long time' (*jiu zai fanlong li, fu de fan ziran*), but conversely, these home-leaving migrant workers enter the cities for a better life but wind up locking themselves into a more inhuman cage."[43]

Lu concludes that, unlike Tao's celebratory tone of homecoming and aspiration toward pastoral landscape, the poems of these migrant-worker poets are full of despair for their lost rural land.

Zheng Xiaoqiong, the migrant-worker poet mentioned above, provides a poetic account as to why they cannot be homebound. In the poem "In an Era" (*Zai shidai zhong*), she grieves over the destruction of the countryside:

> ...this is the 21st century/this is a dusty machine/the felled lychee forest/they fell down, the courtyard turned into rubble, the ruins of the earth/the vast land is scorched by industrial flames, and it's a heap, alas![44] (my translation)

From this excerpt, the villages and homes where the farmers have lived for generations are being transformed into a wasteland. The reason for this is not negligent stewardship as on a farm but rather the brutal industrial uprooting of the population and way of life, as shown in the image of a landscape "scorched by industrial flames." Zheng continues:

> Collapsed in an instant, old traditions accumulated for thousands of years/Falling deeply, the excavator sticks out a huge iron sawtooth/deep from the earth they cut the umbilical cord that connects me to my ancestors/the ancestors walk into the depth of darkness/What else are we looking forward to?[45] (my translation)

Here the theme of demolition in the name of agricultural modernity is further connected to her severance with the past, ancestors, and the earth that connects her as through an "umbilical cord." Lu's calling them "a lost generation" is echoed in Zheng's line "What else are we looking forward to?" after "the ancestors walk into the depth of darkness." The theme of *qian* (retreating and hiding), as discussed in Tao Yuanming's name and Laozi's chapter, is evoked in the poet's sense of alienation as a result of the receding past. What she is expressing is not simply the destruction of the rural countryside, but of the whole agrarian traditional way of life, memory, and values. Hence, this sense of ancestral disconnection adds to another layer of homelessness as the migrant workers who have gone to the township have become "homeless" in both the physical and spiritual or affective sense.

Highly skeptical of the West and the modern agricultural production mode and urbanization that can bring forth both human and ecological well-being, Lu writes:

> this kind of development model that relies on advanced science and technology and industrial management to obtain maximum economic benefits has already been solemnly questioned in Western developed countries.[46] (my translation)

In Lu's view, China, as an agricultural civilization-state, cannot uncritically follow the West, but instead should return to its own cultural roots to seek alternative modernity solutions. Thinking of history in cyclical terms, Lu contends that the future cannot do away with the past, which means that an industrial civilization cannot sever itself from agricultural civilization and that urbanization cannot destroy pastoral traditions. In order to articulate the hidden spirit of Tao Yuanming in contemporary Chinese culture, Lu turns to Derrida's notion of specter as a form of methodology to postulate his own specter theory. Considering specter as people's embedded cultural unconsciousness, Lu writes: "If 'specter' is considered to be similar to, or identical with, 'tradition' or 'spiritual legacy,' it is not necessarily the dead past or simply passive; in fact, it is alive and has a mission."[47]

2.4 The Peach Blossom Spring: Cultural Imagination and Beyond

Tao Yuanming's "Peach Blossom Spring" remains very much alive in the East Asian cultural sphere. Peach Blossom Spring refers to "arcadia" in a time of political instability. According to the story, a fisherman enters a crevasse and traverses its narrow passage, accidentally discovering a bountiful farming village with people of all ages and animals of every kind. The people in the village lead a harmonious existence with nature and are unaware of the dynastic changes happening for centuries. Upon departure, the villagers advise the fisherman not to disclose this place to the outside world so as to remain "hidden."

Peach Blossom Spring is the premodern imagination of a *Shangri-la* that has endured through the ages as an idyllic haven in the East Asian imagination. For example, in Akira Kurosawa's (1910–1998) magic realist film *Dreams* (1990), one of the dream sequences, "Village of the Watermills," reenacts the narrative of Tao's "Peach Blossom Spring." It tells of a young man who blunders into a peaceful, stream-laden village. An elderly wise man tells him that the people of his village, in choosing spiritual health over convenience, decided to forsake the polluting influence of modern technology and return to a happier, cleaner society. At the end, a funeral procession for an old woman takes place in the village, and instead of mourning, the people celebrate joyfully at the proper end to a good life—a motif echoing Zhuangzi's celebration of his wife's death.

The image of Peach Blossom Spring is also present in Hayao Miyazaki's anime *Spirited Away* (2001). In the film, a well-hidden deserted "ghost town" that centers on a fancy bathhouse like a 5-star hotel is separated from the human realm by a river and can only be reached through a narrow tunnel. Miyazaki renders a contemporary twist to the archaic narrative of Peach Blossom Spring, where the spirits in *Spirited Away* flesh out the supernatural undertone to the villagers in Peach Blossom Spring. The bathhouse, which symbolizes hypercapitalist arcadia, is also a site of purification for spirits and nature, as seen when the foul-smelling river god is cleansed of all the industrial debris and pollutants. As a postmodern, feminist, and environmentalist allegory, the teenage female protagonist Sen or Ogino Chihiro, like the fisherman who blunders into Peach Blossom Spring, is advised to leave behind the fantastic world before she returns to humanity. Both *Dreams* and *Spirited Away* demonstrate the way in which the image of Peach Blossom Spring continues to shape the imagination of the East Asian cultural sphere, not only with its Shinto variation, but also charged with post-industrial sensibility.

Though Lu's evocation of Peach Blossom Spring or the re-assertion of a Daoist way of life is often deemed overtly idealistic, if not hypocritical, by the critics who subscribe to the Western linear concept of modernity, Lu's evocation of Tao Yuanming and Peach Blossom Spring in the current stage of China's economic development finds alliances beyond literary circles. Many architects and social entrepreneurs have already been searching for a postmodern possibility that embraces, resurrects, and celebrates the premodern mode of being-in-the-world and aesthetic sensibility. Take architecture as an example. The landscape design of the Miho Art Museum in Japan by the renowned Chinese architect I.M. Pei (1917–2019) also evokes the image of Peach Blossom Spring. Located southeast of Kyoto in Shiga Prefecture, the museum is "hidden" in a hilly and forested landscape to evoke a sense of obscurity, hiddenness, or liminality as mentioned above. To get to the museum, visitors have to walk through a dim tunnel before they reach the open space. Here the Miho Art Museum can be deemed an example of a postmodern building that combines traditional aesthetic and ecological concepts with modern architectural technology.

Social entrepreneurs have also begun to create small-scale eco-farms or villages, with several collaborative projects being initiated. The first example is the Danping Village project in Sichuan in the wake of the Wenchuan earthquake. This rebuilding project is a collaborative project between local governments and the China Project at the Centre for Process Studies at Claremont Graduate University, California, founded by John B. Cobb Jr. and David Ray Griffin. It is a center dedicated to helping facilitate the transition to a post-industrial ecological society. The new community is now not only energy-efficient with a sustainable economy combining organic agriculture, tourism, and handcraft industries, but schools and libraries have also been built with the "aim of reviving traditional Chinese civic consciousness, practices and morality."[48] Another two examples are the Sunshine Ecovillage and the Hua Tao Ecovillage, which are communities built by environmentally minded entrepreneurs who incorporate traditional Chinese culture, including Tao's life philosophy of honorable poverty, simple life, and idleness.[49] These eco-communities, inspired by Daoism or Tao Yuanming, carry out Lu's aspiration of the lifestyle mentioned above: "high quality of life with low material consumption."[50] To create sustainable, harmonious post-GDP, Lu believes that Laozi's philosophy and Tao's espousal of a simple life are still essential for creating a high quality, low-carbon lifestyle.

Another Daoist concept, "knowing the bright but cleaving to the dark,"[51] is just as crucial. As suggested in his pen name "*qian*," which means "hidden" or to "cleave," *qian* can be viewed as necessary values in the current climate of political corruption and material hyperconsumption; it denotes a gesture of rejecting a frenzy of material and worldly possessions and a return to a simpler way of life. The above-mentioned eco-communities also materialize one more quality Lu admires in Tao's lifestyle, that is, the art of idling. The Chinese character for *xian* or "idling" (閒) is an associative compound character that consists of "door" (home) and "moon" (night), conveying the idea of "staying home, doing nothing after the night begins." *Xian* or "idling/doing nothing" is considered a higher mental or affective state. It is the opposite of the character for "busy" (忙), which contains a "heart/mind" radical with a phonetic character of "death," thus rendering an interpretation of "business" as in a state of mental/emotional death. "Idling/doing nothing" is also regarded as the source of creativity and life, from which "the myriad things take shape and rise to activity" (*Daodejing*, Chapter 16). According to Laozi, the myriad things will eventually return to the state of idleness or non-action. Here the ancient philosopher uses a plant metaphor based on his observation of the natural world to illustrate this "eternal law" or "constancy" (*chang*): "I watch them [the myriad things] fall back to their repose. Like vegetation that luxuriantly grows, but returns to the root (soil) from which it springs."[52] Laozi warns that failing to find or understand this organic "eternal law" is to court disaster" (*buzhi chang, wangzuo xiong*).[53] In keeping with the idea of "cleaving to the dark," it returns to the state of idleness, repose, or non-action. This is essential for both recuperation and survival, not only for humans, but also for the planet as a whole. Just as the land needs to lie fallow to recuperate, the same also applies to humans' laboring activities. Being busy leaves no room for emptying the mind, which is when it comes alive and bursts forth innovative ideas that are needed for building a post-industrial ecological civilization.

As the most vocal voice of China's anti-modernist, spiritual ecology, Lu has created a body of scholarship exemplifying China's native development of ecocriticism. Placing Lu in the contexts of post-Mao ideology, fast-track economic reform, and urbanization, as well as the postcolonial quest for subjectivity, provide a key to understanding why Lu considers spiritual ecology and the evocation of "Tao Yuanming spirit" as being imperative—a position that is quite at odds with state ideology, his contemporaries, and later-generation ecocritics.

2.5 Conclusion

As China's economy begins to slow down, both the Chinese people and the government are preparing for a transformation of the lived environment into a sustainable one, as seen in efforts such as cutting carbon emissions and adopting renewable resources like solar energy on a massive scale. According *China Daily*, China aims to increase forest coverage to 23% of its total landmass by the end of the decade, and this year China will plant new forests covering an area roughly the size of Ireland.[54] Pan Jiahua, director of the Institute for Urban and Environmental Studies at the Chinese Academy of Social Sciences, argues for what he calls "ecological prosperity" (*shengtai fanrong*), which is based on the traditional Chinese vision of "unison between heaven and humanity" (*tianren heyi*). Pan urges people to reject a utilitarianism that "aims at the maximization of profit, and ultimately places its priority on the accumulation of capital for the few."[55]

Westerners might not appreciate just how extraordinary it is for a social scientist like Pan, and even the CCP itself, to find common cause with a thinker like Lu. However, Lu's evocation of a premodern poet and pastoral traditions now resonates quite powerfully with many Chinese social scientists, state officials, activists, and avant-garde artists who hope to make his vision a reality. Critics of Lu and his ideas, both Western and Chinese, might say that he is asking people to return to the past. In reality, Lu has understood that for people to live in a proper future, they will have to decisively turn their back away from the one they are living in now.

Notes

1. Dillinger, "Biggest Contributors To Global Warming In The World By Country."
2. Petras, "China: Rise, Fall and Re-Emergence as a Global Power."
3. Wang, He, and Fan, "The Ecological Civilization Debate in China."
4. Ibid.
5. Ibid.
6. Ibid.
7. Ibid.
8. Jacques, *When China Rules the World*, 196–197.
9. Lu, *The Ecological Era and Classical Chinese Naturalism*, ix. The term *Shangri-la* is a British colonial invention that originated in the novel *Lost Horizon* (1933). This term then was picked up by imaginative entrepreneurs in Yunnan who adopted the term and turned Zhongdian County (中甸县) into Shangri-la (香格里拉). So *Shangri-la* has always been associated with a kind of non-Western utopia.
10. It first appeared in a paper "Literary Ecocriticism" ("*Wenxue de shengtaixue piping*") in the *Journal of Foreign Literary Criticism*.
11. Marinelli, "How to Build a 'Beautiful China' in the Anthropocene," 371.
12. The Man and the Biosphere (MAB) Programme is an intergovernmental and interdisciplinary research program. It was launched by UNESCO in 1971 to tackle the population, resources, and environment problems facing the whole world. For more information, see Anon, "The Chinese National Committee for MAB."
13. Huang Zongying and Wang Meng, for example, founded the China Environmental Literature Research Association in 1991. In 1992, this association changed its name to The Promotion Association of Chinese Environmental Culture. This association heralds the onset of ecological literature in China. The literary magazine *Green Leaves* was launched in 1992.
14. Sima Yunjie, "On the Study of Ecological Wenyixue," 12.
15. Zhang, "*Wenyi shengtai xue*," 69–74.

16. The term "cultural ecology" was coined by American anthropologist Julian Sturd (1902–1972). See "The Concept and Method of Cultural Ecology," 30–42.
17. Zhang, "*Wenyi shengtai xue*," 69–70.
18. Ibid., 73.
19. Glotfelty and Fromm, *The Ecocriticism Reader*, xviii.
20. Ibid., 73.
21. Xia, "China as a 'Civilization-State'," 43–47.
22. Lu, *The Ecological Era and Classical Chinese Naturalism*, 94.
23. Ibid.
24. Lu, "Why do contemporary Chinese poets only look Westward."
25. Wei, "Thoreau and I."
26. Ibid.
27. Seto. "What Should We Understand about Urbanization in China?"
28. Lu, *Tao Yuanming de youling*, 264.
29. Liu, *Corporate China 2.0*, xxii.
30. There were still 30.46 million rural people living below the national poverty line at the end of 2017, according to the National Bureau of Statistics. "China brings nearly 13 mln people out of poverty in 2017," *XinhuaNet*.
31. Duara, *The Crisis of Global* Modernity, 3–4.
32. Lu, *Shengtai piping de kongjian*, 45.
33. Slovic, "Landmarks in Chinese Ecocriticism and Environmental Literature."
34. For more on Lu's spiritual ecology, see Wei, "Chinese Ecocriticism in the Last Ten Years," 540–541.
35. Morton, *The Ecological Thinking*, 4.
36. Ibid., 2.
37. Heise, "Comparative Literature and the Environmental Humanities."
38. Anon, "Lu Shuyuan jiaoshou huo 2018 nian 'Kebu gongtong fuzhi jiang' fabiao ganyan."
39. Yang, "Migrant Poets are an Emerging Power."
40. Zheng, *Zheng Xiaoqiong's Blog*.
41. Lu, *Tao Yuanming de youling*, 266–267.
42. Ibid., 265.
43. Ibid.
44. Zheng, *Zheng Xiaoqiong*, 137.
45. Ibid.
46. Lu, *Tao Yuanming de youling*, 268.
47. Lu, *The Ecological Era and Classical Chinese Naturalism*, 121.
48. Wang, "The Deep Convergence between Constructive Postmodernism and Chinese Marxism," 87.
49. Cui, "Hua Dao China: Innovation for Chinese Civilization," 146–151.
50. Lu, *Shengtai piping de kongjian*, 98–102.
51. Laozi, Chapter 28.

52. Lin Yutang's 1948 translation, Terebess Asia Online, https://terebess.hu/english/tao/yutang.html
53. Ibid.
54. World Economic Forum, "China's new forests will be the size of Ireland," https://www.weforum.org/agenda/2018/01/china-is-creating-new-forests-the-size-of-ireland/
55. Marinelli, "How to Build a 'Beautiful China' in the Anthropocene," 380–381.

Bibliography

Anon. 2013. "The Chinese National Committee for MAB." *ABDC Net: Asian Biodiversity Conservation and Database Network*, October 23. http://www.abcdn.org/member/136. Accessed 30 Aug 2018.

———. 2018. "Lu Shuyuan jiaoshou huo 2018 nian 'Kebu gongtong fuzhi jiang, fabiao ganyan'" 鲁樞元教授获 2018 年 「柯布共同福祉奖, 发表感言」 ("Professor Lu Shuyuan was awarded the 'Cobb Common Welfare Award' in 2018"). *International Daily News*, May 17. http://www.chinesetoday.com/zh/article/1228483. Accessed 18 Sept 2018.

Bird, Mike. 2014. "China Just Overtook the US as The World's Largest Economy." *Business Insider*, October 8. http://www.businessinsider.com/china-overtakes-us-as-worlds-largest-economy-2014-10. Accessed 4 Mar 2018.

Chen, Guili. 1999. *Huaihe de jinggao* 淮河的警告 (*The Warning of the Huai River*). Beijing: Remmin wenxue chubanshe.

Cui, Woody. 2018. "Hua Dao China: Innovation for Chinese Civilization." In *Ecovillages Around the World: 20 Regenerative Designs for Sustainable Communities*, ed. Frederica Miller, 146–151. Rochester: Findhorn Press.

Dillinger, Jessica. "Biggest Contributors to Global Warming in the World by Country." *Worldatlas*. https://www.worldatlas.com/articles/biggest-contributors-to-global-warming-in-the-world.html. Accessed 30 Aug 2018.

Duara, Prasenjit. 2015. *The Crisis of Global Modernity: Asian Traditions and a Sustainable Future*. Cambridge: Cambridge University Press.

"Ecological Civilization Is Meaningful to China." 2012. *ChinaDaily.com*, November 19. http://www.chinadaily.com.cn/opinion/2012-11/19/content_15942603.htm. Accessed 30 Aug 2018.

Glotfelty, Cheryll, and Harold Fromm. 1996. *The Ecocriticism Reader: Landmarks in Literary Ecology*. Athens: University of Georgia Press.

Heise, Ursula K. 2014. "Comparative Literature and the Environmental Humanities." *State of the Discipline Support*, 9 Mar. https://stateofthediscipline.acla.org/entry/comparative-literature-andenvironmentalhumanities. Accessed 30 Nov. 2018.

Jacques, Martin. 2009. *When China Rules the World: The End of the Western World and the Birth of a New Global Order*. New York: Penguin Press.

Laozi. *Daodejing*. Trans. Lin, Yutang. *Terebess Asia Online*, https://terebess.hu/english/tao/yutang.html. Accessed 30 Aug 2018.

Lent, Jeremy. 2018. "What Does China's 'Ecological Civilization' Mean for Humanity's Future?" *Pattern of Meaning: Exploring the Patterns of Meaning That Shape Our World*, February 8. https://patternsofmeaning.com/2018/02/08/what-does-chinas-ecological-civilization-mean-for-humanitys-future/. Accessed 30 Aug 2018.

Liu, Junxian. 2010. "China Is the Place Mostly Likely to Achieve Ecological Civilization: An Interview with Constructive Postmodern Thinker John Cobb." *Journal of China Executive Leadership Academy Pudong* 3: 5–10.

Lu, Shuyuan 鲁枢元. 2000. *Shengtai wenyi xue* 生态文艺学(*The Studies of Ecological Literature and the Arts*). Haikou shi: Hainan University Press.

———. 2006. *Shengtai piping de kongjian* 生态批评的空间 (*The Space for Ecological Criticism*). Shanghai: East China Normal University Press.

———. 2011. *Tao Yuanming de Youling* 陶渊明的幽灵 (*The Specter of Tao Yuanming*). Shanghai: Shanghai wenyi chubanshe.

———. 2017. *The Ecological Era and Classical Chinese Naturalism: A Case Study of Tao Yuanming* (A translation of *Tao Yuanming de youling*). Trans. Xiangchun Meng. Singapore: Springer.

Marinelli, Maurizio. 2018. "How to Build a 'Beautiful China' in the Anthropocene." *Journal of Chinese Political Science* 23: 365–386.

Morton, Timothy. 2012. *The Ecological Thought*. New York: Harvard University Press.

Oswald, James. "China Turns to Ecology in Search of 'Civilization'" *Asian Currents/Asian Studies Association of Australia*, http://asaa.asn.au/china-turns-to-ecology-in-search-of-civilisation/. Accessed 30 Aug 2018.

Petras, James. 2012. "China: Rise, Fall and Re-Emergence as a Global Power." *Global Research: Center for Research on Globalization*, March 7. https://www.globalresearch.ca/china-rise-fall-and-re-emergence-as-a-global-power-2/29644. Accessed 30 Aug 2018

Rabinovitch, Simon. 2012. "Pollution Protests Force China U-Turn." *Financial Times*, October 28. https://www.ft.com/content/788bc3c4-20c9-11e2-babb-00144feabdc0. Accessed 30 Aug 2018.

Rapoza, Kenneth. 2013. "Chinas Pres Xi: GDP No Longer the Measure of Success." *Forbes*, 1 July. https://www.forbes.com/sites/kenrapoza/2013/07/01/chinas-pres-xi-gdp-no-longer-the-measure-of-success/#3cc83d518dbd. Accessed 30 Aug 2018.

Seto, Karen C. 2013. "What Should We Understand About Urbanization in China?" *Yale Insights*, November 1. https://insights.som.yale.edu/insights/what-should-we-understand-about-urbanization-in-china. Accessed 30 Aug 2018.

Sima, Yunjie 司马云杰. 1986. "Lun wenyi shengtai yanjiu" 论文艺生态研究 ("On the Study of Ecological Wenyixue"). *Wenxue pinglun* 3: 12.

Slovic, Scott. 2012. "Landmarks in Chinese Ecocriticism and Environmental Literature: The Emergence of a New Ecological Civilization." *Chinese Social Sciences Net*, October 26. http://english.cssn.cn/research/literature/201312/t20131203_892490.shtml. Accessed 30 Aug 2018.

Steward, Julian. 1955. "The Concept and Method of Cultural Ecology." In *Theory of Culture Change*, 30–42. Urbana: University of Illinois Press.

Wang, Zhihe. 2016. "The Deep Convergence Between Constructive Postmodernism and Chinese Marxism." In *Recent Advances in the Creation of a Process-Based Worldview: Human Life in Process*, ed. Łukasz Lamża and Jakub Dziadkowiec, 75–93. Newcastle upon Tyne: Cambridge Scholars Publishing.

Wang, Zhihe, He, Huili, and Fan, Meijun. 2014. "The Ecological Civilization Debate in China: The Role of Ecological Marxism and Constructive Postmodernism—Beyond the Predicament of Legislation." *Monthly Review: An Independent Socialist Magazine*, November 1. https://monthlyreview.org/2014/11/01/the-ecological-civilization-debate-in-china/. Accessed 30 Aug 2018.

Wei, Qingqi. 2006. "Opening Ecocriticism's Sino-American Dialogue: An Interview with Lawrence Buell." *Tamkang Review* 36 (3): 75–90.

———. 2014. "Chinese Ecocriticism in the Last Ten Years." In *The Oxford Handbook of Ecocriticism*, ed. Greg Garrard. New York: Oxford University Press.

Wei, An. 2017."Suoluo yu wo" 梭罗与我 ("Thoreau and I"). *Chinadialogue*, May 26. https://www.chinadialogue.net/culture/9823-Thoreau-and-I/ch. Accessed 30 Aug 2018.

World Economic Forum. 2018. "China's New Forests Will Be the Size of Ireland," January 12. https://www.weforum.org/agenda/2018/01/china-is-creating-new-forests-the-size-of-ireland/. Accessed 30 Aug 2018.

World Trade Organization. 2005. "Developing Countries' Goods Trade Share Surges to 50-year Peak." WTO: 2005 Press Release, April 15. https://www.wto.org/english/news_e/pres05_e/pr401_e.htm. Accessed 30 Aug 2018.

Xia, Guang. 2014. "China as a 'Civilization-State': A Historical and Comparative Interpretation." *Procedia – Social and Behavioral Sciences* 140: 43–47.

XinhuaNet. 2018. "China Brings Nearly 13 mln People out of Poverty in 2017." *XinhuaNet*, February 2. http://www.xinhuanet.com/english/2018-02/01/c_136942195.htm. Accessed 30 Aug 2018.

Xu, Gang. 1988. *Famuzhe, xinglai!* 伐木者醒来 (*Wake up, Woodcutters!*). Changchun shi: Jilin renmin chubanshe.

Yang, Lian. 2014. "Nongmin gong shiren shi yizhi xinxing de wenxue liliang" 农民工诗人是一支兴的文学力量 ("Migrant-Worker Poets are an Emerging Literary Power"). *China News*, October 23. http://www.chinanews.com/cul/2014/10-23/6707630.shtml. Accessed 30 Aug 2018.

Zhang, Songkui 张松樾. 1987. "Wenyi shengtai xue: yimen yunyu zhong de xinxueke" 文艺生态学:一门孕育中的新学科 ("The Studies of Literary and Artistic Ecology"). *Yishu guangjiao* 4: 69–74.

Zheng, Xiaoqiong 郑小琼. 2008. *Zheng Xiaoqiong shixuan* 郑小琼诗选 (*Poetry Selection of Zheng Xiaoqiong*). Guanzhou shi: Huancheng chubanshe.

———. *Zheng Xiaoqiong's Blog*. http://blog.sina.com.cn/xiaoqiong81. Accessed 30 Aug 2018.

CHAPTER 3

The Nakedness of Hope: Solastalgia and Soliphilia in the Writings of Yu Yue, Zhang Binglin, and Liang Shuming

Stephen Roddy

Dr. James Hansen's 1988 testimony to the US House of Representatives brought unprecedented American public attention to the specter of anthropogenic climate change.[1] Yet, as we now know from ExxonMobil documents dating back to 1957, he was hardly the first eminent scientist to conclude that industrial carbon emissions could measurably alter the earth's biosphere.[2] Almost exactly a century before him, in fact, the Swedish chemist Svante Arrhenius (1859–1927) predicted that rising levels of atmospheric CO_2 due to industrialization would eventually lead to milder winters, and thus (he hopefully predicted) greater crop yields, in northern latitudes.[3] Needless to say, Hansen's far more sober assessment of the impact of increased greenhouse gases on the web of life has since been borne out by the symptoms of imminent ecological collapse that are becoming more evident by the year, indeed by the month, in this second decade of the twenty-first century.

S. Roddy (✉)
Department of Modern and Classical Languages, University of San Francisco, San Francisco, CA, USA

C. Chang (ed.), *Chinese Environmental Humanities*, Chinese Literature and Culture in the World, https://doi.org/10.1007/978-3-030-18634-0_3

Around 1885, just before Arrhenius began to formulate his theory of the greenhouse effect, a prominent Chinese scholar expressed unease over what he had been able to glean from various sources about the rapidly accelerating use of fossil fuel in the West. From the garden-dotted, canal-laced city of Suzhou located approximately 100 km west of Shanghai, the classicist Yu Yue (1821–1907) almost uncannily foreshadowed Hansen's warnings of a century later. He predicted that the unprecedented rise in coal consumption during the industrial age had the potential to irrevocably harm the Earth, perhaps even to the point of turning it into a dead planet:

> Today, people of certain other countries…exhaust what heaven and earth have provided for our daily needs…I have heard that in those countries, they burn coal immoderately…We can extrapolate from [our experience with] this one substance to understand the principle that if we use things without limit, the day will come, no doubt, when all will be exhausted. Heaven and earth may be vast, but even they are not sufficient to supply such inordinate demands. Should this happen, the beautiful flower of our earthly plenitude will shrivel and die, leaving the world as a barren clump of dirt, without the capacity to sustain human or any other form of life.[4]

Yu may not have foreseen the exact nature of climate change, but he presciently identifies fossil fuel use as the most salient symptom of the West's failure to curb its self-destructive appetites. Nor was Yu alone; others among his contemporaries also bemoan, in poetic form, the banes of coal mining, industrial-scale production, and mass consumption. Gao Huiran (fl. 1890), a poet from Xizhi (near Taipei), decried the despoliation of the verdant hillsides of northern Taiwan by the expansion of mining operations: "Since coal mines have opened one after another, countless tea plantations have become fouled and overgrown."[5] And the Shanghai resident Yuan Zuzhi (1827–1898), a grandson of the poet Yuan Mei (1716–1797), lamented that the Western-style profligacy that was taking root in the treaty port had cast a pall over the nation and even the world as a whole:

> A million bolts of precious silk
> Cannot buy up all of the western products in this springtime town
> When will we ever go back to simpler, more authentic ways?
> The clear-eyed bystander is wracked with sorrow. (Author's Note: This tiny patch of land—the most wasteful and extravagant in all of China—brings dread to anyone concerned with the fate of the world.)[6]

Yu, Gao, and Yuan share palpable unease over the implications of accelerating industrialization, resource depletion, and urbanization in late-nineteenth-century China. Following Yu's death in 1907, however, objections to the development policies of the late-Qing court and its successor regimes, the Republic of China, 1912–1949 and the People's Republic of China, 1949–, grew fainter. As younger generations of intellectuals came to accept the presence of modern industries, and to actively advocate for their further expansion, Yu's Cassandra-like warnings receded from the public sphere.

Nevertheless, among the small number of skeptics who remained wary of the societal and environmental ramifications of industrialization, two figures in particular stand out for their warnings about the still relatively small but unsettling changes already apparent to Yu. These were the renowned scholar-philosophers Zhang Binglin (1868–1936), and Liang Shuming (1893–1988). Both men continued well into the twentieth century to express varying degrees of unease toward the transformations they observed firsthand, whether along Euro-American capitalist lines, or based on the socialist model pioneered in the USSR. Their dubiousness toward modern industrial production—primarily its cultural and sociopolitical implications, but extending to the natural environment, as well—echoes the aforementioned words of Yu Yue, not only during the early decades of the twentieth century, but even as late as Liang Shuming's final interviews of the 1970s–1980s.[7]

Read together, Yu's, Zhang's, and Liang's respective criticisms of industrial civilization collectively demonstrate a profound disquiet. I will argue below that their emotionally charged writings prefigure solastalgia, a psychosocial condition that according to its leading theorist, the Australian philosopher Glenn Albrecht, is a "form of homesickness one experiences when one is still at 'home.'" It is "a feeling of dislocation; of being undermined by forces that destroy the potential for solace to be derived from the immediate and given."[8] In 2010, Albrecht coined a matching notion, soliphilia, as a kind of antidote to the despair and dislocation of solastalgia. It is a compensatory solidarity "needed between people to overcome the alienation and disempowerment present in contemporary political decision-making. Soliphilia introduces the notion of political commitment to the saving of loved home environments at all scales, from the local to the global."[9] By exploring the parallel currents of despair and hope in the thought of these three exemplary figures, I will suggest how their writings of a century or so ago might enable us to more fully face the "normative dimensions of current environmental practices" that have led us well along the path toward the oblivion that Yu Yue feared might very well be our long-term fate.[10] By retrieving the words of these three men from their marginalized positions in the discourses of both environmental humanities and Chinese ecocriticism, thereby adding their voices to those of others within this volume attempting to "environ from the margins," we might uncover valuable clues to understanding the pathologies that continue to paralyze much of the world's humanity.[11]

3.1 The Evil Twins of Materialism and Militarism

Yu Yue was a consistently strident critic of all things occidental, but this aversion was more nuanced than some of the more doctrinaire forms of nativism voiced by his contemporaries such as Weng Tonghe (1830–1904) or Woren (1804–1871).[12] Although generally opposed to the reformists Kang Youwei (1858–1927) and Liang Qichao's (1873–1929) plan for a constitutional monarchy, he remained open to learning from Japan's hybrid political institutions that emulated the imperial German state while retaining core Confucian values.[13] The danger in the current-day Chinese proposals for reform, he averred, lay in their generally uncritical acceptance of the materialistic worldview that he believed lay at the heart of Euro-American civilization, and which was evident in the technological and military aggressiveness of foreign businesses and governments.[14] He was especially eloquent in denouncing the military technologies that the West and, later, Japan deployed in their respective quests for domination over various world regions, including in China.

To Yu, industrialization and militarism were in fact closely intertwined.[15] He spent his entire adult life either directly or indirectly under the threat of foreign assault: first, during the Opium War which forced him to flee from the British assault of Ningbo in 1841; then, the French and British invasions of 1858–1860 and 1878–1881, the Sino-Japanese War of 1894–1895, and the Boxer Rebellion and its suppression by the eight powers of 1900. He was acutely conscious of never being entirely safe from the ever-more lethal weapons deployed in these conflicts. In the twilight of his life, the Russo-Japanese War of 1904–1905 evoked an especially wrenching *cri de coeur* about the waste and destruction wrought by mechanized warfare: "Japanese and Russians are vainly fighting for territory, creating huge killing fields in rural areas and in cities, bringing great suffering to millions of living beings, and wasting vast amounts of money."[16] This despair reaches a crescendo in his final deathbed poem, where he singles out Kang Youwei, using a pun for Kang meaning "chaff," as the object of his undying ire:

> What vague resentment lurks around me still?
> I have not swept away the loose, stray chaff
> Only when I am placed in a seven-foot plain coffin beneath three feet of soil
> Will I have finished this lengthy letter of remonstrance[17]

On the cusp of the 1911 revolution, with disasters following one after another, and reformists' attempts to resuscitate the empire only further muddying the waters, Yu saw little hope of a decent future, or of any future at all. The depths of this despondency are evident in many of his writings of this period, but his metaphor of wastage left by the attempts at implanting Western institutions and ideas, read against his fears for the survival of life itself, prefigures the desolation toward loss of place that lies at the heart of solastalgia in Albrecht's formulation.[18] For Zhang Binglin and Liang Shuming, too, the benefits of industrial civilization were outweighed by its costs to the human and the natural environments, but each sought human solidarity, resistance, and adaptation—in other words, something akin to soliphilia—and through them, the revival of hope.

3.2 Revolution, Not Evolution!

The revolutionary Zhang Binglin has been called one of the two most influential thinkers of the early twentieth century (the other being Hu Shih, 1891–1962), a leading practitioner of modern classical scholarship who also played a significant role in the early reception of Western philosophy. Zhang studied Confucian classical philology under Yu Yue at the Gujing Jingshe, a Confucian academy in Hangzhou that Yu headed for thirty-one years (1868–1899). The imprint of Yu's oft-cited text of philological scholarship, *Gushu yiyi juli* (*Examples of Doubts About Ancient Writings*), is evident not only in Zhang's methodological approach to classical studies, but also in the writings of his associates in the Guocui (National Essence) Movement, Liu Shipei (1884–1919), Deng Shi (1877–1951), and Huang Jie (1873–1935).[19] This intellectual debt to his teacher weakened after Zhang turned to anti-Manchu activism in the 1890s, however, and led Yu on his part to cut off all contact with his former pupil.[20] A faint but discernible condescension is evident in Zhang's 1908 obituary for Yu, published in the movement journal *Guocuibao*.[21]

Zhang later praised the work of the philologist Sun Yirang (1848–1908) over all other contemporary scholars, thus implicitly belittling Yu.[22] Yet, although Zhang came under the sway of Kant, Hegel, Schopenhauer, and other Western thinkers, selectively employing their ideas to pioneer a syncretic approach similar to that of Wang Guowei (1877–1927) and other contemporary philosophers, his critique of the economic and political dimensions of Western modernity remained indebted to the views of Yu Yue, even in the language and metaphors he employed.

Insisting that classical Confucian learning remain at the core of the educational curriculum, and rejecting the adoption of vernacular speech as the basis for the written language, in his later years Zhang came to be labeled as a "stubborn" (*wangupai*) holdover by both the liberals and radicals who came of age during the May Fourth Movement.[23] His skepticism toward the reforms of the 1920s was grounded not only in his commitment to classical Confucianism but also in his hostility toward the Western educational and other civic institutions on which these reforms were largely modeled. Complementing this rejection was his advocacy of a pan-Asian cultural and political dialog among Buddhists, Hindus, and followers of other religions of South and East Asia. Harkening back to the Buddhist pilgrims Yijing (635–713) and Xuanzang (602–664), who exemplified the cosmopolitan openness of the Tang Dynasty, Zhang urged his fellow Chinese scholars to study contemporary Indian languages to assist in promoting solidarity between China, India, and the rest of Asia as a counterweight to European cultural and economic imperialism.[24] In both the material and the spiritual realms, Asian cultures' common heritage could serve as the glue that might help to thwart Western domination of the region, in all of its political, economic, and cultural ramifications.

This promotion of pan-Asian solidarity is also evident in various comments to the effect that Buddhism and other spiritual traditions of Asia are free from and thus superior to the unsophisticated, idolatrous, and (to his mind) patently false anthropocentrism of European Christianity.[25] Even though cultic worship of Kali, Krishna, and the pantheon of other Indian deities had continued to flourish down to the present day, he considered such popular religious practices to have been fully transcended in the sophisticated doctrines of the Vedanta School.[26] China, too, had managed to overcome the theism that he believed had inspired the early Confucian and Taoist notion of *tian* or "Heaven." Echoing some of the Sino-Babylonian theories then circulating in Japan and Europe, and also among some Chinese at the turn of the twentieth century, he (rather improbably) traces *tian* to what he calls the Zoroastrian deity "Tihuan"(actually the abbreviated form of the Chinese name for Indra, Shitihuanyin).[27] Chinese thinkers had succeeded in moving far away from such primitive beliefs by reinterpreting *tian* in more abstract, philosophical terms, he thought.

Moreover, he argued that the historical and doctrinal affinities between the major religious and philosophical traditions of Asia—Buddhism, Zoroastrianism, Hinduism, Confucianism, and Daoism—were evident in references within all of these creeds' foundational texts to the centrality of the mountains located in and around the Tibetan plateau, the geographical heart of Asia. He quotes the Han Dynasty dictionary *Shuowen jiezi* (ca. 100) to the effect that "the Middle Kingdom lies to the southeast of the Kunlun [Mountains]."[28] This statement demonstrates that even before Buddhism's entry into China, Chinese writers recognized that the world was oriented around this earthly "pivot" (*shuniu*), which corresponds to both the Buddhist Mt. Meru and the Zoroastrian Mt. Agumi. Christianity's physical and cultural distance from these points of reference makes it alien to the Asian identity shared by otherwise seemingly disparate traditions and their adherents.

Zhang's arguments for an Asia-centric alternative to Euro-American modernity led him to accord Buddhism a critical role in addressing the contemporary secular needs of Asian nations. Far from being a mere historical relic, it remained fully relevant to the present and deserved both civic and official support for its revival. He praised the China Inner Learning Academy (Zhina Neixue yuan, 1922–1952) of Nanjing as an exemplary model of what Buddhist lay movements (which he thought better suited to this task than the clergy) were capable of achieving. The Academy's simultaneous promotion of Buddhist practice and doctrinal scholarship brings "control over the mind" (*zhixin*), which could assuage the physical and spiritual turmoil of the present age.[29] Recent events in Japan were also living proof of the continuing vitality and relevance of Buddhism, he thought, especially the re-emergence of Chan (Zen) doctrines there. Even though China had certainly suffered from the excesses of Chan in the late Ming, he conceded, Chan ideals could inject dynamism in addressing worldly problems, such as how the syncretic philosopher Wang Yangming applied its lessons during his suppression of the revolt of Prince Ning (1519–1521). The successes of the Meiji era reformers, many of whom saw themselves as followers of Wang Yangming's doctrines (Yōmeigaku), illustrated the power of Chan teachings.[30] If a state religion were ever deemed necessary for China (a point of strength for Europe, as

Liang Shuming later stressed), Zhang favored elevating not Chan, however, but the Consciousness-Only School (*Weishi zong*) founded by the Tang pilgrim Xuanzang to this position.[31] Perhaps he did so in recognition of its association with this most well-traveled and arguably most cosmopolitan, pan-Asian figure among all of the Chinese Buddhist saints.

It was less Christianity, however, than some of the more recent ideological justifications for various unpalatable socioeconomic and political practices, which posed the greatest existential threats to the Sagely Culture (*shengwen*) of China. As he put it in his obituary for Yu Yue, Western ideas had infected China like maggots voraciously devouring flesh and blood, yet without replacing the indigenous cultural heritage with anything of value. They only knew how to kill their host, as parasites are wont to do. And this infestation was especially virulent because it had taken the form of *gu*, the shamanistic insect-derived poisons thought to inflict physical, mental, and spiritual harm. Zhang identifies the gu of the present as four Western concepts that had assumed a leading place in the contemporary thought of both East and West: conventions or axioms (*gongli*), evolution (*jinhua*), nature (*ziran*), and materialism (*weiwu*). Together, these four have fatally clouded the minds of those Chinese who foolishly seek accommodation with the West.[32]

Among these four poisonous ideas, Zhang devotes by far the most attention to evolution. Following the publication and wide dissemination of translations of the works of Charles Darwin and Herbert Spencer by Yan Fu (1854–1921), Social Darwinism gained a strong following among intellectuals who came of age around the turn of the twentieth century. It was under such influence, for example, that Lu Xun called for dismantling much of the Chinese social and cultural legacy, which he and others viewed as impediments to nurturing the vitality necessary to compete in the now worldwide struggle for survival among nations and peoples.[33] Needless to say, Social Darwinism provided a justification for capitalism that continues to function, albeit in disguised forms, as the ideological underpinnings for the neoliberal global order down to the present day.[34] Zhang's identification of its conceptual flaws thus remains relevant, especially since he anticipated some of its ramifications that became more fully evident in the Nazi Holocaust, forced sterilization campaigns, and recently, neoliberal economic policies.[35]

While it may be useful to explaining the organic world, Zhang says, the application to human society of the theory of biological evolution is mistaken, since this has had the unintended effect of turning humans into beasts: "Unending evolution is also a means of expanding [people's] animalistic nature."[36] For any self-respecting Confucian—or just about any Chinese, for that matter—such a result was morally abhorrent, he thought.[37] Zhang's insight extends beyond this relatively obvious one, however, to link social utilitarianism to the emergence of industrial society, and the combination of labor exploitation and environmental damage that had ensued:

> When people seek to evolve, they must serve steam-powered machines, and once they commit to serving such machines, then they must first search for the coal [with which to power them]. What humans truly need is not this. Instead of bringing suffering upon themselves in the bowels [of the earth] in their search for deferred gratification, it would be better for them to gather firewood and reeds, to sow and to reap, and thus to fill their stomachs in such a way that they can still enjoy leisure for other things...If there are those who wish to find true happiness, then they should try to sever the desire for endless gratification that has taken root in humankind. This will naturally return us to tranquility and stability.[38]

He goes on to discuss the relationship of Social Darwinism to the recently coined neologism "labor" (*laodong*). "Motion" (*dong*) is natural, but "labor" (*lao*) is emphatically not, he thought, and the latter had enabled contemporary apologists for the socioeconomic hierarchy to brainwash ordinary people into believing that most humans are destined to devote their lives to obediently serving the privileged classes:

> [Labor] is at odds with human nature, yet such thinkers force this upon naïve, uneducated people by deploying the maxim, "this is simply in your nature." Having done this, these thinkers then court the powerful, providing another maxim with which to delude the lowly: "To submit to the powerful is in your nature."[39]

It all boils down to the deceptive belief that evolution can be applied to human relations because it arises out of "nature," a word that in Zhang's formulation had mostly lost its hoary Taoist associations with spontaneity and absence of artifice. Despotic political and economic forces endeavor to force populations to accept this, because they can then condemn any recalcitrant forces that dare to challenge their authority as "deviant" (*feiyi*). Although this sort of indoctrination was also true of Confucian ideological proponents in ages past, modernity had imposed a much less stable, much more socially and morally destructive paradigm that Zhang lamented was wreaking havoc in both the human and the natural environments. And thus, human beings had to free themselves from the subterranean prisons reinforced by these newly devised ideological chains and rise back up to their natural habitat on the surface of the earth if they were ever to overcome the nefarious forces pushing them all toward self-immolation.

3.3 Life, the Supreme Good

Zhang's fondness for metaphors such as poisonous insects and maggots in describing Western ideas or practices aligns with the vituperative tone of much of his writing on these and other related topics. Liang Shuming, by contrast, did not make a bogeyman out of evolution, nor of the West as a whole, and found much to praise in some of his Western contemporaries' ideas. But more so than either Yu or Zhang, he consistently searched for alternatives to what he, too, regarded as the enslavement of the working class, and the impoverishment of the peasantry, by new forms of capitalist exploitation that originated in the West. Having come of age during the 1911 Republican revolution, when the incremental reforms of the late Qing gave way to more radical sociopolitical experiments, his own career as a journalist, academic, and social reformer testifies to the rapid transformation of intellectual roles during the early twentieth century. Even while pursuing these newfound vocational opportunities, however, he sought alternatives to Western liberalism, and later Marxist revolution, by restoring traditional values and ways of life to the greatest extent possible. And while sympathetic to socialist ideas like the abolition of private property (Liang befriended Mao Zedong in 1918 when both worked at Beida and remained his interlocutor until 1953), he rejected class struggle, which he regarded as wholly foreign and even inimical to Chinese experience. Liang argued that only by returning to the roots of Chinese social traditions, and rejecting the "Western trickery" that had deceived his countrymen, could the nation begin to restore both its physical and its spiritual health.[40] To him, that meant reviving the indigenous values of selflessness, reciprocity, and interpersonal trust, which had been eroded by the debilitating impacts of Western imperialism and the severing of the previously strong links between urban and rural society.

Although the abolition of the examination system in 1905 freed Liang and his peers from the classical curriculum taught in institutions such as Yu Yue's Confucian academy, for the rest of his life he remained fully engaged in and committed to the Confucian philosophical bedrock that underlay the old order. His writings contrast with those of the philologically trained Zhang Binglin, however, since they are cast in a colloquial prose easily accessible to even moderately educated readers. He himself only completed formal education through middle school, but his books and public lectures won him a series of prestigious academic posts and fame as a public figure during the 1920s–1940s. Though less overtly hostile to the West than either Yu or Zhang, he too rejected the application to China of Western models of development then being championed by liberal intellectuals such as the geologist and mining engineer Ding Wenjiang (1887–1936) and Hu Shih. Moreover, like Zhang Binglin, Liang immersed himself in Buddhist practices, observing a strict vegetarian diet and spiritual exercises over most of his life, and referring to his Buddhist faith as a significant aspect of his thought in many of his writings.[41]

While, as noted above, Zhang attacked the concept of "nature" (*ziran*) as deployed in the pseudo-scientific Social Darwinist discourse discussed above, Liang retrieved the positive values of "nature" as human interconnectedness with the ten-thousand things, and elevated it along with "life" (*shengming*) as the guiding lights of his thinking:

> In my way of thinking, the fundamental concepts are life and nature. I view the universe as a living entity. All things take nature as their leading principle. I share this with [Henri] Bergson and [John] Dewey [whom Liang respected above other contemporary western thinkers], both of whom were deeply inspired by the idea of life, perhaps because they were influenced by their studies in the life sciences while in the process of formulating their philosophies.[42]

As the foil for this overarching ideal of organicism, Liang denounced the "mechanical character" (*jixiexing*, a metaphor often employed by Yu and Zhang, as well) of Western societies, and against which both Dewey and Bergson had rebelled. Western medicine, for example, has "a tendency to view the human body as a composite of discrete parts, and thus to treat illnesses as if this were no different from repairing a machine. Chinese medicine, at least that which could be called learning, is permeated by the concept of life, and this is precisely what distinguishes it from western medicine."[43] Indeed, Western science as a whole regard the human mind itself "mechanically," unlike in China and India where the mind is deployed toward fully crafting the "life of the self" (*ziji shengming*) instead of exhausting its powers in the service of mathematical formulae or other "external" pursuits.[44]

Liang concedes that the tendency of Chinese medicine and other indigenous practices to examine problems holistically, without subjecting them to analytical deconstruction, had resulted in the prevalence of a metaphysical obscurantism (*xuanxue*) in this and other fields of practical endeavor. This meant that fundamental needs like feeding and housing a rising population have been neglected. In his magnum opus, *The Essential Meaning of Chinese Culture* (1949), Liang notes that by focusing their intellectual energies on contemplative pursuits, and not developing more scientifically based agriculture or other practical disciplines of knowledge, traditional literati had failed to improve the sustenance (*yang*) of society. Yet, in spite of these shortcomings, the literati's ideals of harmonious, peaceful social relations, and indeed what Liang elsewhere refers to as the overwhelmingly ethical character of Chinese culture as a whole,[45] had contributed immeasurably to humanity as a whole:

> Fundamentally, what is to be prized in the ancient Chinese [for their contribution to] the life of humanity, is how they acquired knowledge through kindness and generosity, how they embraced questions of peace and tranquility within their hearts, and then, through the difficult process of living them out in practice and fully realizing their inner natures, they came to a deeply personal understanding, so that the effort of self-cultivation overshadowed virtually all else in their lives.[46]

Offsetting his glowing account of the Chinese cultural legacy in this text are some considerably harsher statements about the "sicknesses" afflicting his nation. Published in the very same month as the founding of the PRC, *Essential Meaning* concludes with some soul-searching commentary over how it had come about that the neo-traditionalist renaissance Liang championed throughout his career was now being supplanted, seemingly with little resistance or consideration for long-term consequences, by the Marxist radicalism toward which he remained wary. Liang once again invokes the figure of "life" in addressing Chinese modernity, but its earlier youthful vigor and spontaneity had degenerated into a spiritless, "mechanical" formalism caused by the debilitating senescence (*shuailao*) to which this ancient civilization had fallen victim.[47] Perhaps these remarks applied equally to their author who, nearing his seventh decade of life, no longer could muster the energy to tackle problems to which he had earlier devoted his considerable talents. Although the book as a whole can still be described as reasonably upbeat, in its last sections this tone gives way to a sense of impotence that is less evident, if not entirely absent, in his earlier writings.

During the peak of his civic activism from the mid-1920s to the mid-1940s, Liang became a major figure in the rural reconstruction movement (*xiangcun jianshe*). Investing considerable time and purpose into various ventures, he both articulated and attempted to implement the ideals of a rural, grassroots-based path to national renewal. This came to full fruition during his stint as the director of the rural reconstruction program based in Zouping County, Shandong, in 1931–1937. During his tenure there, Zouping made demonstrable progress in raising literacy rates, improving agricultural efficiency, lowering infant mortality and communicable diseases, and achieving yet other markers of material and social betterment. By 1936, its model had been adopted by over half of the prefectures in Shandong, and was on the verge of being implemented across the entire province, when the looming threat of Japanese invasion brought it to a halt. Once the Sino-Japanese War began in earnest, Liang was forced to evacuate into the interior, and in the unsettled conditions that followed the war, he—like James Yen (Yan Yangchu, 1893–1990) and other rural reformers—was unable to fully realize the goals of the movement before the communist revolution swept them all aside.

Liang considered the rural villages to have borne by far the greatest brunt of the scourge of urban civilization that had been adopted under Western tutelage. Their political and economic disempowerment at the hands of the elites of Shanghai and other Westernized cities had deprived rural villagers of the means of improving their lives. Hence, to Liang, it was of the utmost importance to restore political autonomy and economic self-determination to rural residents, and to prevent their exploitation by predatory urban elites who had no interest in the concerns or welfare of the villages:

> Politics takes the village as the basis. If the root of all rights rests in the [rural] folk, and political power is exercised by these masses, then the power to rule will begin in the village, power cannot concentrate, and classes cannot form. The organization of the nation's villages will follow a single pattern, and though power is dispersed, this will not harm our national unity.[48]

As the scholar Ma Yong notes, Liang saw this enterprise first and foremost as cultural work, necessary to counter the insidious influence of Europe, the US, and the USSR, all of which were providing inappropriate models for rural development.[49] Although he urged the abolition of private ownership of land, and thought bourgeois capitalism "outdated," he also regarded the Russian method of collectivization of agriculture as "mechanical," and thus unnatural.[50]

Ultimately, the core of rural construction lay in spiritual renewal, through the revival of the values of reciprocity and selflessness that had been eroded by Western individualism. And this required a tempering of and control over materialism:

> What does it mean to say that people's lives improve? The improvement of human life means not considering the enjoyment of good fortune as a worthy end, but instead worrying that one's own actions might fail to conform to moral principles. It is as the ancients said: "While eating do not seek satiety, in dwelling do not seek comfort, and be sensitive to events and careful in speech; if you act thusly, you will be correct in your attainment of the Way."[51]

In this assessment of the corrosive effects of Western civilization—materially but also spiritually—on the rural landscape, we can detect continuities with Yu Yue as well as with Zhang Binglin. Yu Yue was the most prescient of these three toward the environmentally destructive potential of both industrial-scale production and warfare. His worries about planetary catastrophe stemmed in part from his conviction that the mechanization of daily life would alienate humanity from the source of its sustenance in the organic world, the "flower" of earth's beauty.[52] Zhang Binglin commented extensively on the Western philosophical and scientific currents that were making inroads into China during his lifetime, and singled out Social Darwinism as the instrument for enforcing subservience to the same sort of rapacious exploitation of natural and human resources that Yu also deplored.

Liang echoes Yu and Zhang, and more fully develops these themes in the economic realm, discussing in some detail the implications of recent technological developments then making inroads into China and across the industrialized world. Writing in the aftermath of World War I, he blames economic dislocations on the effects of industrial mechanization, and the spiritual vacuity that was its ultimate cause. "The machine is truly the demon of the modern world. But it is the direct consequence of the Westerner's attitude toward human life. On the one hand, he demands material happiness, and seeks to utilize nature to conquer nature. On the other hand, he uses his rational, analytical brain to produce science, and combines these together to invent machines."[53] It was only a matter of time before society collapsed under the onslaught of a Darwinian economic system that "violates human nature—that is, humanity [*ren*]—to the point where people can no longer endure it...Everyone has had their vitality ripped to shreds by the Westerner. His unnatural attitude, his machines, his desiccated, deadened lifestyle all spring from this source."[54]

In addition to Bergson and Dewey, Liang gives credit to some other Western thinkers, notably Rudolf Christoph Eucken (1846–1926), for having seen the light and advocated renewal through *Geistesleben* (spiritual life). Jean Jacques Rousseau also receives Liang's qualified approbation for understanding that Western thinkers "without exception strangle, dissipate, murder, corrupt, or bring chaos to life," thereby harming its vitality. Confucius had had the wisdom to foresee what was now finally coming to pass, nearly two and a half millennia after his death. He had surpassed everyone, not just Western but other Eastern thinkers as well, in being able to fulfill the needs of life.[55]

3.4 Reviving Hope, Nakedly

Solastalgia and soliphilia were coined both in and for the early twenty-first century, a time of unfolding ecological disruption amidst accelerating climate change. While the scale of environmental destruction may have been relatively smaller a century ago, the psychosocial trauma of that era and its specifically ecological dimensions are evident in each of the thinkers I have surveyed. All three exhibit signs of disorientation and despair over the rapid changes of their times that I have identified as akin to Albrecht's solastalgia, but among them, Liang Shuming articulates the most comprehensive proposals to re-establish social solidarity that could turn despair into a springboard for repairing the injuries to the political, moral, and physical landscape. And thus, surveying the arc of approximately half a century from Yu Yue's essays of the 1880s to Liang Shuming's works of the 1940s, we are tempted to discern the nascence of soliphilia amidst the arid wasteland of solastalgia.

But any hopeful narrative such as the one I have just constructed may not be justified, at least not according to Zhang Binglin. After identifying evolution as one of the intellectual poisons of his age, Zhang went on to craft his own version of this concept in an essay entitled "Ubhotabhāga

Evolutionary Theory" (*jufen jinhualun*).[56] As in his earlier discussion of the topic, he readily admits that evolution in the natural world is both logical and indisputable, but that its application to human society has been misguided. Here, however, Zhang humanizes the concept, so to speak, by pairing it with the Buddhist ideal of *ubhotabhaga-vimutto* (*jufen jietuo*), or liberation achieved alternatively through the complementary paths of insight (Skt. *pañña*) or faith (Skt. *saddhā*).[57] Arguing that evolution can magnify either good or evil, either strength or weakness, he then explains that while the human and the animal domains are equally subject to forces of competition that encourage efficiency, the final expression of this process might just as well be of greater suffering or of moral and physical decline, as of greater flourishing or more orderly societies. In the human realm, this is most certainly the case—the pursuit of fame and fortune through institutions like the examination system is one example of a perversely masochistic evolutionary vector, he notes sardonically.

Zhang affirms modern biology's assignment of humankind to the mammalian class, but he distinguishes us from all other fellow mammals for our vulnerability, our "nakedness" (*luo*). In ancient texts such as the *Classic of Ritual* (*Dadaili*, ca. 400 BCE), *Classic of Mountains and Seas* (*Shanhaijing*, ca. 400–200 BCE), and Wang Chong's (27–100) *Balanced Discussions* (*Lun Heng*), humanity was classed with earthworms and other primitive life forms as the "naked creatures" (*luochong*) that possess neither feathers, nor fur, nor scales to protect them from the elements. Zhang makes a feint toward this hoary idea, but then focuses on the intelligence that has enabled humans to overcome our physical weakness and project our power over the earth in dramatic forms. Our very vulnerability has evolved hand in hand with our intelligence, resulting in the proclivity to turn against one another in internecine, fratricidal conflicts and exploitation, such as in the current cycle of imperialist wars. Zhang is pessimistic that humanity will ever extricate itself from these vicious habits, even in the best of times. And in our own twenty-first-century moment, it is even more difficult to avoid lapsing at least episodically into solastalgic despair toward our collective futures, however hard we may try to buoy our spirits through the comfort that is still possible to find in our natural surroundings.

If the writings of these three men still have any power to inspire the present-day naked creatures of this earth, so that we might still reach equilibrium with the ever more daunting environment around us, it might be by demonstrating the importance of acknowledging the distinct possibility of failure, and the need to face this naked truth squarely and unflinchingly.

As Ian Angus has put it, "[t]he first step is to tell the truth—about the danger we face, about its causes, and about the measures that must be taken to turn back the threat. In a time of universal deceit, telling the truth is a revolutionary act."[58] Both Zhang the revolutionary and Liang the social reformer would certainly agree that it is only when reality is fully confronted that it is possible for solutions to be discovered, and Yu went even farther by daring to contemplate the end of life itself. All of them clearly saw that a merely intellectual nod to this reality falls far short of the kind of profound self-awareness and moral strength that are needed—now more desperately than ever—to slow down and eventually to halt our depredation of the biosphere. Let us plumb the depths of our grief for the harms we have wrought upon this world, and as Liang Shuming put it, come to an intensely personal understanding of ourselves and our fellow human beings so that we might begin to gain the tranquility we so obviously still lack.

Notes

1. "Global Warming Has Begun, Expert Tells Senate" *New York Times* (June 24, 1988).
2. "Brannan: Exxonmobil On Notice."
3. Svante Arrhenius, *Words in the Making*.
4. Yu Yue, "San dayou lun," in *Binmengji*, vol. 6: 10a (*Chunzaitang quanji*, 1910, vol. 4). Robert B. Marks argues that an obscure Chinese scholar-official anticipated, by several decades, Europeans' recognition of species extinction. In an official gazetteer of a county in Guangdong, this official describes the disappearance of various animals in Guangdong, implicitly recognizing the possibility of anthropogenic reasons for this. See Marks, "'People Said Extinction Was Not Possible': Two Thousand Years of Environmental Change in South Chin," in Hornborg, McNeil, and Martinez-Alier, ed., *Rethinking Environmental History*, 41–59.
5. Gong Xianzong, ed., *Taiwan zhuzhici sanbai shou*, 126.
6. Yuan Zuzhi, "Hubei zhuzhici," 2808.
7. Liang Shuming, *Wo sheng you ya yuan wujin*, 351–3; Guy Alitto, *The Last Confucian: Liang Shu-ming and the Chinese Dilemma of Modernity*, 335–50.
8. Albrecht, Glenn, "Solastalgia," 95–98; http://www.psychoterratica.com/solastalgia.html
9. Albrecht, Glenn, "Psychoterratic Conditions in a Scientific and Technological World" in Peter H. Kahn and Patricia Hasbach, eds., *Ecopsychology*, 241–65; http://www.psychoterratica.com/soliphilia.html

10. Timothy Morton expresses some of this emotional complexity in his description of the "ecological thought": "It has to do with love, loss, despair, and compassion. It has to do with depression and psychosis. It has to do with capitalism and with what might exist after capitalism. It has to do with amazement, open-mindedness, and wonder. It has to do with doubt, confusion, and skepticism. It has to do with concepts of space and time. It has to do with delight, beauty, ugliness, disgust, irony, and pain." See Morton, *The Ecological Thought*, 3.
11. The editors of the inaugural issue of *Environmental Humanities* follow the Australian philosopher Val Plumwood's identification of two central tasks for the environmental humanities: to "resituate the human within the environment, and to resituate nonhumans within cultural and ethical domains," 3. This essay similarly attempts to "resituate" intellectual and cultural history within a broadly environmental framework.
12. Fairbank, John K. ed., *The Cambridge History of China*, 233ff.
13. See Yu Yue, "Siyi nianbiao xu" in *Chunzaitang zawen xu*, vol. 2, 27–28.
14. See, for example, Yu's "Ziqiang lun," in *Binmengji*, vol. 6, pp. 5–8; also see "Xinzhai Ding Gong jiazhuan," in *Chunzaitang zawen 6 bian*, vol. 2, 10a–15a.
15. Yu Yue, "Zhishuo xia," in *Binmengji*, vol. 2, 1–7.
16. Yu Yue, "Mi bing yi," in *Binmengji*, vol. 2: 16ab.
17. Yu Yue, "Linzhong hen" in *Chunzaitang quanshi*, vol. 33: 14a.
18. Yu Yue, *Chunzaitang shibian*, vol. 20, 3.
19. Huang Jie, in particular, echoes Yu Yue in his condemnation of the Ming Christian Xu Guangqi's (1562–1633) advocacy of firearm manufacturing. See Huang, "Xu Guangqi zhuan" in *Guocui xuebao* #19 (1906), 11a.
20. Viren Murthy, *The Political Philosophy of Zhang Taiyan*, 83ff.
21. Zhang Binglin, "Yu xiansheng zhuan" in *Guocui xuebao* #44 (1908), 2a.
22. Zhang Binglin, "Sun Yirang zhuan" in *Guocui xuebao* #44 (1908), 3a.
23. Zhang Binglin, *Guoxue gailun*, 134.
24. Zhang Binglin, "Zhina Yindu lianhe zhi fa," in Zhang, *Taiyan wenlu chubian*, 40–41.
25. Zhang Binglin, "Wushen lun" in *Taiyan wenlu chubian*, 262.
26. Ibid., 260.
27. Sun Jiang and Hu Minghui, "Did the Yellow Emperor Come from Babylonia?" in Minghui Hu and Johann Elverskog, eds., *Cosmopolitanism in China, 1600–1950*, 221–261.
28. Zhang, "Da Tiezheng" in *Taiyan wenlu chubian*, 71–74.
29. Zhang, "Zhina neixin xueshe" in *Zhang Taiyan ji*, 64.
30. Ibid.
31. Zhang Binglin, "Jianli zongjiao lun" in *Taiyan wenlu chubian*, 355.

32. Zhang Binglin, "Sihuo lun" in *Zhang Taiyan ji*, 87. Zhang makes it clear in this passage that he is referring to the modern scientific meaning of the term "nature," and not to the Daoist notion of "[things] as they are" (also *ziran*).
33. See Shen Jinyao, *Lu Xun zawen shixue yanjiu*, 45ff.
34. Needless to say, Darwin's own understanding of evolution was far more nuanced than that of some of his interpreters. He reportedly adopted the phrase "survival of the fittest" reluctantly on the advice of a publisher. According to Timothy Morton,

    > [s]trict Darwinism is profoundly anti-teleological (Marx liked it for this reason). Individuals and species don't abstractly 'want' to survive to preserve their form: only macromolecular replicators 'want' that...A vast profusion of gender and sex performances can arise. As far as evolution goes, they can stay that way. Thinking otherwise is 'adaptationism.' (See Morton, 98)

35. For a brief overview of the relationship between Nazi racial theory and Social Darwinism, see "Victims of the Nazi Era: Nazi Racial Theories," in *Holocaust Encyclopedia*. On neoliberalism's debt to Social Darwinism, see Simon Springer, *Violent Neoliberalism*, 98ff.
36. Zhang Binglin, "Sihuo lun," 92.
37. Zhang followed the standard Confucian practice in maintaining a strict separation between the human and animal domains, the latter representing bestial natures toward which humankind must remain perpetually vigilant. See Michael Nylan, *The Five Confucian Classics*, 9.
38. Ibid., 94.
39. Ibid., 89. Zhang disregards some of the more positive examples of *lao* in classical texts, probably because here he is addressing its imposition on the common folk, rather than its potential for spurring heroic actions or fostering unusual talents (e.g. in *Mencius* 12.15).
40. Liang Shuming, "Cunzhi zhi ziyou" quoted in Li Qingyu, *Zai chushi yu rushi zhi jian*, 272.
41. Li Qingyu, 42, 241. Liang noted late in life that while he had at first been impressed by Zhang Binglin's Buddhist practice, he eventually concluded that Zhang's knowledge of this subject was in fact relatively superficial.
42. Liang Shuming, "Zhongxi xueshu zhi butong," in *Chushi yu rushi*, 145.
43. Ibid., 147.
44. Ibid., 147.
45. Liang Shuming, *Zhongguo wenhua zhi mingyun*, 210.
46. Liang Shuming, *Zhongguo wenhua yaoyi*, 273.
47. Ibid., 280.
48. Liang Shuming, *Xiangcun jianshe lilun*, 287.

49. Ma Yong, *Sixiang qiren Liang Shuming*, 184.
50. Alitto, 214.
51. Liang Shuming, *Xiangcun jianshe lilun*, 72.
52. Yu Yue, "Wang Ganchen Gezhi guwei xu," in *Chunzaitang zawen liubian*, vol. 7, 21.
53. Liang Shuming, *Zhongxi wenhua ji qi zhexue*, 162.
54. Ibid., 165.
55. Ibid., 180.
56. Zhang Binglin, "Jufen jinhua lun," in *Taiyan wenlu chubian*, 49–56.
57. Nyanatiloka and Nyanaponika, *Buddhist Dictionary*, 200.
58. Angus, Ian, "The Myth of 'Environmental Catastrophism.'"

Bibliography

Albrecht, Glenn. 2007. "Solastalgia: The Distress Caused by Environmental Change." *Australasian Psychiatry* 15 (1): 95–98.

———. 2012. "Psychoterratic Conditions in a Scientific and Technological World." In *Ecopsychology: Science, Totems, and the Technological Species*, ed. Peter H. Kahn and Patricia Hasbach, 241–265. Cambridge, MA: MIT Press.

———. http://www.psychoterratica.com/soliphilia.html. Accessed 3 Aug 2016.

———. http://www.psychoterratica.com/solastalgia.html. Accessed 3 Aug 2016.

Alitto, Guy. 1986. *The Last Confucian: Liang Shu-ming and the Chinese Dilemma of Modernity*. 2nd ed. Berkeley: University of California Press.

Angus, Ian. 2013. "The Myth of 'Environmental Catastrophism'." *Monthly Review* 65 (4). http://monthlyreview.org/2013/09/01/mythenvironmental-Catastrophism. Accessed 22 Aug 2016.

Arrhenius, Svante. 1908. *Worlds in the Making*. New York: Harper.

"Brannan: Exxonmobil On Notice." *Smoke and Fumes* (Center for International Environmental Law). https://www.smokeandfumes.org/fumes/moments/5. Accessed 3 Aug 2016.

Fairbank, John K., ed. 1978. *The Cambridge History of China: Late Ch'ing, 1800–1911*. Vol. 11. Cambridge: Cambridge University Press.

Gong Xianzong 龚显宗, ed. 2013. *Taiwan zhuzhici sanbai shou* 台湾竹枝词三百首 (*Three Hundred Bamboo Branch Poems of Taiwan*). Taipei: Jin'an. *Holocaust Encyclopedia*. https://www.ushmm.org/wlc/en/article.php?ModuleId=10007457. Accessed 2 Sept 2016.

Huang Jie 黄杰. 1906. "Xu Guangqi zhuan 徐光启传" (*A Biography of Xu Guangqi*). *Guocui xuebao* 19: 2–23.

Li Qingyu 李庆余. 2015. *Zai chushi yu rushi zhi jian: Liang Shuming xiansheng dui foxue de lijie yu dingwei* 在出事与入世之间: 梁漱溟先生对佛学的理解与定位 (*Between Disengagement and Engagement: Mr. Liang Shuming's Understanding and Placement of Buddhism*). Taipei: Xuesheng shuju.

Liang Shuming 梁漱溟. 1971. *Xiangcun jianshe lilun* 乡村建设理论 (*On Rural Reconstruction*). Taipei: Wenjing.

———. 2006. *Zhongxi wenhua ji qi zhexue* 中西文化及其哲学 (*The Culture and Philosophy of China and the West*). Shanghai: Shanghai Renmin chubanshe.

———. 2010. *Zhongguo wenhua zhi mingyun* 中国文化之命运 (*The Fate of Chinese Culture*). Shanghai: Zhongxin chubanshe.

———. 2011*Chushi yu rushi: Liang Shuming suibi* 出世与入世: 梁漱溟随笔 (*Disengagement and Engagement: Essays by Liang Shuming*). Beijing: Beijing daxue chubanshe.

———. 2013. *Wo sheng you ya yuan wujin: Liang Shuming zishu wenlu* 我生有涯愿无尽: 梁漱溟自述文录 (*Though My Life Will End, There Is No Limit to My Hopes: Liang Shuming's Writings About Himself*). Shanghai: Renmin chubanshe.

Ma Yong 马勇. 2008. *Sixiang qiren Liang Shuming* 思想奇人梁漱溟 (*Liang Shuming, a Remarkable Thinker*). Beijing: Beijing daxue chubanshe.

Marks, Robert. 2007. "'People Said Extinction Was Not Possible': Two Thousand Years of Environmental Change in South China." In *Rethinking Environmental History: World System Change and Global Environmental Change*, ed. Alf Hornborg, J.R. McNeil, and Joan Martinez-Alier, 41–59. Lanham: Altamira Press.

Murthy, Viren. 2011. *The Political Philosophy of Zhang Taiyan: The Resistance of Consciousness*. Leiden: Brill.

Nyanatiloka and Nyanaponika. 2004. *Buddhist Dictionary: Manual of Buddhist Terms and Doctrines*. Kandy: Buddhist Society.

Nylan, Michael. 2001. *The Five Confucian Classics*. New Haven: Yale UP.

Springer, Simon. 2015. *Violent Neoliberalism: Development, Discourse, and Dispossession in Cambodia*. New York: Palgrave Macmillan.

Sun, Jiang, and Minghui Hu. 2016. "Did the Yellow Emperor Come from Babylonia?" In *Cosmopolitanism in China, 1600–1950*, ed. Minghui Hu and Johann Elverskog. New York: Cambria Press.

Wang Liqi 王利器, Wang Shenzhi 王慎之, and Wang Zijin 王子今, ed. 2006. *Lidai zhuzhici* 历代竹枝词 (*Bamboo Branch Lyrics Through the Ages*). Xi'an: Shaanxi renmin chubanshe.

Yu Yue 俞樾. 1910. *Chunzaitang quanji* 春在堂全集 (*The Complete Works of Yu Yue*). N.p.

Zhang Binglin 章炳麟. 1991. *Taiyan wenlu chubian* 太炎文录初编 (*Zhang Binglin's Writings*), vol. 1. Shanghai: Shanghai shudian.

———. 1995. *Zhang Taiyan ji* 章太炎集 (*Collected Works of Zhang Binglin*). Beijing: Zhongguo shehui kexue.

———. 1997. *Guoxue gailun* 国学概论 (*General Discourse on National Studies*). Shanghai: Shanghai guji chubanshe.

CHAPTER 4

Blurred Centers/Margins: Ethnobotanical Healing in Writings by Ethnic Minority Women in China

Dong Isbister, Xiumei Pu, and Stephen Rachman

In this chapter, we draw on "Herbs Living in the Body" (2012) by the Tujia writer Chen Danling and "Snow Lotus" (2012) by the Hui writer Mao Mei to articulate how ethnobotanical healing allows us to address questions of margins and centers in post-socialist China.[1] We suggest that the ethnobotanical—the ways in which the vegetal world has been linked with cultural and group identity—enables discursive practices associated with therapeutics and healing that are not inherently connected with perceptions and

The authors' names are arranged alphabetically by last name. The authors contributed equally to the chapter.

D. Isbister (✉)
University of Wisconsin-Platteville, Platteville, WI, USA

X. Pu
Westminster College, Salt Lake City, UT, USA

S. Rachman
Michigan State University, East Lansing, MI, USA

C. Chang (ed.), *Chinese Environmental Humanities*,
Chinese Literature and Culture in the World,
https://doi.org/10.1007/978-3-030-18634-0_4

configurations of margins and centers. However, the ethnobotanical offers a distinct way in which it informs our understanding of margins and centers in geopolitical terms and beyond. As we see it, the ethnobotanical serves as a source of connection among the environmental, corporeal, and spiritual, and is central to processes of physical or spiritual healing, especially forms of recovery from diseases, injuries, deaths, and the effects of environmental degradation. We contend that ethnobotanical healing takes place on a continuum of multidimensional relationships and derives from culturally specific perceptions of the interconnectedness between humans, the vegetal world, and culture, and thus not only destabilizes conventional geopolitical constructions of margins and centers but also moves beyond them into an imaginary of cosmovisions. "Cosmovisions," Joni Adamson suggests, see the cosmos and biosphere as "a cosmos of relations that speak to complex entanglements of the human with the more-than-human that must be creatively and thoughtfully negotiated" (2). Cosmovisions point toward the web of life, a complex and open system where relationships among humans and the more-than-human, where boundaries of margins and centers are often slippery, shifting, and cannot be understood exclusively in geopolitical terms, even though the geopolitical is undeniably one component of this complex and open system.

The analysis of geopolitical margins and centers in the context of ethnic minorities in China has produced a familiar pattern of thought. The centralized power structures established in metropolitan and megalopolitan areas dominate smaller entities that have less political or economic power within urban centers or smaller entities that typically exist in rural peripheries. In this way, it is almost a default to view ethnic minorities as marginal—smaller in population, lesser in power, influence, and voice. Nevertheless, it may come as a surprise that the body of literature by ethnic minority women writers under consideration in this chapter does not express geopolitical relations in this default imaginary. For these writers, the center is often not some far-off metropolis or megalopolis, but something closer to home that is directly part of their own culture and deeply rooted in the environment, such as a plant or an herb. In other words, what is most central to their voices is something natural and local. In what follows, we will explore how two ethnic minority accounts of ethnobotanical healing alter what one tends to perceive as central and marginal. By focusing on the healing nexus between plants and humans, we render visible that the conventional geopolitical power structures of center and margin are altered. What might appear marginal from one perspective turns out to be central, and what seems central is marginal. More importantly, the central and the marginal, like the *yin* and *yang* energy flows, are dynamic and in a constant state of flux. A given situation in a given time and space may be simultaneously central, marginal, and a turning point.

Our perceptions of margins and centers diverge from the kinds of oppositional theoretical frameworks usually found in existing scholarship concerning questions of geopolitical margins and centers pertinent to ethnic minorities in China. A significant number of scholars of Chinese ethnic minority studies in China and the United States display a tendency to use the lens of centrism or elitism (such as Euro-American ethnocentrism, Han ethnocentrism, and Han elitism) in their articulations of margins and centers.

In his article "Triple Oppressions of Chinese Ethnic Minority Literary Studies" (2005), Cao Shunqing points out that the construction of ethnic minority literary history, classification of genres, and criticism of ethnic minority literature has been under triple oppressions of the three hegemonic discourses of Euro-American ethnocentrism, Han ethnocentrism, and Han elitism. He argues that using established standards in these three discourses to understand, classify, and evaluate Chinese ethnic minority literature has led to its misunderstanding, distortion, silencing, erasure, or appropriation. Cao calls for scrutiny of the triple oppressions in order to create a healthy ethnic literary ecology that simultaneously allows for the coexistence of differences, complementarity, and hybridity. Other scholars in China share Cao's perceptions (Li 2007; Xu 2009). As such, for many years now, the marginalization argument has gained visibility and prevalence in Chinese literary circles.

The marginalization argument in China, however, is not without its critics. In "Understanding from other perspectives: the marginalization of the research on Ethnic Minority Literature" (2014), Li Yuntuan calls this approach into question, positing that arguments about marginalization require careful historical contextualization. Although it is true that marginalization happens within the parameters of constructing an account of the general history of Chinese literature, it is also important to acknowledge the complexities and challenges of writing Chinese ethnic literary histories—a practice that formally began in 1958.[2] Li identifies three factors contributing to the marginalization of Chinese ethnic literary histories within the general history of Chinese literature: (1) the uneven development of literature among the fifty-six ethnic groups, (2) language differences, and (3) literary standards in terms of content and periodization.

However, Li's main point is that there is an entirely different way to approach the question of marginalization. Rather than placing and understanding Chinese ethnic minority literary history within the general history, which logically lends to the perception of marginalization, it is more realistic and meaningful to disentangle the two and recognize the coexistence and plurality of literary trajectories and histories among different ethnic groups.

Unlike Cao and Li, whose inquiries are grounded in ethnic minority literary history, Ralph Litzinger examines the ways in which cultural politics in post-socialist China have created a margin within the margin, the de facto (if not intended) dyad of center and margin between ethnic minority intellectuals and the *zhishifenzi* social class who themselves are marginalized in China's polity. The *zhishifenzi* social class that Litzinger scrutinizes refers to a particular group of Chinese intellectuals including "academics, novelists, government officials, and policy-makers, all of whom are typically assumed to be located in Beijing, Shanghai, Guangdong, and other metropolitan centers" (26). This group of cultural and power elites has gained visibility in Europe and the United States for their participation in the Tiananmen Square Protests of 1989. While criticizing the Cultural Revolution, they also doubt "the ability of Western theory to question the existing ways in which China is perceived and understood" (35). As a way out, some seek to construct new notions of Chinese cultural identity and national subjectivity in the new global order. Litzinger argues that, although this elite group of Chinese intellectuals have opened up cultural spaces in which competing Chinese realities and situations are produced and contested, they have also engendered erasures of "the politics of cultural, economic, and political struggle in many of China's minority regions" (36). Litzinger proposes a politics of "writing an alternative history of postsocialism by way of the ethnic margins," asking "how does one rip the concept of marginality away from the writings and ruminations of the Han intellectual elite and extend it into a broader and more heterogeneous field of cultural and ethnic struggle?" (36).

U.S. anthropologist Susan D. Blum approaches questions of margins and centers through a comparative and wide-ranging lens. In a review essay of seven ethnographic studies from U.S. academic presses published around the turn of the twenty-first century, Blum observes that Western anthropologists and Chinese ethnic minorities themselves actually differ in their attitudes toward the perceived center (the nation-state).[3] Whereas Western anthropologists are typically highly critical of the power disparity between what has been assumed to be central (the nation-state and its modernization policies) and marginal (ethnic minorities), Chinese scholars and members of ethnic minorities themselves often express differing attitudes. Blum notes that the rise of global ethnic studies in the 1990s has raised questions about the extent to which the models of ethnic relations from one national context might be applied to the ethnic relations of another nation. She cautions that "ethnicity was in the air worldwide, but it was not easily understood by simple extrapolation from familiar American experience" (1299). Blum's arguments underscore our contention that the emergence of ethnicity in China—down to the ways that ethnic groups were identified and assigned—has a very different history than that found in the United States and generalizing along lines that use the "American experience" as a default reference point runs risks of simplification and distortion.

Blum further suggests that "Ethnicity is not the only important marker of identity. It intersects in complex ways with class and locality" (1301). One might add gender and the environment to this intersection and our sense of the ethnobotanical is precisely one of these mixed markers in the exploration and expression of ethnic minority identities within specific localities. In an attempt to account for the vogue of center/margin analysis, Blum speculates that "Perhaps this tension between center and periphery, the local and the national, operates in China in ways that intrigue Westerners who have a different relationship to their own localism" (1301). Blum explains that Western scholars of Chinese ethnic cultures have "so internalized our own biological notion of race that when confronted with a non-biological system [as one finds in China], we must point out, over and over, its irrationality" (1302). So, while centralized state power and its policies are an ongoing part of the story of how minorities in China express and understand themselves, it is just one piece of the puzzle, and Western preoccupations with polarities of margins and centers is very much a matter of Western and, often, U.S. concerns more than a matter of how these are experienced in China by minority groups themselves.

While we acknowledge a perceived boundary between center and margin as portrayed by these authors discussed above, we suggest an analytical framework in which the body of environmental literature analyzed in this chapter actually destabilizes these now familiar conceptions of margins and centers. Given that these works were written against a backdrop of China's rapid political and economic rise and all the transformations and dislocations within China's political boundaries as a result of that rise, it is inevitable that the voices of the ethnic minority women writers will reflect these shifts on scales both large and small and how centers and margins have been altered or blurred.

Where one locates center and margin depends upon location and interest. In "Eternal Spirits," for example, the Zhuang author Cen Xianqing reflects on her visit to the rock paintings in Hua Shan, elicits from her a cry of the heart at the recognition that, through her modern Chinese existence in the city, she has felt disconnected from her ancestral home and the long history of her people. Only through an imaginative act of recovery, one that we describe as "ecomemory," can she express a longing for a lost centrality for the Zhuang people—the landscape and the natural and cultural histories inscribed on it.[4] In another example, "The Little Reindeer," by the Evenki writer Dekeli, she describes the disconnection of this Northern Chinese reindeer-herding ethnic group from its roots after being resettled near a city as part of the state government mandated ecological migration program in 2003. In these instances, one senses tensions among several central and marginal forces. First, there is perhaps the overarching sense of the centralized state forcing its authority on ethnic minority groups. The Evenki situation seemingly reflects a paradigmatic form that Ralph Litzinger identifies in colonial or provincial contexts. Litzinger writes:

> Culture, in both colonial and early modernist anthropology, was thus almost always situated in the realm of custom, festival, and ritual, all of which were seen to be outside the historical problematic of Western modernity. Moreover, these various modes of knowing and classifying the other were often linked to practices of development that promised to bring civility, education, modernity, and progress to the non-European periphery. (44–45)

The relocation of the Evenki from their traditional homelands for the purposes of, in part, a state ecological migration program that has impacted many ethnic groups, including the largest Han group,[5] appears to replay this power dynamic on a domestic level, but to see it this way risks ignoring many complexities of ecological migration in China. For one thing, the relocation of the Evenki also involved other ethnic groups in the same region, so any binary understanding of the state ecological migration strategies is limited. Similarly, the Zhuang example does not so easily conform to the same paradigm. The forces of migration and the deep past of an ethnic minority are certainly in play, but the borderline between center and periphery remains somewhat obscure.

Furthermore, because ethnicity is a given and thus central to the narration in these texts, the conventionally conceptualized dyad of margin (ethnic minorities) and center (the majority Han culture) is often less important than the colloquy of multiple voices that have constructed culturally specific centralities for themselves. More importantly, to place this body of literature by ethnic minority women writers at a site of marginality ignores the complex interplay of ethnic histories in China. Several considerations are worth noting here: (1) close interethnic relationships have historically been established on many different levels in China; (2) the same ethnic group can also be widely distributed across the country, such as the Hui people, and have diverse connections with other ethnic groups and the natural environment; and (3) interethnic mix is common among different ethnic groups. The same ethnic group, if living in different regions in China, can also have close genetic relationships with the other groups in the same geographical location. For example, "the Chaoshan She population had close genetic relationships with Chaoshan Hakka, Chaoshanese, and southern minorities, especially the Tujia and Gaoshan nationalities" (Liu et al. 289). Given these considerations, we suggest that classifying ethnic minority groups as a single and homogenous entity over and against the largest Han group cannot do justice to a wide range of ethnic diversity, differences, and dynamics. Each ethnic group is culturally, linguistically, historically, and/or religiously established with its own genealogy and simultaneously remains diverse within its community and complexly integrated with other ethnic groups.

We might take up the two ethnic groups of the authors we are discussing to illustrate this complexity. Chen Danling, author of "Herbs Living in the Body," lives in Guizhou, a province with a significant Tujia population. The majority of the Tujia people—they refer themselves to Bizika (native dwellers)—live in the Wuling Mountain Ranges in Hunan and Hubei provinces in Central China. Some also live in Guizhou, Guangdong, Zhejiang, Sichuan, and Fujian provinces in Southern China. The Tujia living in remote mountainous areas speak Tujia, a Tibeto-Burman branch of Sino-Tibetan languages, whereas the majority speak local Chinese dialects, or Miao, an ethnic language spoken by the Miao people. The Tujia people practice Taoism, ancestor worship, and shamanism. Their religious totems include the turtle and the white tiger, which is believed to be the Tujia ancestor. From this sketch, it is evident that even within this so-called unified ethnic minority, there is great linguistic, religious, and geographical diversity.

Similarly, Mao Mei, the Hui author of "Snow Lotus," is part of a complex and diverse ethnic community, and like the Tujia, the Hui are internally diverse—that is to say, without respect to the Han majority. The Hui people are "the descendants of Persians, Central Asians, and Arab traders on their paternal side, but they are also the descendants of the Chinese on their maternal side" (Chang 77). They are unequivocally Muslims living in almost every region in China, from the Northwest to the Southwest, and from the Southeast to the Northeast. While religion connects the Hui people irrespective of their whereabouts, they are "far more integrated into mainstream Chinese life than those ethnic minorities living in China's borderlands" (Beech 2014). They speak Mandarin Chinese, some other ethnic languages, or local Chinese dialects, which means the Hui people do not have an easily identifiable ethnic spoken language of their own, even though some Persian or Arabic expressions are still used in religious practices or daily conversations. Unlike the Uyghur, the other ethnic minority group in China that is predominantly Muslim, the Hui people do not have their own written language.[6]

The diversity and complexity of these ethnic groups give a strong indication that a center/margin discourse is equally complex, and presuming that national political binaries are the most relevant might be misguided. That is to say, the blurring of centers and margins has produced an array of shifting binary relations of equal importance, be that of the human/nonhuman, state/ethnic minorities, nature/culture, the wild/domesticated, or rural/urban. It is incumbent upon us to do justice and pay attention to all of them. To think of this body of literature in light of center/margin is only useful when we examine the ways in which it blurs the center/margin boundary, not by moving ethnic experiences in relation to the environment to the center or front of narration, but by challenging the conventionally perceived binary of center/margin through articulations of ethnobotanical healing.

4.1 Allegorical Marginality: Herbs as Agents of Healing

Chen Danling's prose "Herbs Living in the Body" recounts the human-vegetal bond established through using local plants, such as *yueyuehong* (roses), *qibulian* (lotus), *huibaolu* (puffball mushroom) to treat pain, wound, or snake bites. The author, a native of Yinjiang Tujia and Miao Autonomous County in Guizhou Province, Southwest China, shares her memories of three incidents in which these plants were used to treat cramps during her menarche initially and regular menstrual cramps afterwards, the snake bites on her mother's right thumb, and her grandfather's foot injury. The prescription for menstrual cramps by Da Gong, the village herbalist who is her grandfather's elder brother, was to have her family grow a rosebush, wait until flowers bloom, pick some flower petals, boil them with two egg yolks, and eat them. When her mother got snakebites on the right thumb one night, one of her aunts found one lotus, chewed it, and applied it to the wound so the venom could be dissolved and her mother's life could be saved. Many years later, puffball mushroom was gathered and applied to her grandfather's injured foot for healing purposes. When telling the three stories, Chen frequently suggests that the plants not merely sacrifice their lives to save human life, but they achieve a form of reincarnated life in those they have saved. In her own case, roses sacrifice a life for a teenage girl by silently living inside her body, and the village naturally becomes "a place where a rose bush and I lived" (102). Recovering from a wound, her mother "lived on" after "the herb was gone" (103). The grandfather's injured foot was cured with puffball mushroom that was "engaged in a life-cycle within him: giving their lives to help him go on living" (103).

Here, we emphasize that the ingestion of healing agents is conceived as radically different from being merely metabolized or entering the bloodstream with pharmaceutical properties. Instead, a symbiosis, or perhaps spiritual symbiosis, occurs in which the herbal agent lives on within the human body, housed or contained, as it were. Perceived as having an alternative form of life, the herbal agent is reincarnated now in animal terms by having a heart and a voice. In this sense, the plants have become the *being* instead of being evaluated simply by their usefulness to humans, and their "non-verbal, ecological, and corporeal voice" is "the manifestation of vegetal presence and human recognition of it through our capacities for taste, smell, touch, and proprioception" (Gagliano et al. xxxiv). In this way, the ethnobotanical, the nonhuman life, expresses itself within the human body, that is to say the healing agent cannot be viewed as an imprisoned object in the body and thus marked as marginal. Instead, the healing agent and the body are unified as one entity, with the medicinal herb now living in the body.

If read from a conventional lens of margins and centers, Chen's prose seems to portray what falls on the margin perceptually, spatially, temporarily, and experientially: an old herbal medicine store standing in a small village, medicinal herbs growing in the corners of vegetable gardens, daisies blooming on stone walls, wooden walls receding to the background (the margin of time, so to speak), and an ordinary village girl experiencing cramps during her menarche. These conventionally perceived margins, however, are the centers where Chen anchors her stories and memories, and where healing happens. In Chen's narration, what stands out in her consciousness are the old medicine cabinet and Da Gong, owner of the village herbal medicine store—symbols and sources of the Tujia ethnobotanical tradition of healing. For her, "whenever people suffer from wounds and pain, they can find a cure in this old herbal medicine" (102). In her heart, "the old herbal medicine store and the herbal scents from Da Gong represent the kind of unique classic makings of Chinese culture. Peaceful, grand, expansive, soul-nourishing" (102).

In Chen's narration, connection with the vegetal world is also transgenerational, transcending time and generation. To her father, his small yard is a beloved world where he grows roses, hyacinth, *shechuangzi* (cnidium), and pumpkin. When his granddaughter was one month old, he planted an orange tree in the yard. He said, "it will grow with my granddaughter, and she will have a long life like the tree" (103). In the planting season, it is in the yard where he demonstrates his love for vegetables, flowers, grass, and the land. In this way, he transmits through his own example a kind of stewardship, as well as his love, to the third generation of the family. Vegetables, grass, flowers, sweats dropping on blades of grass, laughter, the grandchildren's hands in his hand—all of these merge to form a picture of a transgenerational relationship with the vegetal world. Year after year, the vegetal world exists in the background, greening the village. It is a world that is simple and hidden but dear to the villagers and has shaped their attitudes to life: patient, honest, peaceful, generous, and easy. In this sense, Chen's narration adds a temporal dimension to disrupt the binary of margins and centers, which is often perceived in terms of space and thus demarcated by spatial boundaries or borders.

Although Chen's prose depicts her personal experience and a family story of healing, it taps into a particular ethnobotanical healing tradition of the Tujia ethnic group whose ancestral home is the Wuling Mountain Range in central China, an area rich in medicinal herbs and unique in medicinal practices. Tujia people's practice of shamanism, ancestor worship, and Taoism has undoubtedly shaped and informed how they connect with plants such as herbs. Put in the context of shamanistic tradition, the herbs, like all other natural earthly entities, actively participate in human well-being, possess spirits, and forge intimate relationships with humans. Taoism holds a similar view of the human-vegetal relationship and aims to foster respectful and harmonious connections for the energy of life. Chen's prose evokes this rich tradition and reflects on the role of medicinal herbs in physical and spiritual healing in changing social and natural environments. The common Tujia practice that Chen describes of using local plants for healing purposes is more than a search for remedy of human ailments (though that is certainly a benefit), it is connected with the shared destiny of the plants themselves. The plants and humans who harvest and use them are bound together in a coevolutionary partnership. Dorothy Abram and Laura Mernoff argue in *Seven Scents: Healing and the Aromatic Imagination* (2017):

> the interrelationship between humans and plants has been a core aspect of their evolutions, each depending on the other for their growth, reproduction, survival, and flourishing. More than simply providing food for humans, plants have also satisfied the human desire for transcendence—that is, to experience worlds of healing, wonder, and divinity beyond the limitations and restrictions of ordinary perception. (11)

The Tujia people's ethnopractice of herbal healing, as Chen explicates in the essay, not only resonates with a call for the reconfiguration of interspecies relationships but also reiterates the Tujia belief that the lives of plants and humans are mutually sustained and contained by varying forms of life. Chen's prose brings this body of knowledge into her literary imagination to create a healing narrative, calling readers to pay attention to the otherwise overlooked vegetal world of healing agents, such as those Chinese roses in the corner of the front yard or medicinal herbs growing on the roadside. For Chen, these herbs are spirits that find their reincarnation into the world, quietly offering their care and companionship and even sacrificing their lives to save another living being. Chen reminds readers that the small herbs effect healing in a hidden and secret way, but it does not mean they do not exist; they often grow in undisturbed corners, but it does not mean they are marginal. They exist in terms of their own intrinsic value as *beings* and have special value to non-vegetal beings. They are, as portrayed in Chen's prose, central to this ethnic author's literary imagination and a cure for "the worm-eaten cabbage that is the human heart" (103).

4.2 Cosmic Transcendence: "Snow Lotus"

While "Herbs Living in the Body" portrays the deep human-vegetal bond in the Tujia culture developed in Southwest China, "Snow Lotus," written by the Hui author Mao Mei, takes readers to a different locale, offering another example of how the human-vegetal bond blurs conventionally perceived center and margin and thus functions as a means to connect with one's geographically specific cultural identity, healing, and sense of belonging. Mao opens her prose with two tales—one of Narcissus in the Greek mythology, who "fell in love with his own reflection in the river and lingered by the water's edge until he ultimately fell in and drowned" (29) and the other of Kazakh lore about punishment. In the Kazakh lore, a fairy, who assumed responsibilities to watch and protect snow lotuses atop the snow-capped peaks, broke the heavenly decree by being talked into giving away a snow lotus to a Kazakh young man and was "condemned to spend eternity on that icy summit," and "her tears came to form, in this way, the vast, sweeping glaciers" on which the snow lotus grows (30). Mao tells her readers the famous story of Narcissus as a caution to "avoid a similar fate on these snowy plains; avoid vanity, avoid my omnipresent image, and avoid these countless glassy surfaces" even if temptations and perceived magnificence of nature are irresistible (29). By evading the narcissistic self, or the narrow self, Mao directs readers' attention to the myriad faces of the self (in the image of the rare Tianshan snow lotus): the ecological, the mythical, the cultural, the poetic, the spiritual, the therapeutic, and the cosmic.[7]

The rare Tianshan snow lotus grows in extreme weather conditions and blooms every three to five years on the northern slope of Glacier No. 1 at the Headwaters of Urumqi River, Tianshan—the Heavenly Mountain—in Xinjiang Uyghur Autonomous Region of Northwestern China. The ethnic minority groups in the region have used it as herbal medicines for many generations to help prolong life and treat medical conditions such as rheumatoid arthritis and menstrual irregularities. The snow lotus, however, has experienced uncontrollable demands from tourists and businesses in post-socialist China, and become scarce to gather over the years. This has inevitably created concerns about its extinction in the near future.

Mao Mei is a native from Changji Hui Autonomous Prefecture in central Xinjiang, where the Tianshan range stretches and twelve ethnic minority groups inhabit, including Daur, Mongolian, Kazak, Tatar, Uyghur, and Xibo. Islam is the dominant religion of six ethnic minority groups, followed by Lamaism, Shamanism, or Christianity for the other groups. These groups have shared natural environment and resources and most of them have established similar connections with the vegetal world around them and, in particular, the Tianshan snow lotus. Therefore, the cultural and spiritual meanings of the snow lotus are polyethnic in the region, and Mao Mei's prose is situated in a larger geo-ethnic context and is cosmically transcendent. The polyethnic nature of the prose complicates the notion of marginality as if there were a clear-cut boundary between the margin and the center. The region in which the Tianshan snow lotus grows is inhabited by a range of ethnically diverse populations, and these groups can be construed as more or less equally marginal from the regional and national political centers of Urumqi and Beijing, to say the least. This region is also religiously diverse to the extent that to single out one religious minority or ethnic group as more marginal than another would be to engage in perhaps an arbitrary form of marginality. Nevertheless, if we were to map this same region ethnobotanically in terms of the habitat of the Tianshan snow lotus, we might observe a less noticeable center. The habitat of the plant as a healing agent connects multiple ethnic groups across differences in language, ethnicity, religion, and ideology. It serves as a center for physical and spiritual healing amidst the politically and economically marginal in this sense. It is here that neither ethnicity nor natural environment is necessarily marginal, as each is simultaneously the central and the marginal in its own right.

The perceived geopolitical margin—the north slope of Glacier No. 1—in Mao Mei's narrative is a place "without any rulers, laws or people" (31). As such, it is actually a poetic place where love stories happen, where Buddhas reside, and where poets find inspiration. The snow lotus, an embodiment of what grows at the margin, in Mao Mei's eyes, is "the epitome of beauty in bloom; the very evidence of its existence supports those in hardship, giving them the strength to pull through and live meaningful lives" (31). By celebrating the spirit of the snow lotus, "a hermit in seclusion – humble, yet high above everything" (31), Mao Mei's prose suggests that the margin has its own value and may even represent a symbolic state of being to which one may actually aspire:

> Only when you are on the same latitude, on the same frequency, and in the same climate as a given thing, can you possibly hear the same thing it hears, see the same thing it sees, take the same form it takes, have the same thought it has, and adopt the same course of action it adopts. (30–31)

4.3 Conclusion

What lies in the heart of "Herbs Living in the Body" and "The Snow Lotus" is a profound sense of healing informed by ethnobotanical knowledge and wisdom. Healing is achieved through the human-vegetal bond that allows the transformation of therapeutic energy among different forms of life. The human-vegetal bond is premised on ethnobotanical beliefs that the vegetal world, like the human world, is intelligent. Healing, built upon this bond, is intertwined with a cosmic or an ecospiritual identity that simultaneously encompasses and moves beyond any single axis of constrained or rigid margins and centers. An ecospiritual identity may be culturally specific in these texts, but it suggests a larger identity that is vital to an increasing global awakening to transcend narrow individualism and anthropocentrism. Through an interpretative analysis of "Herbs Living in the Body" and "Snow Lotus," we propose to conceptualize a discourse that moves beyond a troubling dichotomous structure of center and margin, which inevitably privileges "the dominating hegemonic cultural system" (Keating 90). We contend that margins and centers cannot be reduced to one of oppositional relationships, and invite readers to consider transformational ways of restoring relationships between what is perceived to be on the margin and/or in the center.

Notes

1. There are fifty-six officially recognized ethnic groups in China, with the Han being the largest group in terms of population and political, economic, and cultural power and influence. "Herbs Living in the Body" and "Snow Lotus" are part of an anthology we started to edit and translate in 2014. It features mostly recent works (poetry, prose, and short stories published between 2012 and 2015) of new, emerging, or established women writers from thirteen ethnic minority groups in China, highlighting those from ethnicities even less visible than those from more noticeable areas, such as Tibet and Taiwan. Altogether, this body of literature gives voice to a panoply of diverse ethnic perspectives on the challenges and opportunities of post-socialist China. We use "post-socialist China" broadly to encompass a range of historical, political, economic, and cultural meanings that have been articulated by historians and theorists of sinology and intellectuals in China since the 1980s. For detailed theoretical discussions of post-socialism, see Dirlik and, also Litzinger.
2. In 1958, the Publicity Department of the Communist Party of China organized a symposium in Beijing to discuss the task of writing ethnic minority literary history and literature.

3. For a list of the seven books and Blum's commentary, see "Margins and Centers: A Decade of Publishing on China's Ethnic Minorities," *The Journal of Asian Studies*. 61.4. November 2002.1287.
4. For a detailed discussion of ecomemory, see "(Re)connecting People and the Land: Ecomemory in Environmental Writings by Ethnic Minority Women Writers in China." *English Language Notes*. 55. 1–2 Spring/Fall 2017. 135–142.
5. Beginning 1983, the ecological migration program in China has undergone three phases and relocated millions of Chinese residents, mostly in the rural area, to address issues of poverty, degrading environment, deforestation, and imbalanced access to resources. The majority of relocated residents are from ethnic minority groups living in different provinces and ethnic autonomous regions. For the Evenki people, the relocation put the reindeer at risk, as they were used to a specific type of moss in the forest and were sick or died as a result of the relocation. The state government adjusted the migration plan and allowed the Evenki people to herd reindeer in the forest by living in the new settlement location and up in the mountains so they can watch their reindeer.
6. Several American scholars who study ethnic minorities in China have published about the complexity of Chinese ethnic languages and the ambiguity of ethnic groups' attitudes to Chinese and Han education. See David Bradley "Language Policy for the Yi." Martin Schonenhal. "Education and Ethnicity among the Liangshan Yi."
7. Snow lotus, a plant usually seen at over 12,000 feet above the sea level, also grows in Tibet, Qinghai, and Sichuan. It is Saussurea involucrata or saussurea laniceps in Latin and *Tagilis* in the Uyghur language meaning "king of the plant kingdom." Because of human-induced disturbances and climate change, the snow lotus has become scarce and is facing threat of extinction.

Bibliography

Abram, Dorothy, and Laura Mernoff. 2017. *Seven Scents: Healing and the Aromatic Imagination*. Philadelphia: Singing Dragon.

Adamson, Joni, and Salma Monani. 2017. "Introduction: Cosmovisions, Ecocriticism, and Indigenous Studies." In *Ecocriticism and Indigenous Studies: Conversations from Earth to Cosmos*, ed. Salma Monani and Joni Adamson. New York: Routledge.

Beech, Hannah. 2014. "If China Is Anti-Islam, Why Are These Chinese Muslims Enjoying a Faith Revival?" http://time.com/3099950/china-muslim-hui-xinjiang-uighur-islam/

Blum, Susan D. 2002. "Margins and Centers: A Decade of Publishing on China's Ethnic Minorities." *The Journal of Asian Studies* 63 (4): 1287–1310.

Cao, Shunqing. 2005. "Triple Oppressions of Chinese Ethnic Minority Literary Studies." *Studies of Ethnic Literature* (3): 5–10.

Cen, Xianqing. "Eternal Spirits." http://www.chinawriter.com.cn/2003/2003-08-14/46290.html

Chang, Hajji Yusuf. 1987. "The Hui (Muslim) Minority in China: An Historical Overview." *Journal of Muslim Minority Affairs* 8 (1): 62–78.

Chen, Danling. 2012. "Herbs Living in the Body." *National Minority Literature* (12): 101–107.

Dekeli, 2011. "The Little Reindeer Story." *National Minority Literature* (3): 78–80.

Dirlik, Arif. 1989. "Postsocialism? Reflections on Socialism with Chinese Characteristics." *Critical Asian Studies* 21 (1): 33–44.

Gagliano, Monica, et al., eds. 2017. *The Language of Plants: Science, Philosophy, Literature*. Minneapolis: University of Minnesota Press.

Keating, AnaLouise. 2013. *Transformation Now!: Toward a Post-Oppositional Politics of Change*. Chicago: University of Illinois Press.

Li, Xiaofeng. 2007. "Multiethnic Literature: Absence in Chinese Literary History." *Studies of Ethnic Literature* (3): 10–16.

Li, Yuntuan. 2014. "Understanding from Other Perspectives: The Marginalization of the Research on Ethnic Minority Literature." *Journal of Guangxi University for Nationalities* (Philosophy and Social Science Edition). 36(5): 174–178.

Litzinger, Ralph A. 2002. "Theorizing Postsocialism: Reflections on the Politics of Marginality in Contemporary China." *The South Atlantic Quarterly* 101 (1): 33–55.

Liu, Shuhui, et al. 2017. "Patrilineal Background of the She Minority Population from Chaoshan Fenghuang Mountain, an Isolated Mountain Region, in China." *Genomics* 109 (3–4): 284–289.

Mao, Mei. 2012. "Snow Lotus." *Literature of the Hui Nationality* (1): 29–31.

Xu, Qichao. 2009. "Absence and Construction: Perspectives on Literary History and Ethnic Minority Literature." *Studies of Ethnic Literature* (2): 65–70.

Yuan, Zerui. 2017. "Analysis on the Eco-Ethics Ideas of She Ethnic Group's Folk Religion." *Journal of Lishui University* 39 (4): 60–65.

CHAPTER 5

From Jiang Rong to Jean-Jacques Annaud: An Ecological Rewrite of *Wolf Totem*

Runlei Zhai

In an interview right before the release of *Wolf Totem* (*Lang tuteng*, 2015) in mainland China, Jean-Jacques Annaud pointed out that he decided to make the novel *Wolf Totem* into a film when he read less than sixty pages of its French version. "The words in the novel made me feel that the author, Jiang Rong, speaks right to my heart."[1] It is widely known that *Wolf Totem* is a 2004 semi-autobiographical novel by Chinese author Lü Jiamin under the pen name Jiang Rong. Based on his experience as an educated youth sent for "re-education" on the grasslands of Inner Mongolia during the Cultural Revolution, Jiang wrote the novel to discuss a series of ecological, social, cultural, and political problems in modern China. To explain why he has the passion to film this Chinese story, Annaud said, "I was really excited when I read this novel. It seems that I was instantly taken to an exotic, epic-like land. I was also moved by the warm and deep emotions between this young man and the little wolf. From a deeper perspective, all these reflect a global topic, which is the balance between man and nature."[2] Obviously, the attention to nature and its

R. Zhai (✉)
Department of Foreign Languages and Literatures, North Carolina State University, Raleigh, NC, USA

C. Chang (ed.), *Chinese Environmental Humanities*,
Chinese Literature and Culture in the World,
https://doi.org/10.1007/978-3-030-18634-0_5

relationship with people on the grasslands works as a common topic between Jiang and Annaud, who could transcend the national, cultural, and language barriers in this book-to-movie adaptation. Both Jiang and Annaud exhibit "the ethics of care at the margins" that Professor Chang Chia-ju advocates in the introductory chapter of this book to call attention to the important aspect of cultural and biological diversity in current era of global homogenization.

When Ren Shanshan, a journalist in China, questioned Annaud for emphasizing the ecological instead of political discussion of the novel, Annaud responded, "The novel and the movie are two different works. The movie faces not only the readers of the original novel but those who have not read it before. Thus, my responsibility is to convey my understanding and impression of the novel. I have to tell my audience not only what the novel is about but also why I like it."[3] Annaud's *Wolf Totem* is not a simple copy of Jiang Rong's novel. It is an ecological rewrite. Although Sheng-mei Ma touches upon Annaud's rewrite in his paper, "Sino-Anglo-Euro Wolf Fan(g)s from Jiang Rong to Annaud," and there are also some Chinese articles discussing the differences between the novel and the movie, none of them answers or succeeds in answering the following questions. First, why does Annaud focus on the ecological theme in this book-to-film adaptation? Second, can the ecological theme function as a bridge in this trans-national and trans-cultural communication? Third, what are gained and lost in this ecological rewrite? This chapter hopes to answer the above questions and argues that a comparative approach is necessary in evaluating the gain and loss of this trans-cultural, trans-lingual, and trans-media ecological rewrite.

5.1 Annaud's Focus on the Ecological Theme

Critics and readers generally agree that Jiang Rong's *Wolf Totem* is a complicated novel. It has more than six hundred pages and describes in great details how Chen Zhen falls in love with wolves on the Mongolian grasslands and how he ruins the life of a little wolf by stealing it from the mother wolf and preventing it from going back to nature. Different from traditional novels, Jiang's *Wolf Totem* is interspersed with commentaries and discussions. Each chapter starts with one or two epigraphs from historical documents and the book concludes with a seventy-page long epilogue in which Jiang underlines the importance of the wolf totem in the form of conversations between Chen Zhen and Yang Ke thirty years later. Sheng-mei Ma believes that all these historical documents and lengthy discussions serve for the purpose of illustrating the differences between "Chinese sheep" and "Mongolian wolf."[4] Han Chinese are criticized for their sheepish passivity while Mongols are celebrated for their wolf-like passion for independence, equality, and freedom. In an interview, Jiang explained that he wrote the book as a self-reflection. "I needed to make Chinese people look inside themselves and see their weaknesses."[5]

Nevertheless, Jiang's nationalistic discussion annoyed many Chinese and non-Chinese critics. Calling it "an extended political metaphor," Jerry Varsava finds the wolf versus sheep metaphor highly problematic.

> At first glance, the wolf—whether of Mongolian or other origin—is not an obvious emblem for the democratic sensibility that Jiang Rong is clearly at pains to portray and promote in the novel, and which he has supported through his political dissidence over the decades. Notionally rapacious and predatory, the lupine community might be more obviously likened symbolically to a fascistic form of politics and governance, with the legendary "alpha male" serially dispatching weaker pretenders to the role of dominator in an age-old melodrama scripted by DNA until his own position is compromised through the ill fortunes of battle and/or the running down of the biological clock.[6]

Although Jiang tries hard to celebrate the wolf as a freedom-lover and pursuer, many critics and readers feel difficulty accepting it since the wolf is also associated with a predator, a cruel, greedy, and aggressive animal. Jiang also praises the wolf as the teacher of Genghis Khan, who is famous for his political and military ambitions. That explains why the German sinologist Wolfgang Kubin calls the novel "fascist" and "a shame for China."[7]

As a French director and a common foreign reader of *Wolf Totem*, Annaud shows no interest in this nationalistic discussion. For Annaud, it does not make sense to argue about whether Mongols are better than Han Chinese or about whether there is a certain superior culture.

> It is not right to judge whether nomadic culture is superior to agrarian culture. They are just different lifestyles. Some people like the nomadic lifestyle that follows herds and camels, while others like to build houses, grow crops, and keep animals in fences. They are two different cultures. Though controversial, they should not be defined by the good-vs-bad dichotomy.[8]

Although the wolf versus sheep metaphor has attracted a lot of Chinese "who feel threatened and oppressed in the Social Darwinism of the post-Mao capitalist market, who wish to transform themselves from underdogs to top wolves,"[9] the nationalistic discussion fails to convince Annaud, who expressed his desire to emphasize the book's different themes from the very first draft of the script.[10]

According to Annaud, the ecological message is the bridge to connect him with the author, Jiang Rong.

> I was approached by the writer, Jiang Rong, who became a wonderful friend. I spent three weeks with him in Inner Mongolia. He knew my work, and some of his friends at the production company also knew my work. They came to me, and I found it was right up my alley. It has been my conviction to find true stories about the environment. I was very excited to see that one of the bestselling books in China was precisely about something that everyone in the West is unaware of— that China has a deep movement that understands the need for the conservation and protection of nature and promotion of environmental issues.[11]

Annaud's words are important in three aspects. First, he attracted Jiang Rong and the production company because of his world-renowned interest in filming stories about the environment. He has made many successful films about the human-nature relationship, such as *The Bear* (1988) and *Two Brothers* (2004). Second, Jiang Rong also attracted Annaud because of the novel's attention to and the care of the marginalized human and nonhuman groups, including nomadic herders, wolves, and the Mongolian grasslands. Annaud believes that the ecological message is not only what the novel is truly about, but is also the reason why he loves the story and feels it a mission to show it to the global audience. As Annaud emphasized in another interview, "I feel a strong affinity with the author, Jiang Rong. Man and nature living harmoniously is what I strongly believe in and strive for. I love the novel since the author and I share the same concept on environmental protection."[12] Last but not least, there was a need for transnational and trans-cultural communication because "the West is unaware" of China's growing ecological consciousness. Annaud's ecological rewrite, in this sense, provided a perfect chance for the West to understand China: the environmental problems it faces and the ecological ethics and aesthetics it embraces.

5.2 The Ecological Theme as a Trans-National and Trans-Cultural Bridge

In "Sino-Anglo-Euro Wolf Fan(g)s from Jiang Rong to Annaud," Sheng-mei Ma describes Annaud's rewrite as a "de-fanging" process, meaning that Annaud cuts Jiang's aggressive ethnocentrism and focuses more on the widely accepted topic of environment.[13] This chapter argues that it is nothing wrong for Annaud to have an ecological rewrite so long as it can truly work as a tool for the ecological discussion among the author, the director, and their global audience. As Jiang Rong admitted in an interview, "The book is so long [544 pages in the English edition]. How could you fit all that into a two-hour film? There's no way you could fit it all in and still write a good screenplay, so some stuff had to be cut."[14] Although there are arguments as to whether *Wolf Totem* could be summarized into an ecological novel, its ecological significance is unneglectable. In his book *Ecocriticism Perspective of Twentieth Century Chinese Literature*, Wu Jingming sings high praise of *Wolf Totem*, listing four contributions the novel has made. First, it reflects ecological wisdom on the grasslands. Second, it shows an ecological vision that all beings are equal. Third, it creates an ecological character, Bilgee. Fourth, it gives serious ecological warnings.[15] He Chengzhou also praises the novel by admitting its role in "calling for people's moral responsibilities for environment and nature."[16] Even the author, Jiang Rong, admitted in an interview that environment is "an important issue in the novel, and one of the main reasons why [he] wrote it."[17] Thus, when Annaud expressed his desire to emphasize the ecological theme in his book-to-film adaptation, he didn't get any negative comments from the author or the production company. He also enjoyed complete freedom from Chinese censors. "What I can say is that I had carte blanche at every level until this day. The movie you see is the same movie I cut."[18]

When the film was released in 2015, it attracted immediate attention. The film grossed US$110.46 million from China, US$8,811,832 from France, and US$125,588,448 worldwide.[19] It won a lot of film awards in China and was celebrated as "the first Chinese eco-film made by a foreign director."[20] Many critics and reviewers speak highly of its ecological message. Stephen Schaefer describes the film as "ultimately a hymn and a plea for ecological harmony."[21] James Kendrick claims that "if the film has a true star, it is the environment itself, which is enormous and daunting, dangerous, yet serene."[22] Dennis Schwarts believes that the film "warns us that an old-fashioned way of life is passing to an encroaching civilization."[23] Most importantly, the author, Jiang Rong, approves and acclaims Annaud's ecological rewrite. "I thought the film did a good job of expressing the book's spirit and its manner. I think Annaud did a very good job of this. He even exceeded my expectations in some respects."[24] Nonetheless, the film also received plenty of negative comments and reviews. Mongolian writer Guo Xuebo criticized both the novel and the movie, claiming that "there's no record of any wolf totem in Mongolian literature or history."[25] Peter Keough in *Boston Globe* summarizes the film as "a dramatic dud, a cut-and-paste mess of plot snippets."[26] *Variety*'s chief Asia film critic, Maggie Lee, also expresses her disappointments, arguing that the film "boils down to a familiar environmentalist allegory that doesn't move or provoke too deeply."[27]

5.3 Gain and Loss in the Ecological Rewrite

What are the criteria to judge Annaud's ecological rewrite? Is it a success or failure in terms of ecological discussion? This chapter contends that a comparative approach is necessary in the analysis of this trans-cultural, trans-lingual, and trans-media ecological rewrite. In "Ecocriticism and Comparative Literature," Chinese scholar Hu Zhihong advocates a comparative approach in today's ecocritical studies, especially Asian ecocritical studies. Hu believes that ecocriticism originated in the West and "must vary in some aspects, whether in content, methodology or ideological foundation to some extent when it emerges in other countries, especially in Non-Western countries whose cultural traditions are different in nature from Western ones."[28] Hence, critics should step over national or cultural boundaries in the understanding of ecological issues. In the analysis of Annaud's rewrite, the multicultural strategies are even more important. Jiang's *Wolf Totem* is written in Chinese and talks about the distinctions between Mongolian culture and Han Chinese culture. Annaud's adapted script was written first in French and later in Mandarin Chinese with English subtitles. It has altogether seven rounds of edits worked out together by Annaud and his writing partner Alain Godard in cooperation with Chinese screen writer Lu Wei and the author Jiang Rong. From book to movie, from the Chinese author to the French director and then the global audience, this rewrite exemplifies the trend of global ecological communication and cooperation, and therefore cannot be truly understood and fairly evaluated without a comparative approach.

5.3.1 The Gain

Aided by the multicultural strategies, this chapter finds that there are both gain and loss in Annaud's ecological rewrite. Compared with the original novel, it has three advantages.

First, it achieves much more impressive visual and audio effects. Although Jiang's novel is captivating in describing the beauty of the Mongolian grasslands and the battles between wolves and gazelles, wolves and horses, and even wolves and human beings, Annaud's movie version is undoubtedly more impressive. Katie Walsh believes that "[the] greatest aspect of *Wolf Totem* is the gorgeous, sweeping cinematography that captures the landscape in breathtaking aerial shots and crystal-clear color."[29] Annaud is an expert in using the cinematic language. He uses long shots and sometimes extreme long shots to visualize the breathtaking beauty of the landscape. He changes to close-ups to emphasize the vitality and power of Mongolian wolves when Chen Zhen sees them for the first time. There are also shots in different angles and slow motions to picture the confrontation between wolves and human beings. Annaud is very particular about showing the real beauty of the Mongolian landscape. He spent a long time scouting the locations in Inner Mongolia and showed extreme patience in picturing the four seasons on the grasslands. Even Jiang Rong admitted that "[one] aspect where this film really succeeded was in finding a piece of untouched grassland. It's so beautiful."[30] Annaud also insisted on using real Mongolian wolves in the film, which was supported by the author Jiang Rong and Chinese producer Wang Weimin. He spent more than three years finding and training real Mongolian wolves in cooperation with Scottish animal trainer Andrew Simpson. The effect is truly stunning when the audience see real wolves running, fighting, and eventually dying on the grasslands and hear the howl of wolves together with the background music by the famous composer James Horner. It is undoubtedly one of the few movies on Chinese eco-screen that could impress the audience with the excitement of finding the beauty of nature and at the same time the shocking sadness of seeing how human power destroys it.

Second, compared with the novel, the film is more direct in its ecological message. Jiang discusses more than ecology in his novel *Wolf Totem*. He opens the novel by showing how wolves trap and kill gazelles like brave, well-trained military soldiers. In order to introduce the wolf versus sheep metaphor, Jiang intersperses the first chapter with recollections of how terrified Chen Zhen is in seeing wolves and how calm and brave

Mongols, including Mongolian women and kids, could be in fighting. As a representative of Han Chinese, Chen Zhen is compared to sheep, easy to be terrified. Mongols, on the contrary, are more like wolves, who are brave and wise. Such a comparison, however, is downplayed in Annaud's rewrite. The film begins by introducing Chen Zhen as a young man fed up with the chaotic life in Beijing during the Cultural Revolution and who hopes to find something new on the Mongolian grasslands. As Chen Zhen says at the beginning of the film, "That was in 1967, the second year of the Cultural Revolution. City students were sent to the countryside. I volunteered. I was among the very first to go. I picked Inner Mongolia..... wherever that was. It was so exciting to escape the hard life we had in Beijing during that troubled period." Instead of the wolf versus sheep metaphor, the central conflict here is between city and countryside, a theme commonly seen in ecological writings and films. Annaud uses the image of crowded buses in Beijing to represent the city and a five-second extreme long shot of the green and vibrant grassland to represent the countryside. It is easy to follow Annaud's story in a chronological timeline as Chen Zhen moves from the city to the grasslands, finds a paradise in that place, and eventually sees the destruction it faces.

It is also easy to get the film's central ecological message: the grassland logic. Although Annaud keeps the stories about Chen Zhen's first encounter with wolves and the thrilling confrontation between wolves and gazelles in the movie version, the purpose is not to show the differences between Han Chinese and Mongols. He uses them to stress the important role wolves play on the grasslands. They are brave, powerful, and wise, and most importantly, they play a key role in maintaining the ecological balance on the grasslands. When Chen Zhen blames wolves for killing "pitiful" and "innocent" gazelles, the Mongolian old man, Bilgee, retorts, "No. Gazelles are the cruel ones. They eat all the grass. Out here, the grass is life. The big life. All else is little life that depends on the big life for survival." The message is clear and direct. It differs a lot from Jiang's novel, which mixes the ecological discussion with cultural and ethnic comparisons. In addition, Jiang's novel is like an academic paper. In order to introduce the grassland logic, Jiang uses a lot of examples written in long paragraphs to show the destruction gazelles may bring to the grasslands. He likes to deliver these ecological messages through long monologues, especially monologues by Bilgee, who is regarded as the most knowledgeable and experienced nomad on the grasslands or in He Chengzhou's words a "native informant" and a "speaking wolf."[31] In Chap. 2, for example, Bilgee gives long monologues on how destructive gazelles could be by running on the grasslands and eating all the grass. He also emphasizes the ignorance of Han Chinese who do not understand the fact that wolves protect the grasslands by controlling the number of gazelles. All these monologues disappear in the movie version. Even the big life versus little life theory is shortened from a page-long monologue into short and easy-to-understand conversations. It is true and also a pity that some detailed illustrations such as why grass is the big life are sacrificed in the movie version. Yet, Annaud's rewrite is undoubtedly more suitable for the big screen with the potentiality of attracting a larger audience, especially those who may not be interested in long and complicated environmental lectures.

Last but not least, Annaud has done a great job in explaining the sky burial, a key element in the understanding of the Mongolian ecological culture. It is generally agreed that the wolf totem is vitally important in both the novel and the movie, since both have *Wolf Totem* as their title. In his book-length study of Jiang Rong's *Wolf Totem*, Li Xiaojiang argues that primal nature and primal freedom are two basic concepts of the wolf totem. Primal nature could be understood in modern terms as the pursuit of equality, struggle, and balance. Primal freedom differs from the concept of "liberty" in the category of political science.

> It is a spirit of primal freedom generated and developed in a state of existence in primal nature, unrelated to consciousness of self, power, and subservience, or civilized order, and still less to political positions of "positive" or "negative." This is an a priori free spontaneity, life possessing and using to the greatest degree the survival space and mode that nature has given, without restriction or transgression.[32]

The problem here is how to express those complicated ideas in the story.

Jiang Rong pointed out in an interview that "one important way in which the totem is expressed is the traditional Mongolian practice of giving dead bodies to the wolves to eat."[33] This traditional Mongolian practice is called the sky burial. In the novel, Jiang gives a very detailed explanation of how the sky burial works.

> For centuries on the Olonbulag, when a herder died, people stripped him naked and tied his body up in a roll of felt, although sometimes they left the corpse clothes so they could forego the felt. Then they loaded the corpse

> onto a cart on which a long board had been laid across the shafts and made secure. In the predawn hours, two senior male members of the family, each hold one end of the board, drove the cart to the sky burial site, where they whipped their horses into a gallop. Inevitably, the deceased bounced out of the cart, and that was the spot where the soul would return to Tengger.[34]

The two relatives would then leave the deceased to the surrounding wolves. If by three days, nothing but bones are left of the corpse, people believe that the soul of the deceased would enter Tengger. Jiang believes that Mongols "do this because they believe wolves return to heaven. So when wolves die, they bring the human souls with them to heaven."[35] The sky burial plays a critical role in transforming the real and virtual wolf to the mythical wolf, and should therefore be emphasized in the story. Yet, in the novel, Jiang introduces the sky burial mainly in the form of commentaries. Long paragraphs are given to the explanations of how the sky burial is practiced on the Olonbulag and why Mongols have deep reverence for Tengger, the wolves, and the souls they entrusted to them. Although Chen Zhen is highly curious, he can only go there after the sky burial and meditate on how that works and what that means.

Annaud is different. Instead of commentaries, he visualizes the sky burial on screen by making radical changes to the plot. In his rewrite, Bilgee's son, Batu, dies in the attempt to protect the herd of horses. Bilgee has to give his son a sky burial, and thus Chen Zhen has a chance to witness the whole process. With the sun rising in the east, the camera follows the cart to a piece of rocky land and zooms in onto the dead body falling from the cart to the ground. As a Han Chinese, Chen Zhen knows nothing about the sky burial and thus comes the following conversation.

> Chen Zhen: A'ba. It's too rocky to bury him here. Let's find another place.
> Bilgee: Mongols don't bury their dead. We grasslanders eat meat all our lives for which we kill many creatures. After we die, we donate our meat back to the grassland.

With all the visual images and the conversation, the audience can easily understand the meaning of the sky burial. One gets a lot from nature and is therefore willing to give himself back to nature when he dies. Here, Annaud does not overemphasize the role wolves play in the sky burial. His interpretation is more from an ecological perspective. The concept of taking and returning nullifies the dichotomized relationship between humans and nature, or humans and wolves. All beings including humans, wolves, and even the grasslands live in an interconnected web. When one only takes but not returns, the connection and the balance will be affected. As Jiang admitted, "The scene was an opportunity to present this tradition without having to lecture on it. This was different from the book. In fact, it exceeded the book in this respect. So I thought Annaud had a good reason to kill off Batu. It didn't distort the original intention of the book."[36]

The sky burial reappears near the end of the movie but this time it is to emphasize the ecological and cultural crises on the grasslands. Although both the novel and the movie mention how crazy people become in using traps, guns, poisons, and bombs to kill wolves, Annaud highlights the destruction by making Bilgee die in an accidental bomb explosion. Seeing Bilgee's sky burial, Chen Zhen says nothing. The audience can only hear the slow tempo background music with occasional cries of a distant eagle. The sky burial near the end of the movie is more like Chen Zhen's silent farewell to Bilgee, to the disappearing wolves and the grasslands. As he says after the burial, "With Bilgee's death, it was the grassland I knew and I loved that was dying. Time had come for me to become again the young city man I had never ceased to be." Compared with Jiang, Annaud has more successfully wedded the sky burial to the plot. It works like a clue to the Mongolian grassland culture and the ecological and cultural crises the director hopes to show on screen.

5.3.2 *The Loss*

Although many critics and reviewers praise Annaud for his emphasis on the ecological interpretation and his contribution to the trans-national and trans-cultural communication, the rewrite has some serious problems in terms of the ecological discussion.

First, it simplifies the complicated conflicts in the novel into a good versus bad battle. In the movie, the new head of production, Bao Shuigui, seems to be the culprit of all disasters. He comes from the farming areas of Inner Mongolia and believes firmly that human beings can conquer and transform nature. Annaud emphasizes Bao's belief in the power of culture by adding two details into the movie. First, Bao introduces Chen Zhen and Yang Ke to Bilgee as city students who can teach Mongolian kids Mandarin. The implication is that the area needs to be civilized. Yet, the civilization Bao hopes to bring is human-centered. For the benefits of human beings, he is willing and happy to do anything to nature. The second detail is when Chen Zhen proposes a scientific experiment to mate wolves with dogs, Bao immediately approves and tells Chen Zhen that he taught the improvement of species at university. In Jiang's novel, Bao has no experience of teaching at all. Annaud adds the detail to highlight Bao's belief that human beings can manipulate nature.

Everything Bao does in the movie is to manipulate nature. He asks people to kill wolves, including their cubs. He uses guns, traps, and poison to eradicate toxic plants, pests, mosquitoes, rats, and wolves. He brings farming people to the swan lake, determined to transform the grasslands into farmlands. All his decisions turn out to be ecologically destructive. As Bilgee explains, "Our soil is too thin. It's not suited for plowing." It is also dangerous to kill wolves because fewer wolves will mean more ground squirrels and rabbits, which will eventually destroy the pastures. The climax is the wolf-extermination campaign, which aims to eliminate all the wolves on the grasslands. When Bao orders people to kill wolves by all means, even by burning the grasslands, Chen Zhen is enraged and says, "The nomads will hate you." Here, the conflict is visualized into a war against Bao. People hate him because he is the destroyer of nature and the violator of the grassland logic. It seems that he is the cause of all troubles, and the grasslands would still be fine if he was not there.

Annaud's rewrite, however, goes against the original novel, which shows in many details the complicated conflicts on the grasslands. In Jiang's novel, Bao is just one of many destroyers. There are long paragraphs describing how outsiders kill swans, rats, and dogs for their meat, and wolves for their pelts. They kill not only the males and the old ones, but the pregnant and the newborn, and most importantly, they feel no shame at all. They are not afraid of animal extinction and even feel proud of killing "enemies" in nature. Jiang is very careful to show the destruction as a group activity. The destroyers include Han Chinese, immigrants and laborers from the farming areas of Inner Mongolia, and some local nomads. They use all kinds of modern technologies such as guns, bombs, and even airplanes to scatter poison. Their destructive force is frightening. As Chen Zhen says, "It had taken only one summer for the people to turn the lovely swan lake area into a graveyard for swans, wild geese, wild ducks, and wolves."[37]

This widespread destruction of nature cannot be fully understood without the social and cultural backgrounds. The story is set during the Cultural Revolution (1966–1976). Many people, including Bao Shungui, are firm followers of Mao Zedong. They love to quote Mao's words such as "man is the primary element" and use these words to guide their life. In *Mao's War Against Nature: Politics and the Environment in Revolutionary China*, Judith Shapiro points out that "Maoist China provides an example of extreme human interference in the natural world in an era in which human relationships were also unusually distorted."[38] Mao advocated that human beings were the master and conqueror of nature. It was popular to see notions such as "humanity's determined triumph over nature" (*rending shengtian*) and "fight Heaven, struggle with the earth, and win a high yield" (*zhantian doudi duogaochan*) in the Mao era (1949–1976). Nature was regarded as a target for transformation and sometimes even an enemy to be destroyed. A famous example is the campaign to "Wipe out the Four Pests" (*chu si hai*) in the Great Leap Forward movement (1958–1962). Believing that rats, sparrows, flies, and mosquitoes were four enemies in nature, Mao started a nationwide campaign to eradicate these species. The wolf-extermination campaign in *Wolf Totem* is just another example of how human beings hope to manipulate nature. Although Bao has done many bad things against nature, he himself can never be the cause of all the disasters. Annaud is in this sense wrong to de-emphasize the social and cultural backgrounds. He is too hasty to conclude that "the Cultural Revolution did not have a great impact on people there" simply because "they lived in a faraway place" and "did not know what actually happened due to the lack of communication tools such as televisions, newspapers, or radios."[39]

In addition to the blind belief in human power, agricultural encroachment is another destructive force on the grasslands. Jiang points out in his novel that China has a long tradition of moving agrarian immigrants to the border areas for the sake of protection and development. Such a tradition started in the Han dynasty (206 BC–220 AD), became a highly important policy in the Qing dynasty (1636–1912), and continued to be practiced in the Mao era (1949–1976), which witnessed a population explosion from 500 million to 900 million. When vast numbers of farming immigrants rushed to Inner Mongolia, they posed huge and irreversible risks to the local environment. As it is described in the novel, these immigrants believe in the agrarian culture and do not show any respect to the grassland logic. They are fast in building sturdy houses, killing local species, and using trucks to transform the grasslands. When the land fails their agrarian dreams, they have no hesitation in leaving. It is stunning to see how destructive these migratory farmers could be. "Most of that grassland had turned into poor, sandy farmland in their time, and it could no longer provide for them. So, like migratory birds, they came out here."[40] It does not take long to see the impact of their destruction. Aided by modern technologies, these migratory immigrants ruin the swan lake area within a summer and turn the green grasslands into sandy farmlands in less than thirty years. It's really sad when Chen Zhen and Yang Ke revisit the grasslands thirty years later only to find the place totally different from their memory.

> On a highway, half covered with dry sand, caravans of trucks equipped with iron cages to transport sheep and cows rumbled toward them, trailing thick columns of yellow dust as they made their way to China proper. They hardly saw a yurt or a herd of horses or cows along the way; every once in a while they spotted a flock of sheep, but they were small and thin, with dirty, tangled black wool. Even the "processed" sheep looked better than those. The two friends nearly gave up on the trip, not wanting the moist, lush grassland in their hearts to be replaced by dry dust.[41]

Unfortunately, all these detailed descriptions are lost in the movie version. Although Annaud touches on agricultural encroachment in the swan lake episode, he fails to show in real pictures how destructive these people could be. The audience can only see the death of a few swans, after which their attention is quickly drawn back to the story between Chen Zhen and his little wolf. In his discussion of the movie version, Jiang expresses his great disappointment in seeing the end of the movie.

> [The] way I imagined it, the end would show the grasslands as a giant desert. In fact, that's why I don't want to go back to the grasslands. The grasslands today are terrifying. When I went back in 1997, the terrain was still there, but it was already in a very bad state. Telephone poles, wireless towers and mining—there was so much. You didn't get the sense anymore of an untouched grassland.[42]

People would not realize the destructive power of agricultural encroachment without the visual contrast between the green grasslands and a giant desert.

Annaud's problem comes mainly from his western perspective. Among many conflicts in the novel, Annaud chooses to focus on the nature versus culture confrontation, which is a common topic in western ecological studies. It becomes also a hot topic on the western screen, particularly the Hollywood screen. Movies like *Bambi* (1942), *The Lorax* (1972, 2012), and *Avatar* (2009) are all stories about the human destruction of nature with the nature versus culture conflict as their major theme. It is understandable when Annaud emphasizes Bao as a representative of culture that destroys nature. Nevertheless, as Hu Zhihong points out, critics should explore the multicultural strategies in contemporary ecological studies, which become increasingly trans-national and trans-cultural.

> On one hand, awareness of comparative literature [requires] that ecocritics should explore the ecological affinities or identities of different cultures or civilizations, i.e., commonalities and similarities. Meanwhile, they should pay more attention to the differences and uniqueness of different cultures and civilizations. Thus it explores the multicultural strategies to deal with ecological degradation.[43]

The ecological degradation in *Wolf Totem* can never be fully understood without proper attention to the widespread belief in human power during the Cultural Revolution or agricultural encroachment that has been threatening and destroying the grasslands throughout Chinese history. By simplifying the conflicts and de-emphasizing the backgrounds, Annaud may make his story easy to understand for his global audience. Nevertheless, the rewrite is reduced to a common ecological allegory that fails to show either the actual causes or the appalling effects of Mongolian ecological degradation.

The second problem with Annaud's rewrite is the portrayal of Chen Zhen and the little wolf. In the novel, Jiang points out clearly that Chen writes *Wolf Totem* out of a deep remorse for what he has done to the cub. He steals it from the mother wolf out of pure curiosity.[44] When the cub grows up, he tries all means to prevent it from going back to nature. He keeps the little wolf famished and parched for the whole day to ensure its return after each walk. To avoid wolf bites, he snips off the tips of its fangs, the weapon a wolf needs to survive on the grasslands. He forces it to migrate with him. When the little wolf refuses, he chains it to the wagon, causing serious wound on its neck and paws. The saddest moment is when the little wolf finds itself unable to respond to the howls of distant wolves due to its injured throat, it goes mad. "Crazed by anxiety, he grew reckless, jumping and running, jerking the chain and the wooden post, oblivious to the possibility of mortal injury."[45] The consequence is of course fatal. As Jiang describes,

> In the light of dawn, they pulled the felt away to see the consequence of the cub's crazed struggle for freedom and longing for his father's love: The infected tooth was now protruding from his mouth; the root had broken when he tore at the felt. He was bleeding, his wounds possibly made worse by the dirty felt. His throat was still bleeding as well, worse than they'd first moved here. The old wound had clearly been reopened.
>
> His eyes bloodshot, the cub kept swallowing the blood, but it was everywhere on their deels, on the felt, inside the pen, a worse sight than when a foal was killed. The blood quickly turned to ice. Chen's knees buckled from the fright and he stammered in a shaky voice. "It's all over. He'll die for sure."[46]

The little wolf's suffering and its final death are like warnings. Nature could be victimized not only by those who hope to control and dominate nature but also by those who are curious about nature. Although Chen loves the little wolf and tries his best to take good care of it, he is after all a victimizer. He is the direct cause of the little wolf's loss of freedom, happiness, and health. The little wolf's tragedy testifies to the arrogance and stupidity of human beings and plays a critical role in Jiang's ecological criticism.

Nevertheless, all the above warnings and critiques are lost in Annaud's rewrite. Instead of a destroyer, Chen Zhen is pictured as the little wolf's caretaker, a forever friend, and even a "servant." As he admits to Bilgee, "I am the one who is his servant. I wait on him like I would a Mongol Prince. I give him my ration of meat. I can't sleep well at night. I worry about him being cold or sick or bitten by dogs or carried away by an eagle. Yang Ke says I am the wolf's slave." When the cub grows up, Chen Zhen is more like a father figure and a teacher. He teaches the little wolf to guard against wolf traps and some surviving skills like swimming. Though the little wolf bites him, his anger is soon replaced by the pride. "Sorry. I am wrong. You are a real Mongolian wolf. I am proud of you." In the movie, Chen Zhen has never done anything to hurt the little wolf, not to mention snipping off the tips of the wolf fangs or killing it to end its torture. His story with the little wolf is more like an explorer discovering the cub, taking good care of it, befriending it, and eventually sending it back to nature. Even the composer, James Horner, distorts the story by naming the third song "An Offering to Tengger/Chen Saves the Last Wolf Pup" in his *Wolf Totem* soundtrack. From a victimizer to a rescuer, Annaud's portrayal of Chen Zhen carries no more guilty feelings. The power of criticism is also affected when people fail to see how easily human beings can go wrong in their relationship with nature.

Similar to Chen Zhen, the image of the little wolf transforms, as well. In the novel, the little wolf is a fighter for freedom. It refuses from the very beginning to eat from a human hand. It feels excited to hear the howls of wolves in the mountains and seeks every single chance to go back to nature. Smart, courageous, and unyielding, the little wolf is ready to sacrifice anything, even its own life, for freedom. Such an image is in accordance with the collective image of Mongolian grasslands wolves, who Bilgee describes to be warriors and even gods. "Wolves want the dignity to earn their own meat. They don't want to be fed. They want to kill at the risk of being killed. They are warriors." Chen Zhen has therefore made a huge mistake by teaching it to fear death and forbidding it from killing other animals. As Bilgee summarizes, "You captured a god to turn it into a slave. What you are doing is an affront to our ancestors and to Tengger!

Yet, it is hard to relate this heroic figure to the little wolf in Annaud's rewrite. Instead of a warrior, the little wolf looks like a little pet in need of human care. It is lovely but powerless and naïve. The movie does not explain why the little wolf refuses to migrate with Chen Zhen and pictures it more like a disobedient kid unwilling to cooperate. "What are you doing? I'm getting angry. You can't stay here, you know. Dogs and eagles will eat you. Understand? Don't you understand? Don't be silly!" The power relation between Chen Zhen and the little wolf is clear. Chen is knowledgeable and powerful while the little wolf knows nothing about the cruelty of nature. It even becomes a troublemaker in the latter part of the story when it bites not only Chen Zhen but Gasmai's little son.

The most striking difference lies in the end. In the novel, Chen kills the little wolf who is no longer able to survive. He chooses to let the wolf "die as if in battle" so that its soul could "complete its cycle."[47] Although there are debates whether Chen does the right thing, the wolf's death plays two important roles in the novel. First, it launched a lifelong assault on Chen's conscience, making him realize what a huge mistake he has made with regard to the grasslands. Second, the little wolf's death completes Chen's search for his own wolf totem and becomes the most important reason for the writing of the book.

> Chen's vacant gaze followed the impish, lifelike figure of the cub's pelt as it danced in the wind.... The cub, like a golden flying dragon, rode the clouds and mist, travelling on snow and wind, soaring happily toward Tengger, to the star Sirius, to the free universe in space, to the place where all the souls of Mongolian wolves that had died in battles over the millennia congregated. At that instant, Chen Zhen believed he saw his very own wolf totem.[48]

In contrast to the little wolf's heroic death, Annaud rewrote the end by sending the little wolf back to nature after Gasmai's son recovers from the wolf bite. According to Jiang, Annaud made the change based on two reasons. First, "[westerners] would not be able to bear this," which means that it is too cruel to show the torture and the killing on screen; second, the release "provides a little warmth and hope to viewers."[49] It is understandable that Annaud hopes to satisfy his global audience, especially western audience, by transforming the theme of destruction to a more popular, acceptable, and humane motif of bringing the lost animal back to nature. Nevertheless, the little wolf in this warm, back-to-nature story, contradicts the heroic, godlike figure in *Wolf Totem*, making it hard for the audience to understand the logical connection between the little wolf and Mongolian ecological and cultural ethics and aesthetics.

5.4 Conclusion

Adapting the 640-page novel into a 120-minute film is no easy task. It becomes even more difficult when the director is non-Chinese and the audience may have totally different social and cultural backgrounds. Instead of a de-fanging process, this chapter concludes that Annaud's rewrite serves as a bridge between different perspectives and cultures. It does a great job in finding and acknowledging the novel's contribution to contemporary ecological discussions and impresses the world by translating Chinese ecological and cultural ethics and aesthetics to the western audience. It emphasizes "the ethics of care" at the marginalized groups such as herders, wolves, and the grasslands, and gives the world a chance to see them on screen and listen to their tragic stories written in 85% Mandarin and some Mongolian and the rest in the wolf language. Yet, it should also be admitted that Annaud's rewrite has problems. It does not show the real causes or the appalling effects of ecological degradation in Inner Mongolia. It also fails to harmonize the western and eastern perspectives in the understanding of Chinese ecological ethics and aesthetics. A comparative approach is therefore important in the evaluation of Annaud's rewrite. It helps to see both the gain and the loss in this trans-national and trans-cultural communication. Although it has some serious problems in adaptation, it is after all a pioneering work in transcending national and cultural boundaries in ecological communication on world screen.

Notes

1. Ren Shanshan, "Faguo daoyan Jean-Jacques Annaud tan dianying langtuteng: zhe shi yibu paigei quanshijie de dianying."
2. Qin Wan, "Jean-Jacques Annaud: changshuo Zhongguo hao, pengyou juede wozai daguanggao."
3. Ren, "Faguo daoyan Jean-Jacques Annaud tan dianying langtuteng: zhe shi yibu paigei quanshijie de dianying."
4. Ma Sheng-mei, "Sino-Anglo-Euro Wolf Fan(g)s from Jiang Rong to Annaud," 87.
5. Jian Rong, "Q&A: 'Unlike the Wolf, Sheep Are Afraid of Freedom."
6. Jerry Varsava, "Jiang Rong's *Wolf Totem*: Toward a Narrative Ecology of the Grassland of Contemporary Inner Mongolia," 286.
7. Wolfgang Kubin, "Deguo hanxue quanwei lingyizhiyan kan xiandangdai Zhongguo wenxue."
8. Qin, "Jean-Jacques Annaud: changshuo Zhongguo hao, pengyou juede wozai daguanggao."
9. Ma, "Sino-Anglo-Euro Wolf Fan(g)s from Jiang Rong to Annaud," 83.
10. Jean-Jacques Annaud, "Shanghai Film Fest: Q&A with Director Jean-Jacques Annaud."
11. Ibid.
12. Tang Yuankai, "*Wolf Totem*: Another *Dances with Wolves*," 69.
13. Ma, "Sino-Anglo-Euro Wolf Fan(g)s from Jiang Rong to Annaud," 92.
14. Jiang Rong, "Q. and A.: Jiang Rong on 'Wolf Totem,' the Novel and Now the Film."
15. Wu Jingming, *Shengtai piping shiye zhong de ershi shiji Zhongguo wenxue*, 394–399.
16. He Chengzhou, "The Wolf Myth and Chinese Environmental Sentimentalism in *Wolf Totem*," 787.
17. Jiang, "Q. and A.: Jiang Rong on 'Wolf Totem,' the Novel and Now the Film."
18. Michael Martina, "French 'Wolf Totem' Director Says China Censors Gave Him 'Carte Blanche.'"
19. "Wolf Totem (2015) Box office Mojo."
20. Qi Shanshan, "Kua wenhua beijing xia jiyu wenxue gaibian de shengtai dianying piping—yi dianying langtuteng weili."
21. Stephen Schaefer, "'Wolf Totem' Leads Pack with Poetic Beauty."
22. James Kendrick, "Wolf Totem."
23. Dennis Schwartz, "Warns Us That an Old-Fashioned Way of Life is Passing to an Encroaching Civilization."
24. Jiang, "Q. and A.: Jiang Rong on 'Wolf Totem,' the Novel and Now the Film."

25. Laura Zhou, "*Wolf Totem*: Writer Blasts Hit Film over 'Fake' Mongolian Culture."
26. Peter Keough, "In 'Wolf Totem' it's the call of the mild."
27. Maggie Lee, "Berlin Film Review: *Wolf Totem*."
28. Hu Zhihong and Hu Tiantian, "The Integration of Ecocriticism and Comparative Literature—A Win-Win Academic Strategy," 105.
29. Katie Walsh, "Movie Review: Gorgeous 'Wolf Totem' Includes Lesson That Still Ring True."
30. Jiang, "Q. and A.: Jiang Rong on 'Wolf Totem,' the Novel and Now the Film."
31. He, "The Wolf Myth and Chinese Environmental Sentimentalism in *Wolf Totem*," 789.
32. Li Xiaojiang, *Wolf Totem and the Post-Mao Utopian: A Chinese Perspective on Contemporary Western Scholarship*, 75–76.
33. Jiang, "Q. and A.: Jiang Rong on 'Wolf Totem,' the Novel and Now the Film."
34. Jiang, *Wolf Totem: A Novel*, 62–63.
35. Jiang, "Q. and A.: Jiang Rong on 'Wolf Totem,' the Novel and Now the Film."
36. Ibid.
37. Jiang, *Wolf Totem: A Novel*, 470.
38. Judith Shapiro, *Mao's War against Nature: Politics and the Environment in Revolutionary China*, 1.
39. Qin, "Jean-Jacques Annaud: changshuo Zhongguo hao, pengyou juede wozai daguanggao."
40. Ibid, 341.
41. Ibid, 508.
42. Jiang, "Q. and A.: Jiang Rong on 'Wolf Totem,' the Novel and Now the Film."
43. Hu and Hu, "The Integration of Ecocriticism and Comparative Literature—A Win-Win Academic Strategy," 105.
44. It has to be pointed out that in the original novel, Jiang Rong uses "it" to refer to the wolf cub. Yet, in the English translation, Howard Goldblastt changes "it" into "he" to show the wolf cub's importance. This chapter chooses to use "it" in the hope of being faithful to the original novel.
45. Jiang, *Wolf Totem: A Novel*, 496–497.
46. Ibid, 497.
47. Ibid, 499.
48. Ibid, 503.
49. Jiang, "Q. and A.: Jiang Rong on 'Wolf Totem,' the Novel and Now the Film."

Bibliography

Annaud, Jean-Jacques. 2012. "Shanghai Film Fest: Q&A with Director Jean-Jacques Annaud." Interview by Jonathan Landreth. *Los Angeles Times*, June 15. https://latimesblogs.latimes.com/movies/2012/06/shanghai-film-fest-qa-wolf-totem-with-director-jean-jacques-annaud.html. Accessed 12 July 2018.

———. 2015. *Wolf Totem* [*Lang tuteng*]. Sony Pictures Home Entertainment.

He, Chengzhou. 2014. "The Wolf Myth and Chinese Environmental Sentimentalism in *Wolf Totem*." *Interdisciplinary Studies in Literature and Environment* 21 (4): 781–800.

Hu, Zhihong, and Tiantian Hu. 2016. "The Integration of Ecocriticism and Comparative Literature—A Win-Win Academic Strategy." *Cultural Studies and Literary Theory [Zhongwai wenhua yu wenlun]* 35: 103–130.

Jiang, Rong. 2008a. "Q&A: 'Unlike the Wolf, Sheep Are Afraid of Freedom.'" Interview by IPS Correspondents. *Inter Press Service*, June 5. www.ipsnews.net/2008/06/qa-unlike-the-wolf-sheep-are-afraid-of-freedom/. Accessed 22 June 2018.

———. 2008b. *Wolf Totem: A Novel*. Trans. Howard Goldblatt. New York: The Penguin Press.

———. 2015. "Q. and A.: Jiang Rong on 'Wolf Totem,' the Novel and Now the Film." Interview by Amy Qin. *The New York Times*, February 26. https://sinosphere.blogs.nytimes.com/2015/02/26/q-and-a-jiang-rong-on-wolf-totem-the-novel-and-now-the-film/?_r=0. Accessed 3 Mar 2018.

Kendrick, James. 2015. "Wolf Totem." *Qnetwork*, December 22. www.qnetwork.com/review/3467. Accessed 2 June 2018.

Keough, Peter. 2015. "In 'Wolf Totem' It's the Call of the Mild." *Boston Globe*, September 10. www.bostonglobe.com/arts/movies/2015/09/10/wolf/pSi7sI9tp3ViNrfHQfnIdM/story.html. Accessed 3 July 2018.

Kubin, Wolfgang. 2006. "Deguo hanxue quanwei lingyizhiyan kan xiandangdai Zhongguo wenxue." Interview by Xin Ping. *DW*, November 26. https://p.dw.com/p/9R8g. Accessed 8 May 2018.

Lee, Maggie. 2015. "Berlin Film Review: *Wolf Totem*." *Variety*, February 10. https://variety.com/2015/film/asia/berlin-film-review-wolf-totem-1201428207/. Accessed 23 May 2018.

Li, Xiaojiang. 2018. *Wolf Totem and the Post-Mao Utopian: A Chinese Perspective on Contemporary Western Scholarship*. Trans. Edward Gunn. Boston: Brill.

Ma, Sheng-mei. 2016. "Sino-Anglo-Euro Wolf Fan(g)s from Jiang Rong to Annaud." *Ecozon@* 7 (1): 76–93.

Martina, Michael. 2015. "French 'Wolf Totem' Director Says China Censors Gave Him 'Carte Blanche.'" *Reuters*, February 1. www.reuters.com/article/us-china-film/french-wolf-totem-director-says-china-censors-gave-him-carte-blanche-idUSKBN0L51J820150201. Accessed 9 June 2018.

Qi, Shanshan. 2017. "Kua wenhua beijing xia jiyu wenxue gaibian de shengtai dianying piping—yi dianying langtuteng weili." *people.cn*, June 20. media.people.com.cn/n1/2017/0620/c412970-29350688.html. Accessed 3 July 2018.

Qin, Wan. "Jean-Jacques Annaud: changshuo Zhongguo hao, pengyou juede wozai da guanggao." *ifeng.com*. ent.ifeng.com/idolnews/daxierenwu/special/dxrw024/. Accessed 15 Mar 2018.

Ren, Shanshan. 2015. "Faguo daoyan Jean-Jacques Annaud tan dianying langtuteng: zhe shi yibu paigei quanshijie de dianying." *China Daily*, February 17. data.people.com.cn/rmrb/20150217/12. Accessed 29 Jun 2018.

Schaefer, Stephen. 2015. "'Wolf Totem' Leads Pack with Poetic Beauty." *Boston Herald*, September 11. www.bostonherald.com/entertainment/movies/movie_reviews/2015/09/wolf_totem_leads_pack_with_poetic_beauty. Accessed 12 June 2018.

Schwartz, Dennis. 2017. "Warns Us That an Old-Fashioned Way of Life Is Passing to an Encroaching Civilization." *Ozus' World Movie Reviews*, January 20. homepages.sover.net/~ozus/wolftotem.html. Accessed 3 June 2018.

Shapiro, Judith. 2001. *Mao's War Against Nature: Politics and the Environment in Revolutionary China*. Cambridge: Cambridge University Press.

Tang, Yuankai. 2015. "*Wolf Totem*: Another Dances with Wolves." *China Today* 2: 68–70.

Varsava, Jerry. 2011. "Jiang Rong's *Wolf Totem*: Toward a Narrative Ecology of the Grassland of Contemporary Inner Mongolia." *Interdisciplinary Studies in Literature and Environment* 18 (2): 283–301.

Walsh, Katie. 2015. "Movie Review: Gorgeous 'Wolf Totem' Includes Lesson That Still Ring True." *Tribune News Service*, September 14. www.seattletimes.com/entertainment/movies/movie-review-gorgeous-wolf-totem-includes-lessons-that-still-ring-true/. Accessed 3 May 2018.

Wolf Totem. (2015). "Box Office Mojo." *Box office Mojo*, December 31, 2018. www.boxofficemojo.com/movies/?page=intl&id=wolftotem.htm. Accessed 1 Jan 2019.

Wu, Jingming 吴景明. 2014. *Shengtai piping shiye zhong de ershi shiji Zhongguo wenxue* 生态批评视野中的二十世纪中国文学 (Ecocriticism Perspective of 20th Century Chinese Literature). Beijing: Zhongguo Shehui Kexue Chubanshe.

Zhou, Laura. 2015. "Wolf Totem: Writer Blasts Hit Film over 'Fake' Mongolian Culture." *South China Morning Post*, February 26. www.scmp.com/news/china/article/1722433/ethnic-chinese-writer-criticises-fake-culture-forced-mongolians-hit-film. Accessed 3 July 2018.

CHAPTER 6

An Ecotranslation Manifesto: On the Translation of Bionyms in Nativist and Nature Writing from Taiwan

Darryl Sterk

This chapter considers the translation of bionyms—plant and animal names—in modern literature, in the context of mass bionym extinction. For just as many species are now threatened with extinction, so are many bionyms. And just as we seek to protect biological diversity, so we should protect bionymic diversity, because it is a part of the cultural diversity on which humanity can draw to solve environmental problems and to live in harmony with the creatures that have been pushed to the periphery by industrialized civilization. As a way of protecting bionym diversity, I present my own bionym translation manifesto: translate the local bionym literally either as a loanword or a calque, with the common bionym in the target language as a complement and the scientific name as a global equivalent. That is at least three bionyms in three languages. My bionym manifesto is therefore polyglot. My polyglot bionym manifesto may seem overly demanding for a nativist writer like Huang Chun-ming, but it suits a nature writer like Wu Ming-yi, who makes a point of naming and renaming individual species in different languages in his works.

D. Sterk (✉)
Department of Translation, Lingnan University, Tuen Mun, Hong Kong

C. Chang (ed.), *Chinese Environmental Humanities*,
Chinese Literature and Culture in the World,
https://doi.org/10.1007/978-3-030-18634-0_6

I begin with Huang's essay "Waiting for the Name of a Flower" by considering one bionym in particular, which I contextualize locally before finding complements for it globally. The discussion of Huang's essay should serve as a primer in the basic biological background I think every ecotranslator should have. In a brief transition, I lay out my manifesto in the context of scholarly literature on ecotranslation. In this context, I argue against any attempt to lay the blame for bionym extinction at Linnaeus's feet. I also argue for a respect for the scientific knowledge that establishes natural equivalence as a foundation for ecotranslation. I then consider the translation of bionyms in Wu Ming-yi's writings, non-fiction and fiction. The conclusion reflects on how the eco in ecotranslation might apply to the translation of bionyms, specifically whether there really is a need for diversity in bionyms analogous to the ecological need for diversity. Looking ahead to future research possibilities, I also reflect on the fact that referring to a biological organism is not the only thing a bionym can do.

6.1 Waiting for the Name of a Flower with Huang Chun-Ming

In his essay "Waiting for the Name of a Flower" (Huang 1989), the Taiwanese nativist writer Huang Chun-ming remembers waiting for the name of a flower. For the longest time nobody who passed by knew the name. Not the schoolboys on bicycles. Not the civil servant on foot. Not the young lady riding a pair of wheels. Finally, an old lady, walking her grandson home, told him that "It's called a riffraff flower, because it serves no useful purpose. It's been called that in these parts for as long as I can remember," which leads Huang to the conclusion that he himself is riffraff, because he spent his time to no useful purpose, waiting for the name of a flower.

Huang's true concern in this essay is his own modernity, where modernity is a style of life that has separated him from the soil his ancestors farmed. He ironizes modern buzzwords by putting them in quotation marks, then applies them to himself. For instance, the "leisure" he obviously enjoys, having nothing better to do than wait for the name of a flower. To the grandmother, whose life has been labor-filled, leisure would sound like indolence. In this encounter between Huang and the old lady, Huang is the modern, even though he is the nativist writer. For a nativist (*hsiang-t'u*) writer like Huang, estrangement from the soil (*t'u*) of one's homeland (*hsiang*) means forgetting the traditional way of life, which the grandma in the story leads.[1] The schoolboys, the civil servant, and the young lady all think Huang is a weirdo, and even if they took him for normal they would all be too busy to take the time to stop and chat. The grandma, by contrast, answers his question and invites him over for dinner, because she has the friendly warmth that Huang associates with the salt of the earth.

But in "Waiting for the Name of a Flower," the earth also seems to mean the natural world. For it is not just that Huang does not know the name of the flower. In fact, he has never noticed this particular kind of wildflower before. The grandmother, on the other hand, is quite familiar. "It's a creeper with thorns," she says, "what grows around people's yards." Of course, from the perspective of science, the old lady's dismissive appellation is less than ideal. Riffraff flower could, after all, apply to any number of flowers for which no use has been found. Surely each species should have its own name, rather than being consigned to the trash heap of nomenclature. Still, her knowledge beats Huang's.

Though he falls short of the old lady in this respect, Huang is not quite the leisured modern who is ignorant of the natural world. He is closer to the soil than the schoolboys, the civil servant, and the young lady, for he displays some awareness of the kind of ecosystem he is visiting. He is standing on an embankment close to the estuary of the Lan-yang River, located in I-lan County in northeastern Taiwan. Standing there, he is aware of the type of plant that is typical to that ecosystem. We sense his awareness because he notices four plants whose names he does know, and which he has seen in this type of terrain before. And this is where it gets complicated. It does not get complicated for the reader of the Mandarin Chinese essay, for whom the names of the plants may not even register, because they are mentioned once, never to be referred to again. But for the Chinese-to-English translator, the four botanical names lead into a maze of plant nomenclature, both common and scientific. The size of the maze depends on the translator's interest and knowledge. Sometimes too little knowledge is a dangerous thing. But too much knowledge can also be a dangerous thing. The translator has to know when to stop acquiring it and make a choice. But before making that choice, let us begin to explore the maze.

The first plant Huang notices, the only one I will have the leisure to discuss, is the horse-whip herb. A horse-whip is a highly branched herb that can grow to about a meter tall with lobed green leaves and spiked mauve flowers. “Horse-whip herb” has been selected, as it were, by Huang from a number of common names by which the plant is locally known. According to the Plants of Taiwan (PoT) database, the first resource any Taiwanese ecotranslator should check, this plant goes by, or has gone by, a number of other names: iron horse-whip, iron fishing-rod, tea-leaf herb, and dragon-tooth herb.[2] The provenance of these Chinese names is not immediately clear, but the Chinese-English ecotranslator can keep the following principle in mind: the Chinese names of a plant listed in the PoT might survey the botanical nomenclature of the entire world. They might come from nineteenth-century Western nature enthusiasts like Robert Swinhoe. They might come from Japan, for Japanese botanists had studied Taiwan’s flora systematically before the Kuomintang arrived in 1945. They might also come from different parts of China, because of the massive influx of “mainlanders” around 1950.

In addition to Mandarin bionyms, the PoT also lists Taiwanese (also called Southern Hokkien) names brought by immigrants to Taiwan from Fukien Province in the eighteenth and nineteenth centuries, and sometimes the Taiwanese duplicates the Mandarin: *bé-pin chháu* is horse-whip herb, *hih bé-pin* is iron horse-whip, *thih tiò-kuann* is iron fishing-rod, and *tê-bí chháu* is tea-leaf herb. The question is: did the Taiwanese come from the Mandarin or is it the other way around? That “tea-leaf herb” in the Chinese is literally “tea-rice herb” suggests that the direction of influence in this case is from Taiwanese to Chinese, because “tea rice” is the Taiwanese term for tea leaf. If the term was originally Chinese it would have been *ch’a-ye ts’ao* not *ch’a-mi ts’ao*, *ye* for leaf, not *mi* (*bí* in Taiwanese). That leaves dragon-tooth herb as the only term mentioned in the previous paragraph without a Taiwanese analog. According to the *National Language Dictionary* (*Kwo-yü Ts’u-tien*), dragon tooth is a word for dumpling. Whatever the relevance of a dumpling to this species of plant, the term dragon-tooth herb is definitely not Taiwanese.

To a lesser extent, the PoT also lists terms in Hakka and in languages spoken by people who have been officially recognized as indigenous since a constitutional amendment in the mid-1990s. In the case of horse-whip herb the PoT reports that the Paiwanese people, who live about a hundred miles south of I-lan, call the plant *marupoyopoyomu*. *Marupoyopoyomu* was recorded in katakana in a 1928 name list of Taiwan’s flora by the Japanese botanist Syuniti Sasaki, and romanized from the katakana by Yang Tsai-yi (1982). At some point the PoT should be updated with the romanizations developed for Taiwan’s indigenous languages by Paul Jen-kuei Li (2004) who has written about plant names in Formosan languages. As I write, teams of indigenous intellectuals all over the country are compiling collections of short articles on plant nomenclature, whose results, ideally, will eventually be incorporated into the PoT. And as we shall see in the second half of this chapter, indigenous bionyms are important in Wu Ming-yi's nature writing. But Huang Chun-ming shows no awareness of them. He selected the term horse-whip herb from a set of Mandarin and Taiwanese synonyms, because that term was the first to come to mind.

One wonders, though, if he managed to select a local term? Is horse-whip herb perhaps unique to Huang's home county of I-lan. No, horse-whip herb appears for the first time in Taiwanese historical records in the mid-eighteenth century in the *Revised Feng-shan County Gazette*, which is to say from the area today called Kao-hsiung, in the southwest of the island.[3] But the term is attested much earlier in China, in a medical treatise called *Further Records of Famous Physicians* (*Ming-yi pie-lu*), dating to the third and fourth centuries CE (Li Jingwei et al. 2004: 178). In other words, there is nothing I-lan about it. The only reason Huang selected it is because it is the name by which he knew it or thought his readers would know it by.

Still, the name by which Huang knew it is evocative. It is not an arbitrary association of sound and sense. For plants arbitrarily associated with a sound, all that the translator can usually do is find the corresponding term, if there is one, in the target language. What other way to translate rose, roughly [ɹoz] in International Phonetic Alphabet symbols, is there besides *mei-kuei* [mei-kʷei]? By contrast, horse-whip herb would seem different under another name, because unlike rose (or *mei-kuei*) it is descriptive, comparing the plant to a riding crop. For readers in Taiwan who can recognize the plant, horse-whip herb also evokes a mental image of the actual plant. The problem in translation is that "horse-whip" on its own cannot possibly evoke an image of the actual plant, only of a horse-whip. Given that the schoolboys, the civil servant, and the young lady do not know the name of the flower for whose name Huang is waiting, they might be as ignorant of "horse-whip herb" in Chinese as the reader of the English translation. This is the problem for Huang's modern readers, in Chinese or in English translation, who are estranged from tradition and from nature. Huang was primarily concerned with educating people about tradition, but readers of his essay might be inspired to learn more about nature. Today, with Google and Wikipedia, his readers in Chinese can look up pictures of the plant, which might be the first step toward noticing it in daily life, developing a personal connection to it, showing concern for it, caring about it.[4]

6.2 Searching for the Name of a Plant with Carl Linnaeus

Readers in English translation might also be able to look up pictures, because "horse-whip herb" is listed on certain websites as the Chinese name of the plant in question. But to make sure the English reader gets the picture, the translator would almost certainly supply a common name in English. The common name—the entrance to a different maze—is either verbena or vervain. Verbena and vervain, it turns out, are cognate. The Latin *verbena* was adopted into French as *vervain*, and both were adopted into English, though their meanings have narrowed considerably. In the Middle Ages, the French term *vervain* was an "herbaceous plant much valued medicinally," according to the Online Etymology Dictionary. In Latin, *verbena* is "leaves or twigs of olive, myrtle, laurel, or other sacred plants employed in religious ceremonies." Verbena and vervain are the most common common names, but verbena goes by, or at least it has gone by, many other names in English, not just vervain but also simpler's joy, where a simpler is a preparer of simples, meaning herbal formulas. It is also called holy herb, bringing in the religious associations of *verbena* and *vervain*. Apparently, it is called ironweed in the United States, because of the high iron content of its roots (Ahmed et al. 2012: 68). An education in the nomenclature of an organism is an education in its natural and cultural ecology.

Alas, these terms are no longer in current use; the reader is likely to have seen them for the first time. As a modern ecotranslator, I feel sad that these local terms, which reflect traditional uses or outstanding properties of the plant, are passing away. I also find poignant Renato Tomei's elegy to traditional nomenclature in her "Multilingual Phytonymy: Ecotranslation and Vernaculars," a chapter in the book *Descriptions, Translations and the Caribbean: From Fruits to Rastafarians*. Tomei studied Caribbean plant terminology, which she terms "phytonymy." According to Tomei's account, Caribbean phytonymy has for the most part lost currency, or has even been forgotten. Tomei not only helps us remember but also proposes a theory of our forgetfulness: we can blame Linnaeus, because "the systematic model of the Linnaean system caused phenomena of loss and language erosion to names of plants in native languages" (2016: 34). In support of this claim, Tomei references *Imperial Eyes: Travel Writing and Transculturation* by Mary Louise Pratt (2007), who begins her account of the imperial gaze with Linnaeus, who sent out "apostles" to all corners of the earth to be his eyes in absentia. Apparently, "imperial eyes" saw a single species and imposed a single name, willfully ignoring the diversity of folk phytonymy.

Linnaeus is being scapegoated. Would Carolus Linnaeus have wanted us all to speak the same language when he went by Carl von Linné in his native Sweden? And in Mandarin he is Lín-nèi.[5] To Linnaeus, the scientific name was just a reference to an objective biological reality that all languages could refer to in their own ways. In the wake of biologists like Charles Darwin and E. O. Wilson, we do not have to understand the objective reality of a species in the same way as Linnaeus did. Without forfeiting objectivity, we can recognize that any species can be in flux, its boundary fuzzy, just as Linnaeus could surely recognize and accept that some designations for flora and fauna denote cultural categories not biological realities. Verbena is a case in point. The Latin *verbena* guided Linnaeus when he named the species in his *Species Plantarum*, published in 1753. To *Verbena* Linnaeus added *officinalis*, a term used for medicinal herbs, to form the binomial species name, hence *Verbena officinalis* L., where L. stands for Linnaeus.[6]

This particular binomial name conveys the cultural significance of this kind of plant, but like any binomial name it positions a particular species in the family tree of life. From general to specific, *Verbena officinalis* L. is a eukaryote, a plant, an angiosperm, a eudicot, and an asterid; it belongs to the order Lamiales, which is to say that technically speaking it is not a grass; it is not a member of the order Poales, which is why I translated the *ts'ao* in *ma-pien ts'ao* as herb and not grass.[7] It also belongs to the family Verbenaceae. The resemblance between the family name Verbenaceae and the genus name *Verbena* is a clue that the genus is the type genus, meaning representative genus, of the family. Furthermore, it turns out that *Verbena officinalis* L. is the type species of the genus *Verbena*. And just as *Verbena* is the type genus of Verbenaceae and *Verbena officinalis* L. of *Verbena*, there is a specific specimen that is the type specimen of the species. It can be viewed online.[8]

A type specimen is a node in the family tree of life, a hook, so to speak, from which to hang the species. A species is a category of individual organisms that might, in theory, freely mate with the type specimen and produce fertile offspring. This definition leaves the boundaries of the species blurry, because one can imagine an individual *B* close enough to the type specimen *A* to reproduce with it and another individual *C* close enough to *B* but too far from *A*. Still, the boundary of a species is not so blurry that we must throw up our hands and give up on the idea. The idea of a species is still a firm enough foundation for ecotranslation.

Not everyone sees it this way, however. To some, Tomei likely included, *Verbena officinalis* L. is simply one perspective on the diversity of life, one motivated by globalist pretensions but which is no more true than the various folk phytonyms. Corresponding folk phytonyms might include more or less of life than *Verbena officinalis* L., and different nomenclatures might divide life up differently. If so, nomenclatures, whether scientific or folk, would strictly speaking be incommensurable, which is to say untranslatable.

It is for the sake of translatability that I would grant the scientific name a special status. A scientific name facilitates what Anthony Pym calls natural equivalence (Pym 2010: 6–23), where natural means pre-existing in nature, equi-valence equal value. Assuming that a scientific name refers to a pre-existent category of nature allows the translator to hypothesize that folk phytonyms from around the world are semantic equivalents. Horse-whip herb and verbena would then be synonyms, different names for the same thing: a fuzzy-boundaried flux. This fuzzy-boundaried flux may seem to possess an essence without the depth of perspective of evolutionary time. With their own quotidian, practical perspective, indigenous peoples may assume that each kind possesses its own essence. On this issue the biologists and the indigenous people would disagree. However, they agree to an impressive degree on classification. Ethnobotanists, who study different peoples' understandings of botany, have found an impressive correspondence between folk nomenclature and scientific consensus at the levels of species and genus (Kaesuk-Yoon 2009: 119–142). It is because this correspondence is so impressive that I venture to propose that the bionyms in different languages can be treated as natural equivalents, in other words as synonyms. Of course the correspondence is not complete. But when folk nomenclature differs it tends to be because the term is functional, denoting a class of types with a similar purpose, not a class of individuals that are related without regard to utility. Verbena and vervain, after all, are used to mean plants used in religious ceremonies, but everyone could recognize that plants used in religious ceremonies might well be classified differently if considered according to their own qualities.

Above I defended Linnaeus against those who would blame him for imperiling local plant names the world over and argued that recognizing the special status of the scientific name does not mean we have to ditch all the wonderful names by which a kind of life has been called. However, while Linnaeus himself was not a bionym imperialist, in the sense of wanting to replace local names with scientific names, his imperial gaze does imply the imperial enterprise. I would not adopt the extreme position that any influence, imperial or otherwise, is automatically wrong, but I am of course well aware that the imperial enterprise poses a threat to local flora and fauna. Introduced foreign species may eventually become invasive, outcompeting local species. The introduction of foreign species was studied most famously in *The Columbian Exchange* (2003) by Alfred Crosby, who paired import with export to claim that there was botanical give and take in the encounter between Columbian Europe and Aztec America. However, Crosby also noted significant casualties of exchange: many species suffered extinction as a result of the visits of Columbus and others.

Crosby's study can be generalized: people move around and bring species with them, species that may displace endemic species in the places they visit. According to the "out-of-Taiwan hypothesis" of Austronesian cultural dispersal, for instance, the indigenous peoples of Taiwan took the pig and the paper mulberry everywhere from Madagascar to Easter Island (Bellwood 1979; Chang et al. 2015). It might seem strange to say so, but the Austronesian introduction of the pig and the paper mulberry to islands big and small from Madagascar to Easter Island seems somehow imperialistic. Even so, it has nothing to do with capitalism; more contemporary examples, by contrast, have everything to do with capitalism. There is, for instance, a "collusion between public and private" (*kuan-shang kou-chieh*) behind the planting of community parks in Taiwan with trendy imported plants—like *Plumeria alba*, named Plumeria after Charles Plumier, who "discovered" it in the 1690s, and frangipani, after a mythical Columbian botanist invented by a late nineteenth-century perfumist (Kettler 2015)—in order to line the pockets of bureaucrats and horticulturalists.[9] While these plants may originally have been extracted by imperialists, they are now exploited by capitalists. Botanical imperialists and capitalists in East Asia are also well studied, people like George Thomas Staunton (Fan 2004) and Ernest Henry Wilson (Spongberg 1990). Such people spread attractive plants around the world, so that everywhere you go you see the same ornamental plants in the local landscaping, just as you tend to see the same weeds in the local landscape. Horse-whip herb is a case in point. It and the other three plants that Huang Chun-ming mentions are weeds.

6.3 Sustaining Bionym Diversity Through Translation

While as translators we can hardly directly address the threat to ecological diversity posed by imperialism and capitalism (or by inadvertent introductions like the zebra mussel into the Great Lakes), we can at least strive to uphold the value of bionymic diversity, which might offer us indirect assistance as we try to solve or mitigate environmental problems. This is in fact my thesis in this chapter, that translators *should* do their part to maintain terminological diversity by balancing the local and the global, the common and the scientific in translations of bionyms. More specifically, the translator should consider translating the local name literally, either as a loanword or as a calque, where a loanword captures the sound, a calque the meaning, of the original. In other words, there are two kinds of literal translation, sonic and semantic. In translation studies, literal translation, whether sonic or semantic, is termed foreignization and regarded as a way of drawing the reader closer to the original. The opposite of foreignization is domestication, which substitutes a term in the target language with which the reader is familiar. If the reader is not familiar with the common bionym, domestication is not an option, but in that case supplying the reader with what was once a common, familiar bionym gives him or her an opportunity to familiarize himself or herself with the lifeform and its nomenclature.

The lifeform is the scientific side of the issue, its nomenclature the cultural side. In the translation studies literature, the translation of bionyms is often regarded mainly as a cultural challenge. The translator is supposed to consider the cultural significance of a plant in the original and the target languages and adapt accordingly (Rydén 1983). It should go without saying that the translator should be sensitive to cultural context, and part of my purpose in this chapter is to argue that the sound and sense of the bionym itself is part of its cultural significance: perhaps vervain was actually used as a makeshift horsewhip back in the day. But I am also arguing that the meaning of a plant in a work of modern literature depends on what exactly it is scientifically. As a kind of specialized translator, the modern ecotranslator therefore has a professional responsibility to know the scientific side of the issue, to appreciate, for instance, that the cherries in Anton Chekhov's cherry orchard were *Prunus cerasus*. It was, in other words, a "sour cherry orchard" (Osimo 2013), a translation that, it just so happens, better fits the tragicomedy of the play. It often happens that a scientific perspective makes a literary translation more effective. Ecotranslators' principles like representing nature as possessing agency if the original work does and trying to hear "the voice of nature" (Badanes and Coisson 2015: 356 *et passim*) have their place in the literary translator's toolkit, but they are no substitute for knowledge of nature, without which any claim to love nature sounds rather hollow.

I should also distinguish my approach to ecotranslation from the school of translation studies calling itself eco-translatology. This school, which developed in China, attempts to apply concepts from biology, not just ecology but also evolution, to explain how terms and translations are selected in competitive markets (Hu 2011). In other words, it is a descriptive theory with a pseudo-scientific vocabulary. Michael Cronin's approach, by contrast, is prescriptive in his recent monograph *Eco-Translation* (2017), in which he exhorts translators to work out of an awareness of the ecological challenges our planet faces today. To Cronin's exhortation, I would like to add my own imperative to anyone who is translating bionyms: pause to do a bit of research and sample the diversity of nomenclature, including scientific name(s) and common names, then translate the diversity. The translation of this diversity by a translator who loves it might encourage readers to treasure it more.

6.4 Translating the Name of a Fig with Wu Ming-Yi

My imperative to translate the diversity may seem overly demanding for a writer like Huang Chun-ming, but with a nature writer like Wu Ming-yi, it is not demanding at all. Huang, I think, was reacting inchoately to the problematic I have been exploring in this chapter: technologies, including the bicycle and the scooter, have separated people from nature, have excluded wilderness from daily life, such that many modern people, including many who profess to love nature, have little familiarity with it or knowledge of it. But since Huang does not really pay that much attention to nature or its nomenclature, the translator who wants to take the opportunity to educate the reader has to supplement. With Wu Ming-yi, by contrast, the educational objective is in the text; Wu often explains the ecology of a species and its nomenclature in several different languages, so that a polyglot ecotranslator can just follow along.

In his early essays, which he published in the early 2000s, Wu tries to educate the audience, for instance by placing a tourist destination in historical ecological perspective. In "The Compass of the Soul," his essay on Mei-nung (Wu 2018b), the hometown of the late nativist writer Chung Li-ho, Wu explains why a single species of butterfly, the lemon migrant (*Catopsilia pomona*), thrives there. Because of the groves of iron-sword trees (*Senna siamea*) that were planted by the Japanese to yield wood for gunstocks.[10] Why iron-sword trees? Because they are fast-growing and produce a hard wood with enough flexibility to contain recoil. That they attracted lemon migrant butterflies was an ironic bonus, ironic because it was unexpected. What shall we do about such ecological ironies? While Huang Chun-ming seems to resist modernity and to wish to restore the good old days, Wu Ming-yi asserts that "any kind of restoration is change." The way is thus forward, toward a better environment for life, both human and non-human. Such an environment would be diverse, for Wu declares a preference for mixed stands of trees that would be home to a wide variety of butterflies.

In educating the reader about environmental variation, Wu can seem a bit demanding, because he gets technical. He considers the status of three varieties of lemon migrant—lemon-yellow, silver-dot, and ochre-dot—with reference to E. O. Wilson's 1953 co-authored academic attack on the notion of subspecies, "The Subspecies Concept and Its Taxonomic Application" (Wilson and Brown 1953). According to this article, any species has a clear definition, the free-interbreeding criterion—that members of a species interbreed freely—but any subdivision based on observed features is arbitrary, because it depends on what features one chooses to or is able to observe (Wilson and Brown 1953: 99). As Wu notes, Wilson later reconsidered his attack and came to accept the utility of the concept of subspecies as a way of referring to distinct populations of a species whose genetic status was in doubt (see Lott 2015, citing Wilson 1992 and 1995). Wu's reflection on Wilson's attack and its reconsideration achieves several aims. It informs the reader that scientific understanding is in flux, not fixed, because the objects of knowledge are constantly changing, and, even at any given point in time, the members of a species display variation that scientists can categorize in different ways. Wu's main thesis in the essay can be expressed in terms of variation: it would be better to see more variation in the surroundings, in both senses of the term "see." In other words, Wu thinks it would be better for the environment to be more diverse and for us to discriminate more diversity.

Wu's predilection for diversity also applies to nomenclature, to all the terms we use to discriminate diversity. Take *Catopsilia pomona*, for instance. To Wu, the common name in English, lemon migrant, "sounds as if a certain color has gone wandering." Wu also notes that the genus name *Catopsilia* (which means smooth underside, because the butterfly has relatively few scales on its wings) is in the PRC termed "mobile scales." He mentions these things because he finds them evocative and educational, and hopes the reader does, too, in addition to appreciating the understated Mandarin bionym "pale-yellow." The only kind of bionym Wu seems not to like is more about personal or imperial glorification than about the description of a plant or an elicitation of its cultural significance. In another essay, "Koxinga" (Wu 2018a), Wu reflects on a subspecies of butterfly named by German insect trader Hans Fruhstorfer after Koxinga (Kok-sìng-iâ in Taiwanese, Kuo-hsing-yeh in Mandarin), the conqueror of Taiwan. He named the subspecies after Koxinga, because Kuo-hsing Township is where the type specimen was taken. The full name is *Euploea tulliolus koxinga* Fruhstorfer. The common term in Chinese is little purplespot, and in English it is called the dwarf crow. Though naming the creature at all is in some sense imperialistic (for what name would the butterflies themselves use?), Wu's perspective is nevertheless anti-imperial. Why should Fruhstorfer, who never set foot in Taiwan, be allowed to name it after the conqueror Koxinga, who never visited the township named after him?

Wu's anti-imperial perspective does not, however, entail an extreme nativism. Wu is only rejecting a certain approach to the naming of lifeforms, not rejecting the lifeforms themselves. Though he may have a special place in his heart for vulnerable endemic species, he is welcoming toward newcomers, whether short-term visitors or naturalized residents.[11] His accommodating attitude is most evident in his novel *The Man with the Compound Eyes* (Wu 2013), in which I counted fifty-six kinds of life, not including human beings. Wu's novel is populated by species that represent Taiwan, like the Taiwan long-armed scarab (*Cheirotonus formosanus* Ohaus) that Toto sees right before he dies (Wu 2013: 286). But these species share the island with numerous naturalized species that can be found all over Asia or all around the world, like the common scarab Alice finds on the dash of her car (Wu 2013: 21).

Wu's attitude to people is similarly welcoming. The character who knows the island flora and fauna best is Atile'i, an aborigine from another island, which is to say a foreigner, one who can recognize all the birds in the vicinity at a glance (Wu 2013: 214). The local indigenous woman Hafay knows the wild food plants almost as well as Atile'i, but their practical indigenous knowledge is complemented by the scientific understanding of Sara, a Norwegian ocean biologist who has come to the island to study the impact of the trash piled on the beach on marine ecology. Indeed, part of the point of the novel is to introduce aspects of Taiwan ecology. Alice asks her friend Ming—the author Wu Ming-yi's avatar in the novel—for advice about what to plant outside her seaside house to attract butterflies (2013: 53). She is asking at once about plant-environment and plant-insect interactions, in other words about the kinds of interaction that create the network of connections that is any ecosystem.

Wu also tries to educate readers about bionyms in his novels, as in the naming of a kind of tree that is symbolic of ecosystemic connection: fig trees. In *The Man with the Compound Eyes*, the *pai-jung* is symbolic of ecological interconnection, because a single tree can, by means of prop roots, form a grove that is actually a single individual. At first I translated *pai-jung* as albino banyan. I could also have rendered it as white fig, because a banyan is a kind of fig, a species in the genus *Ficus*. I chose albino banyan simply because the rarity of albino made it sound more poetic to my ear and because I thought *jung* was banyan. In other words, I did not know at the time that a banyan is a kind of fig. (Actually, any fig that starts out as an epiphyte, a plant that grows on another plant, can be called a banyan.)

Now how should white fig or albino banyan be categorized in terms of the translation strategies mentioned above, sonic and semantic, literal and free, foreignizing and domesticating? Both ignore the sound of *pai-jung*, and are therefore semantic. White and albino are both literal translations of the sense of the descriptive part of the bionym and are foreignizing. Fig and banyan are arguably literal translations of the sense of the purely biological part of the bionym, but they are familiar. The only way to foreignize in this case is to treat *jung* as a loanword and translate the sound.

In the translation of this bionym, I educated myself about the cultural and scientific background, and considered various literal options, to try to live up to my manifesto. But ultimately, my manifesto is more a flexible guideline than an iron rule, and in this case I did not foreignize, because it did not make the most literary sense. I looked up the scientific name, *Ficus benjamina* L. According to the *Flora of North America*, Linnaeus was thinking about its economic use: "The name probably refers to the supposed relation of the plant to the source of a resin or benzoin procured from the Orient in antiquity."[12] *Benjamina* is the feminine form of

Benjamin (to agree with the gender of *Ficus*). This is not to say this Benjamin is a girl. It is not the name of a person at all, but rather the name of a resin. Benjamin is actually a domestication of the Arabic for the substance, *lubān jāwī*, meaning frankincense from Java. This etymology, though fascinating, did not turn out to be important to my translation: I went with the most common name in English, weeping fig. It is so called because its aerial roots dangle to such an extent it appears to weep. That its aerial roots dangle is why the species is called "hanging fig" (*seoi*^{4}-*jung*^{4}) in Hong Kong. I opted for weeping fig for several reasons. First, because the weeping fig is a reasonably common house and hedge plant, so that English readers might have some chance of knowing it by that name. Secondly, because weeping associates much more readily with death than the color white (which is in the Mandarin bionym "white fig" because it is the color of the wood). This tree is associated with death because one of its branches fell on the character Anu's son Lien's head, killing him instantly. And Lien's death is associated with the theme of interconnectedness by the notion that in some sense, Lien (whose name I should have spelled Vilian, the official Bunun spelling) lives on in the whole of which he was a distinct part when he was alive.

Out of a commitment to nomenclatural diversity, Wu names the tree again. The tree is named once more in the novel because it grows in a Bunun indigenous community, the village of which Anu is the chief. This tree has a Bunun name, Vavakalun, a single word that, as explained in the novel, means a tree that can walk.[13] As the narrative explains, the Bunun people once took them as territorial markers, but then they noticed that the markers had a tendency to move. Out of my own commitment to diversity, and out of sheer curiosity, I wondered how Vavakalun was translated into other languages in other translations of the novel. In Gwennaël Gaffric's French, it was "*les arbres marchant*" (the trees that walk) and "*figuiers pleureurs*" (crying figs). In other words, Gaffric did as I did. In Pavlína Krámská's Czech, it is "*strom, který dokáže chodit*" (the tree that is able to walk) and "*bílá fíkovník drobnolistý*," where *bílá* means white, *fíkovník* fig, and *drobnolistý* tiny. In other words, Krámská added the white in the common name in Mandarin to the common name for *Ficus benjamina* in Czech, adding a foreignizing touch.

In two cases in *The Stolen Bicycle* (Wu 2017), Wu merely refers to "giant *jung* trees," a genus-level designation which I translated as banyan. The first case is at a temple, the Hall of Holy Mother Matsu, where the narrator's father takes a rest. This, Wu reports to me by email, is an "upright banyan," an Indian laurel fig. The second is the massive banyan tree that served as a fortress for Li K'e and his men against a Japanese onslaught in northern Burma. It is obviously reminiscent of the Vavakalun, a figure for the interconnected natural world, for the tree is home not just to humans but also to a multiplicity of flora and fauna on which the humans depend. According to Wu Ming-yi (again, by email), this tree in northern Burma may well be a Vavakalun, a white weeping fig that can walk. It is apparently a real tree, the setting for a real battle in northern Burma in 1945. One wonders why Wu did not name these two trees more specifically than *jung*. Probably because the characters only knew it on that level. At any rate, since Wu generalized, I decided to generalize in my translation, too.

In the naming of another species of fig tree in *The Stolen Bicycle*, however, Wu insists on specificity. The first time this tree appears it is called the birdcrap banyan in Taiwanese (Wu 2017: 18). Starting from the age of eight, the narrator grew just like a birdcrap banyan. In a bracketed in-text note, Wu informs us that the birdcrap banyan is called the sparrow banyan in Mandarin. It is a tree whose figs attract sparrows, with an obvious connection to crap. The scientific name is *Ficus superba*, *superba* because it is a keystone species, a species that is an ecosystem unto itself. In English, it is commonly called a "deciduous fig," because it sheds its leaves in the spring. Here, in the first chapter, I translated it as "birdcrap banyan," with "sparrow banyan" in brackets, reserving "deciduous fig" (Wu 2017: 376) for my translator's note at the end of the novel. In the fifth chapter, however, the deciduous fig makes another appearance, as the *yono* (Wu 2017: 141), which in Wu's Chinese text is treated as a loanword and glossed as a sacred tree of the Tsou people. An online Tsou dictionary identifies the *yono* as a sparrow banyan. It may seem ironic that the birdcrap banyan of the Taiwanese is the sacred tree of the Tsou, but as Julia Kristeva argued, shit has something in common with the sacred. Perhaps the banyan is a key to the interpretation of Wu's novels, in which he is trying to probe "the border of [his] condition as a living being" (Kristeva 1982: 3), to get readers to think beyond themselves.

6.5 Conclusion

At the beginning of the first chapter of *The Stolen Bicycle*, Wu's narrator declares a passion for bicycles and their nomenclature. After listing the word for bicycle in half a dozen of the world's languages, he states that when it comes to bicycles he is a "polyglot" (Wu 2017: 8). In this chapter I have tried to show that when it comes to bionyms Wu tends to be trilingual, and I have argued that the ecotranslator of bionyms should be polyglot. With my translator's note to *The Stolen Bicycle* there are four translations of *Ficus superba*, and with the French and Czech translations of *The Man with the Compound Eyes*, there are five translations of *Ficus benjamina*. Three or four is already an impressive number, when editors and readers usually prefer one. Wu and his translators challenge us to learn various different bionyms. One purpose behind their challenge is to maintain terminological diversity. Diversity in ecology is insurance against cataclysms that might wipe out all the individuals in a species if they are too genetically similar, which might in turn disturb an ecosystem. Having titled my chapter ecotranslation and argued that Wu Ming-yi educates the reader about ecosystems, I wonder now if there is a bionymic ecology analogous to natural ecology, wherein it would behoove us to maintain nomenclatural diversity, just as diversity in nature is the source of the resilience of nature.

The analogy should not be pressed too far, because words are not alive in the same way as organisms are, and they do not depend on one another the way that organisms depend on one another. But like organisms in an ecosystem, words are related to each other systematically by virtue of language. Just as there are different ecosystems in nature, each language is a different ecosystem, where the words relate to each other in different ways and to the world in different ways. Each literary work, too, is like an ecosystem that might let us see new connections in the world, including the world of nature, in potentially enlightening ways. In Arnhem Land in Australia, the word for a white apple tree is *bokorn*, as is the name of the fish that eats its fruits when they fall into billabongs. "Once we go over to calling the *bokorn* fish a 'spangled grunter,'" comments Richard Evans, "and the *bokorn* tree a 'white apple,' our words no longer deliver the ecological link between them" (2009: 22). It is these kinds of links that form any living system, links we might not perceive if we do not have the right words. It is these same kinds of links that a literary work is composed of, or that any attempt to make meaning of the world is composed of. It is impoverishing to translate in a way that fails to deliver these links, which educate us about nature in its complexities and the complexities of our relationships to it. And if translation delivers these links it is enriching, because it adds spangled grunter, which tells us something new about the *bokung*, as does white apple, as do *Ficus superba*, birdcrap banyan, sparrow banyan, deciduous fig, and *yono* about the tree they collectively name.

Indigenous peoples of Arnhem Land, Taiwan, and all over the world have something else to teach us about nature and how we can relate to it in their practices of naming. In the scientific worldview, plants and animals and minerals are objects of knowledge for a scientist to describe objectively, not in terms of his or her relation to them. But like it or not, the scientist has some relationship to what he or she studies, and the scientist's relation to nature often plays into a culturally specific attitude in which plants, animals, and minerals are mere objects, resources to be exploited. As Wu Ming-yi, along with anthropologists of human-nature relations like Philippe Descola (2013), remind us, indigenous peoples relate to the natural world in other ways. Descola is aware that to speak of "the indigenous worldview" is an oversimplification; his four-term typology of naturalism, animism, analogism, and totemism (2013: 233) is a reminder of the diversity of attitudes to nature in indigenous cultures. Yet we naturalists, who tend to treat everything non-human as an object (2013: 392), can recognize that the animists, the analogists and the totemists have something in common: they treat animals and plants and even non-living things as persons.

I therefore call on modern ecotranslators to test the personal approach by using bionyms in new ways. Scientists use bionyms to refer to species, as in "This is a weeping fig." Of course, indigenous people can do the same, as in "This is a Vavakalun." But we know from our everyday social experience that there is another way to use a name: to address someone in various ways, by way of greeting, praising, commanding, disparaging, and so forth. In social interaction, we use someone's proper name. We sometimes assign proper names to non-human creatures, like the General Sherman tree, but we might do so more widely. Alternatively, we might use categorical names to address individual members of a category, as in "Hello, weeping fig." To do so is to ascribe something like personhood to the organism in question. To ascribe something like personhood to a non-human organism transcends the naturalist worldview in which we humans are subjects in an object world that we can understand the better to manipulate it. Trying to imagine the human-plant or human-animal encounter interpersonally will at least make us more sensitive translators. Ecotranslators who try this perspective will be more likely to notice when their authors adopt it.

It stands to reason that nature writers are more likely to adopt it. Huang Chun-ming never treats the flower whose name he waits for as a person, but Wu Ming-yi's narrative does occasionally ascribe personhood to animals, plants, and non-living things, referring to "the tricks the wind and the trees play" (Wu 2017: 147). And Anu in *The Man with the Compound Eyes* addresses the spirit of the mountain to show his respect (2013: 259). In his essays, Wu Ming-yi at least once addresses a butterfly, "tiger butterfly of mine" (Wu 2015: 17). When we are prepared to address plants as people, we might well start to hear what Badanes and Coisson (2015) call the voice of nature.

Meanwhile, we are *still* waiting for another name besides riffraff flower for the wild, blooming beauty Huang Chun-ming waited by.

Notes

1. Etymologically, nativism suggests birth, and was first used in 1845 of an anti-immigrant, American-by-birth movement in the United States. In literary studies, it refers to literary movements that have reacted in various ways against modernity. One such movement is the nativist movement of Taiwan, where nativist translates into *hsiang-t'u*, literally "rural (or homeland) soil." For Taiwan's nativism and its links to German *heimat* or homeland literature, see Lupke 2016.

   A note on Romanization: with a view to preserving diversity in the spelling of Chinese languages, I have adopted a modified Wade-Giles Romanization for Mandarin. For Taiwanese I follow the recommended Romanization of Taiwan's Ministry of Education.
2. http://tai2.ntu.edu.tw/PlantInfo/species-name.php?code=521%20014%2001%200
3. This gazette is the 164th text in the *T'ai-wan wen-hsien ts'ung-k'an* (1961), searchable at Academia Sinica at http://hanji.sinica.edu.tw/index.html
4. For an anthropological perspective on personal connections to plants and animals, with examples of common names from the indigenous language Taroko (also known as Truku), see Simon 2017, Accessed 2018.
5. See Cook 2010 for a defense of Linnaeus against the charge of linguistic imperialism.
6. *Verbena officinalis* L. only has one scientific name, but others go by many names, because there may be superseded terms that remain in use as synonyms. There are, for instance, twelve synonyms for *Senna siamea*.
7. I have referred to https://en.wikipedia.org/wiki/Verbena_officinalis, which lists the sources for this representation of the place of this species in the tree of life.
8. The type specimen, technically the lectotype (so called because it was not chosen by Linnaeus, the namer of the species, himself), is located in the

George Clifford Herbarium, now in the Natural History Museum in London. The database can be searched by inputting the genus and species names: http://www.nhm.ac.uk/research-curation/scientific-resources/collections/botanical-collections/clifford-herbarium/database/index.dsml

9. For the use of the term collusion, and for the specific examples, I am indebted to Jerome Su, the Chairman of Bookman Books and my mentor in all things botanical.
10. Note that I supply no common English name, because none exist, unless cassod or kassod seem common to enough people. This is another common issue of ecotranslation, that the places where the species is common are not English-speaking.
11. For a history of American bionativism, in which foreign species become symbolic of foreign peoples, see Coates 2006. For a critique of the bionativism in Taiwan, see the documentary *Exotic Exoticism* (Liu Chi-hsiung 2007).
12. http://www.efloras.org/florataxon.aspx?flora_id=1&taxon_id=200006350
13. The morphology of the word Vavakalun seems to be *va-vakal-un*, where *va-* is probably "habitual reduplication" (Rik De Busser, private communication), *vakal* the leg of an animal, and *-un* a passive focus marker, probably indicating that the noun is derived from a verb.

Bibliography

Ahmed, Dildar, et al. 2012. "Comparative Study of Antibacterial Activity and Mineral Contents of Various Parts of Verbena officinalis Linn." *Asian Journal of Chemistry* 24 (1): 68–72.

Badenes, Guillermo, and Josefina Coisson. 2015. "Ecotranslation: A Journey into the Wild Through the Road Less Travelled." *European Scientific Journal* (Special Edn.) November: 356–368.

Bellwood, Peter. 1979. *Man's Conquest of the Pacific: The Prehistory of Southeast Asia and Oceania*. Oxford: Oxford University Press.

Chang, Chi-shan, et al. 2015. "A Holistic Picture of Austronesian Migrations Revealed by Phylogeography of Pacific Paper Mulberry." *Proceedings of the National Academy of Sciences of the United States of America* 112 (44): 13537–13542.

Coates, Peter. 2006. *American Perceptions of Immigrant and Invasive Species: Strangers on the Land*. Berkeley/Los Angeles: University of California Press.

Cook, Alexandra. 2010. "Linnaeus and Chinese Plants: A Test of the Linguistic Imperialism Thesis." *Notes and Records of the Royal Society* 64 (2): 121–138.

Cronin, Michael. 2017. *Eco-Translation*. New York: Routledge.

Crosby, Alfred. 2003. *The Columbian Exchange*. Santa Barbara: Greenwood.

Descola, Philippe. 2013. *Beyond Nature and Culture*. Trans. Janet Lloyd. Chicago: University of Chicago Press.

Evans, Nicholas. 2009. *Dying Words: Endangered Languages and What They Have to Tell Us*. Oxford: Wiley-Blackwell.

Fan, Fa-ti. 2004. *British Naturalists in Qing China: Science, Empire and Cultural Encounter*. Cambridge: Harvard University Press.

Hu, Gengshen. 2011. "Eco-Translatology: Research Foci and Theoretical Tenets." *Chinese Translators Journal* 2: 5–9.

Huang, Chun-ming. 1989. "Teng-tai yi tuo hua teh ming-tzu." In *Teng-tai yi tuo hua teh ming-tzu*, 45–51. Taipei: Huang-kuan.

Kaesuk-Yoon, Carol. 2009. *Naming Nature: The Clash Between Instinct and Science*. New York: W. W. Norton & Company.

Kettler, Andrew. 2015. "Making the Synthetic Epic." *The Senses and Society* 10 (1): 5–25.

Kristeva, Julia. 1982. *Powers of Horror: An Essay on Abjection*. Trans. Leon S. Roudiez. New York: Columbia University Press.

Li, Paul Jen-kuei. 2004. "Some Plant Names in Formosan Languages." In *Selected Papers on Formosan Languages*, vol. 1, 767–803. Taipei: Institute of Linguistics, Academia Sinica.

Li Jingwei, et al. eds. *Grand Chinese Medical Dictionary* (in Chinese). Beijing: Renmin Weisheng chubanshe, 2004.

Lott, Tom. 2015. "Edward O. Wilson's Recantation on Subspecies." *Cascabel: A Natural History Blog (sensu lato)*, February 19. http://cascabel.typepad.com/cascabel/2015/02/edward-o-wilsons-recantation-on-subspecies.html. Accessed 19 Oct 2017.

Lupke, Christopher. 2016. "Nativism and Localism in Taiwanese Literature." In *The Columbia Companion to Modern Chinese Literature*, ed. Kirk Denton, 258–266. New York: Columbia University Press.

Osimo, Bruno. 2013. *Literary Translation and Terminological Precision: Chekhov and His Short Stories*. Amazon Digital Services LLC.

Pratt, Mary Louise. 2007. *Imperial Eyes: Travel Writing and Transculturation*. Routledge.

Pym, Anthony. 2010. *Exploring Translation Theories*. New York: Routledge.

Rydén, Mats. 1983. "On the Translation of Plant Names in Literary Texts." *Shakespeare Translation* 9: 7–17.

Simon, Scott. 2017. "What Is a Species?" https://maptia.com/scottsimon/stories/what-is-a-species. Accessed 18 Sep 2018.

Spongberg, Stephen. 1990. *A Reunion of Trees: The Discovery of Exotic Plants and Their Introduction into North American and European Landscapes*. Cambridge: Harvard University Press.

Tomei, Renato. 2016. "Multilingual Phytonymy: Ecotranslation and Vernaculars." In *Descriptions, Translations and the Caribbean: From Fruits to Rastafarians*, ed. Rosanna Masiola and Renato Tomei, 33–70. Palgrave Macmillan.

Wilson, E.O. 1992. *The Diversity of Life*. Cambridge: Harvard University Press.

———. 1995. *Naturalist*. New York: Warner Books.

Wilson, E.O., and W.L. Brown Jr. 1953. "The Subspecies Concept and Its Taxonomic Applications." *Systematic Zoology* 2: 97–111.

Wu, Ming-yi. 2013. *The Man with the Compound Eyes*. Trans. Darryl Sterk. London: Harvill Secker.

———. 2015. "Death Is a Tiger Butterfly." Trans. Darryl Sterk. *Pathlight*, Spring, 9–19.

———. 2017. *The Stolen Bicycle*. Trans. Darryl Sterk. Melbourne: Text.

———. 2018a. "*Koxinga*." Trans. Darryl Sterk. Santa Barbara: Taiwan Literature English Translation Series. 41: 111–120.

———. 2018b. "*The Compass of the Soul*." Trans. Darryl Sterk. Santa Barbara: Taiwan Literature English Translation Series. 41: 121–132.

Yang, Tsai-yi. 1982. *A List of Plants in Taiwan*. Tai-pei: T'ien-jan shu-shih.

SECTION II

Chinese Ecocinema and Ecomedia Studies

CHAPTER 7

Worms in the Anthropocene: The Multispecies World in Xu Bing's *Silkworm Series*

Kiu-wai Chu

In a visual art project titled *Dear Climate*, a group of American artists and scholars designed "a collection of agitprop posters and meditative audio experiences that help you meet, befriend, and become climate change."[1] These posters are designed to promote environmental awareness, and particularly, multispecies co-existence in the Anthropocene age, as reflected in posters with slogans such as "BIODIVERSITY IS THE LAND OF OPPORTUNITY"; "PARTY WITH THE BEES IN THE BUTTERFLY BUSH"; "MEET THE BEETLES"; and "SPEND QUALITY TIME WITH AN INSECT." Human beings encounter insects on a daily basis and yet rarely give them much attention. When they are noticed, they are often exterminated or turned into economic resources for human consumption and utilization. What does it mean to spend quality time with an insect? What is the role of art in cultivating such an attitude in the contemporary world?

K. Chu (✉)
School of Humanities, Nanyang Technological University,
Jurong West, Singapore

C. Chang (ed.), *Chinese Environmental Humanities*,
Chinese Literature and Culture in the World,
https://doi.org/10.1007/978-3-030-18634-0_7

More and more artists have "generated multimedia installations and performative interventions to bring attention to animals, plants, fungi, and others at the periphery of anthropocentric worlds" in recent years (van Dooren et al., 9). As such, we are beginning to see the power of art in shifting our focus toward the non-human world, including that of the worms. However, in a world perceived generally in anthropocentric terms, human beings continue to occupy a role in the center while the rest (animals, insects, plants, and all non-human beings) are at the margins. Because of that, is, responding to the theme of this book, "practicing environing at the margins" a celebration and glorification of humans' role in shaping the environment and the external world? In other words, are we celebrating anthropocentrism? This chapter juxtaposes two ecocritical approaches, namely multispecies studies and Anthropocene criticism, to explore the dynamics between human and non-human agencies in the present world that drives the debates in anthropocentrism. Focusing on contemporary Chinese artist Xu Bing's *Silkworm Series* (1994–2014) and several other art installations that have made use of living insects, it explores how art facilitates ecocritical reflections upon our relationships with the non-human world.

This chapter also illustrates how Xu's *Silkworm Book* is a fitting metaphor of the multispecies condition in the Anthropocene, in which non-human beings are increasingly registering their material presence and agency to impact the world, while at the same time human beings continue to play a role in shaping it. By adopting living insects in art performances and carefully monitoring their movements and activities during the period of exhibition, Xu Bing demonstrates spending "quality time" with insects by "performing" special care toward them, as a way to rethink questions centering on multispecies co-existence and worldmaking in the age we call Anthropocene. This chapter also argues the "artfulness" of *Silkworm Book* as an ecological art lies in the tension and balancing of non-human agencies and human interventions. It presents us an open-ended platform for debates about the role of human beings in the present world (Fig. 7.1).

Fig. 7.1 "Spend quality time with an insect." *Dear Climate*. (Courtesy of *Dear Climate*)

7.1 Worms in Contemporary Art

British artist Damien Hirst has been known for his animal and insect-themed art installations since the early 1990s. As critics point out, most of his works "grapple with the seemingly alchemical, violent transformation of life to death" (Broglio, 1). Many of his controversial pieces involve whole or dissected animals, living insects, and other non-human creatures. From "A Hundred Years" (1990); "A Thousand Years" (1990); to "Let's Eat Outdoors Today" (1990–91; 2011), Hirst created a series of installations with maggots and flies. In the confined vitrine that was divided into two halves, we see one half with trays of maggots being fed slices of raw meat. The thousands of maggots hatch into flies and pass through the circular hole to the other side of the vitrine, where they gather over the leftovers of food and drinks laid on a table. Before long, the flies are electrocuted by the large insect-o-cutor hanging above the table. Hirst's "In and Out of Love" exhibited in Tate Modern in London in 2012 contained two windowless rooms swirling with live butterflies. Viewers were led to observe them close up and as they flew, fed, and rested. However, Hirst was condemned by animal rights activist groups, because it was later revealed that over 9000 butterflies died during the 23 weeks the exhibition was open.[2]

In the Chinese art scene, several artists have also experimented in creating art works with living insects, including Huang Yong Ping's controversial work "Theatre of the World." First exhibited in Germany in 1993, the installation consists of a transparent dome-shaped construction where hundreds of reptiles and insects are placed within. During the several months-long exhibition, the creatures battled against one another, some were eaten, and others died of exhaustion. They were then replaced by new creatures to keep things in action. The insects that were trapped inside the dome structure, which included spiders, scorpions, crickets, cockroaches, black beetles, stick insects, and centipedes, were mostly killed and eaten by the lizards, toads, and snakes that were placed within the confined battlefield. In 2017, "Theatre of the World" was planned to be a highlighted piece in the Guggenheim Museum's exhibition "Art and China after 1989: Theater of the World" in New York. However, animal rights activists started an online petition and garnered over 800,000 signatures in two weeks to criticize Huang's piece, together with two other video art works including Xu Bing's *A Case Study of Transference* (1994) and Peng Yu and Sun Yuan's *Dogs that Cannot Tough Each Other* (2003), for animal cruelty. Within a few days, the museum pulled the concerned three works in response.

Huang Yong Ping in fact was not the only Chinese artist who has experimented using live insects in art works. In the 1990s, Xu Bing's *Silkworm Series* and Cai Guo-Qiang's *The New York Earthworm Room* also created a significant stir in the international art scene. The two works are seen as a more humane and ethical use and involvement of insects in art installations.

Xu Bing's *Silkworm Series* first started in 1994, four years after he moved from China to the United States. In his *American Silkworm Series* exhibited in Boston and New York between 1994 and 1998, Xu placed silkworms under different physical environments, ranging from mulberry branches, television sets, and books, and let them move, eat, grow and transform, and spin their silken cocoons, with regular human monitoring and interferences throughout the exhibition period. China has over

2000 years of history in raising silkworms for silk production. Because breeding similar kinds of silkworms has proven to be difficult in the United States, for his art installations, Xu Bing imported hundreds of live silkworms needed from Linhai, Zhejiang, which is a region most famous for its silk production. The *Silkworm Series* entailed both installations as well as performances during which hundreds of live silkworm moths laid eggs and spun their silk over everyday objects, varying from mulberry branches, books, newspapers, and photographs, to computers and videotapes.

Xu revived his *Silkworm Series* with a two-part installation in summer 2014, exhibited in *It Begins with Metamorphosis: Xu Bing* at The Asia Society Hong Kong. In the first installation, *The Opening* (开幕式), silkworms were placed in a large vase of mulberry branches. Over the three-month period of the exhibition, the silkworms ate up the leaves until only dead branches remained, and clusters of cocoons were seen nestled within (Figs. 7.2 and 7.3). In juxtaposition with this was the second installation,

Fig. 7.2 Xu Bing, *The Opening*, 2014. Silkworms, mulberry plant, and ceramic vase. Collection of the artist. Exhibited in *It Begins with Metamorphosis: Xu Bing*, Asia Society Hong Kong Center. 2014. (Photo taken mid-May 2014)

Fig. 7.3 Xu Bing, *The Opening* (Close up). Silkworms, mulberry plant, and ceramic vase. Collection of the artist. Exhibited in *It Begins with Metamorphosis: Xu Bing*, Asia Society Hong Kong Center. 2014. (Photo taken mid-May 2014)

Fig. 7.4 Xu Bing, *Silkworm Book*, 2014. Silkworms, silk, and book. L39.5 × W39.5 cm. Collection of the artist. Exhibited in *It Begins with Metamorphosis: Xu Bing*, Asia Society Hong Kong Center. 2014. (Photo taken early-May 2014)

Silkworm Book (蚕书), where silkworms were placed on a hardcover edition of Franz Kafka's *Metamorphosis*, which was neatly nailed on a raised wooden platform. Similar to previous works in the *Silkworm Series*, the book provided a physical environment for the silkworms to continuously move around its surface, creating silken strands that gradually obscured the text until the entire book surface was covered up in silk and insect droppings, as the silkworms slowly died and were replaced (Figs. 7.4 and 7.5).

Prior to creating the *Silkworm Series*, Xu established himself as a calligrapher and artist in the international art field by inventing unreadable pseudo-Chinese characters that both confuse and amuse Chinese and foreign viewers, but uniting them "in a kind of egalitarianism of induced illiteracy" (Silbergeld 2006:20). From the late 1980s until the 1990s, he produced his definitive work *Book from the Sky* (天书), dedicating nearly three years to carve over 4000 of these reinvented Chinese characters into wood blocks. Xu's dedication as an artist is reflected in his subsequent work, *Silkworm Series*, in which he "found himself identifying with the way the silkworms diligently and slowly went to work, ignoring the world outside" (Koon, 19).

Fig. 7.5 Xu Bing, *Silkworm Book*, 2014. Silkworms, silk, and book. L39.5 × W39.5 cm. Collection of the artist. Exhibited in *It Begins with Metamorphosis: Xu Bing*, Asia Society Hong Kong Center. 2014. (Photo taken mid-May 2014)

With the gradual consumption of the mulberry branches as well as the pages of *Metamorphosis* being covered by white silk and stains, viewers are reminded of the short lifecycles of the silkworms. Despite the worms' lifespan that lasts only for a few weeks, what Xu Bing has done as an artist is to magnify their lifecycles in a human-controlled, semi-natural environment, allowing the silkworms to do what they naturally do and letting the ecological processes take place in a quiet manner for several weeks.

A joint exhibition took place at Bard College in New York in 1998 for both Xu Bing's *Silkworm Series* (蚕系列。开幕式) and Cai Guo-Qiang's *The New York Earthworm Room* (纽约蚯蚓室). The exhibition was titled "When Heaven Meets Earth," which implies the conditions of the worms used in the two installations: Xu Bing's silkworms spreading over and covering different objects, and Cai Guo-Qiang's earthworms burrowing under the soil. Cai's work was inspired by American artist Walter De Maria's renowned installation *The New York Earth Room*, which has been exhibited since 1977. De Maria transported 250 cubic yards of earth to fill up the ground of a New York art gallery, turning nature into an indoor earth sculpture in which no plant or living matter could live or grow, to ensure the piece will stay unchanged since it was first installed in the late 1970s. The work urges people to ponder on human interaction with, and interference of, nature, at a time when people are seeing a rapid deterioration of natural environment in the real world.

Creating a cross-cultural dialogue with De Maria, Cai replicated his installation, but reversed the concept by reintroducing living organism, in particular earthworms, into the replicated earth room. As a result, the earthworms aerated the soil and nourished it with organic matters, and grass began to grow. The work, as curator Zhang Zhaohui points out, turned De Maria's "almost ossified earth into a laboratory of earth amelioration, and restores its natural condition by means of the earthworms' instinctive puncturing inside the dirt. The ameliorated earth begins to nurture life; green comes back to the ground. [...] Thus the work returns us to the fundamental issue that Cai Guo-Qiang constantly addresses: the relation of heaven, earth, and human beings" (Zhang Z., 42, 43).

The silkworms and earthworms in Xu and Cai's art installations do what they naturally do: create silk; and aerate and nourish the earth. However, it is by doing what they do that the installations and performances are made possible. As the worms are placed not in a natural environment but within a confined human-controlled space, the two pieces are turned into fitting metaphors for the Anthropocene, where our world is inhabited by multiple species, and even tiny creatures like silkworms and earthworms are all playing their roles in shaping not just the physical environment but also human culture and civilization as well as our understanding of the world over the course of history.

Leaving China in the politically unsettling 1980s, Cai Guo-Qiang and Xu Bing's early experience as young artists in the West had not been easy. Like most diasporic subjects, they constantly found themselves at the margins of two worlds. And yet, by finding their ways in practicing art at the margins, they soon established their reputations as cross-cultural artists whose works explore the connections, fusions and confusions, as well as conflicts and collisions, between the East and the West. Both artists use worms to symbolize their own transformation and metamorphosis, from artists rooted deeply in their Chinese cultural origins to cosmopolitan artists whose works address larger issues at a time of globalization. They are "playing the part of the cultural broker, bringing something of China to the United States, and taking something from the United States back to China, bridging the gap" (Zhang, 38). Silbergeld describes the cross-cultural dimensions of Chinese as developing their "signature styles based on borrowing from China and the West, transforming these into something neither East nor West, not just Chinese, not not Chinese. Their art creates and operates in a world of its own, on a space between" (Silbergeld 2011:195). Along similar lines, Wiseman (2011) sees Xu and Cai's works as "subversive" by positioning themselves in between the global and local (4). While scholars have been fascinated by the artists' transcultural experience in China and the world, they do not usually mention how Xu and Cai's works are actually doing more than just to negotiate between the East and the West; they too demonstrate both physically and metaphorically human/non-human encounters in the long history of China and the world. They are not just art practices at the margins of China and the West. They are, at the same time, art practices of environing that negotiate between the human and non-human worlds.

Focusing on Xu's art installation series exhibited in Hong Kong in 2014, particularly the *Silkworm Book*, the following section adopts two different approaches in recent environmental humanities, multispecies studies and Anthropocene criticism, to examine how the art work serves as a metaphor to reflect upon how human history and culture mingle with non-human presence and influence, thus suggesting a rethinking of human civilization, particularly from the non-Western scope and perspective, in the age of Anthropocene.

7.2 Silkworm Books and the Ecological Art

Throughout Xu Bing's *Silkworm Series* that has been exhibited multiple times over two decades, books have played a major role in the exhibition. Since 1994, Xu Bing has created several versions of the Silkworm Books. In the original installation, *American Silkworm Series Part I: Silkworm Book*, silkworms were placed over a book with blank pages, thus instilling new meanings to a book without content. In 2001, in *Silkworm Chairman Mao Red Book*, Xu replicated the installation with the famous *Quotations of Chairman Mao Tse-tung*, the Little Red Book that was read by the entire nation of China during the 1960s and 1970s. In the more recent

Hong Kong installation in 2014, Franz Kafka's *Metamorphosis* was used. Exhibited in different countries at different times and with the varied cultural significance of the books used, the Silkworm Books have generated and conveyed different political, cultural, and ideological messages. It is with such cosmopolitan perspectives that Xu Bing has become known as a cross-cultural diasporic Chinese artist, and his works are considered to be a bridge between Chinese and Western societies. At the same time, the Silkworm Books also encourage us to adjust our perception of the world in non-anthropocentric ways. Through their engagement with insects, the Silkworm Books can be regarded as ecological art that represent not just *cross-cultural* but *cross-species* encounters in the Anthropocene epoch.[3]

In *The Ecological Thought*, Morton suggests that "Ecological art, and the ecological-ness of all art, isn't just *about* something (trees, mountains, animals, pollution, and so forth), Ecological art *is* something. Art is ecological insofar as it is made from materials and exists in the world" (Morton, *The Ecological Thought*, 11). What Morton advocates is to expand our scope beyond representations and semiotic analysis of an artwork and to shift toward the quality of art and "artfulness" (Manning 2015), which we may find in the subject's material presence and agencies as artwork. In Xu Bing's silkworm installations, this would mean thinking beyond what the installation represents metaphorically, or how silkworms are represented in the work, and focusing on the material significance of the presence of these live silkworms.

Manning, on the other hand, adopts a Bergsonian approach to defining eco-art and "artfulness," by emphasizing relationality and the notion of perpetual becoming:

> [t]he work *art* in German (*die Art*) continues today to carry one of the earliest meanings of the term: 'manner' or 'mode.' In the early thirteenth century, art was still connected to this qualifying notion, attuned less to an object than a skill or craft of learning. A way of learning. [...] Art, understood along these terms, is not yet about an object, about a form, or a content. It is still on its way, in its manner of becoming. (Manning, 45)

To her, "[a]rtfulness is always more than human" (Manning, 72). She sees the quality of art in "activating fields of relation that are environmental or ecological in scales of intermixings that may include the human but don't depend on it" (ibid.).

Placing together a book, numerous living silkworms, the white silk they produce, and human monitoring efforts as a multimedia art installation, Xu Bing created in *Silkworm Book* a vibrant art assemblage that relies on both materiality and relationality/interactions among the different components of the piece. However, taking into consideration interspecies ethics and the notion of "care" for the non-human world, can the Silkworm Books still be defined clearly as an ecological art?

Anthropologists, animal studies experts, and ecocritics have proposed various theoretical frameworks in recent years for examining the interconnectedness and entanglement of human and non-human species, so as to re-evaluate human and non-human agencies and to challenge our long-established understanding of human societies and cultures. The emergent framework of "multispecies ethnography" developed by anthropologists in the past decade centers on "how a multitude of (nonhuman) organisms' livelihoods shape and are shaped by political, economic, and cultural forces" (Kirksey and Helmreich, 545) and highlights the "situated connectivities that bind us (humans and nonhumans) into multi-species communities" (549). As they argue, "relationships also have histories. Beyond a static ecological exchange, like the energy circuits mapped by early ecologists, organisms are situated within deep, entangled histories. And so, beyond mere survival, particular lifeways in all their resplendent diversity emerge from interwoven patterns of living and dying, of being and becoming, in a larger world" (van Dooren et al. 2016:2).

Scholars within the field of multispecies studies have emphasized "cross-species kinships" and "interspecies ethics" among various animals, plants, and other species (Haraway 2004, 2016; Rose 2011). Similar to other kinships, the ethics of interspecies kinships is built on the gesture of "care" (van Dooren 2014). While it may be a bit far-fetched to talk about caring for the worms that most urban people in the present world would hardly have contact with, by adopting living insects in art performances and carefully monitoring their movements and activities during the period of exhibition, Xu Bing's Silkworm installations illustrate "performing" special care toward the insects as a way of spending "quality time" with them and help us rethink various questions centering on multispecies co-existence in the Anthropocene. Through these "performances of care," viewers are enabled to see the miniature of a world co-habited and shared by human and non-human beings. As van Dooren suggests, "caring is always a practice of worlding" (van Dooren 2014:294). By highlighting care in human/non-human interactions, we are "remak[ing] ourselves, our practices and our world" (ibid.). Entering the Anthropocene era, theories of cosmopolitanism no longer focus only on humanity as a shared, universal concept but also on "the urgency and challenges of developing a contemporary vision of planetary identity that would be accountable to socioeconomic, racial, sexual, cultural, linguistic, religious, and other kinds of difference" (Heise 2016:225). Building upon her previous work on eco-cosmopolitanism—an ecologically oriented, deterritorialized form of cosmopolitanism that envisions human and non-human beings co-existing in imagined communities on a planetary scale (Heise 2008:61)—Heise refines her focus on interspecies relations in her conceptualization of "multispecies cosmopolitanism," which "aims broadly at reconstructing the understanding of 'human' communities as in reality conglomerates of human and nonhuman species that shape each other" (Heise 2016:195). As a multispecies work of art, Xu's *Silkworm Series* embraces and reflects such an ecological worldview and enables us to rethink in artistic terms what performance of care could mean.

In order to understand what makes the artwork ecological, it is crucial to understand how it cultivates (interspecies) care and attentiveness toward the non-human world.

> [R]ather than simply celebrating multispecies mingling—a basic fact of life [,] [...] work in multispecies studies is concerned with the cultivation of what we have called arts of attentiveness. This attentiveness is a two-part proposition: both a practice of getting to know another in their intimate particularity [...] and, at the same time, a practice of learning how one might better respond to another, might work to cultivate worlds of mutual flourishing, [...] In short, the arts of attentiveness remind us that knowing and living are deeply entangled and that paying attention can and should be the basis for crafting better possibilities for shared life. (van Dooren et al., 17)

Responding to Van Dooren, Kirksey, and Münster's call for "cultivating arts of attentiveness" for the "worlds of (multispecies) mutual flourishing," the Silkworm Books reflect ecocritically on the role of human beings in shaping the more-than-human world, while on the other hand, how non-human beings shape human culture and civilization. In doing so, they reflect the power dynamics between humans/non-humans that contributes to the promotion of multispecies cosmopolitanism that increasingly defines our world today.

7.3 Non-human Agencies in the Multispecies World

From tiny maggots maturing into adult flies and eventually getting electrocuted by a huge fly-zapping machine, Damien Hirst's *Let's Eat Outdoors Today* (1990–91) shows us the moments of life and death of thousands of insects. Huang Yong Ping's *Theatre of the World*, on the other hand, shows us the fights among the numerous reptiles and insects within a dome installation, and their injuries and deaths are being put on display. What is common between the two works is how they highlight the transition of non-human beings from living to dead, and the turning of their deaths into visual spectacles. Both works have been criticized by animal right advocates for their unethical treatments of non-human lives. In contrast to Hirst and Huang's installations, Xu Bing's *Silkworm Series* avoids the intense and dramatic spectacles generated by the torture and extermination of insects and presents what seems a more ethical kind of performance.

According to Sharon Chan, the co-curator of Xu Bing's exhibition in Hong Kong, Xu has given specific instructions in preparation of the *Silkworm Book*: "Select silkworms of the fourth or fifth stage of development. Feed them with dry mulberry leaves three times a day until their appetites decrease. Prepare a book, nail it to a raised wooden platform and let the silkworms spin the silk. Clean the droppings" (Chan, 103). Compared to Hirst and Huang's installations in which insect exterminations became art spectacles for public viewing, the silkworms used for Xu's art installations are handled with better care and in more ethical ways. It shows us how worms live, act, and what they can do as living beings.

Unlike Hirst's maggots that transform into flies in front of the viewers' eyes, Xu Bing did not show his silkworms metamorphosizing into moths. Instead, in the 2014 installation the worms were placed on top of Kafka's *Metamorphosis.* By doing so, the art piece makes symbolic, literary reference to the story of Gregor Samsa morphing into a giant insect and allows it to shape our perception of the live worms used in the piece in imaginative ways. While the silkworms do not metamorphosize in front of our eyes, their presence and activities transform the piece of installation gradually and present us with a multispecies art assemblage that "emerges as an intertwined flux of material and discursive forces ... a material-semiotic network of human and nonhuman agents incessantly generating the world's embodiments and events" (Iovino and Oppermann, 3). Within this material-semiotic network, our focus is shifted from the subjectivity of individual lives to the non-human agencies (of silkworms, silk, and the book) within the art assemblage as a whole. We are led to observe the agential dynamics among human and non-human beings within it.

Differing from Huang and Hirst's insect installations, Xu Bing's *Silkworm Series* focuses on the slow, quiet, and undramatic process and encourages us to turn our attention from spectacles of life and death to the non-human agency of silkworms and how their quiet presence alters and shapes the exhibition environments over time. Versus the trend of individualism in Chinese art scene,[4] Xu Bing and Cai Guo-Qiang have reduced their own presence in their works of the *Silkworm Series* and Earthworm Room, making worms and their non-human agency central to the work, as a response to the call for addressing broader ecological issues in art. To be eco-conscious, artists often have to "downgrade the ego in order to upgrade the eco component of their creative practices" (Weintraub, 45). As Wang suggests, "What makes the Chinese use of animal images distinct is not so much what they stand for as how they stand in relation to each other. In other words, it is the way they all add up and work in concert that matters most. The web of relationships to which they are integrated is the key to Chinese animal 'symbolism'" (Wang, vii). Loaded with various "performers," the Silkworm Books present a series of art installations that highlight such a web of relationships: "from Xu Bing [and his assistants] as the invisible silkworm farmer[s] at the preparatory stage, to silkworms as animals that carry cultural connotations of material production, and the book as the symbol of cultural production, as well as another site of creativity" (Chan, 103). In addition to that, the spectators coming for the exhibition and other dynamic elements, such as the mulberry leaves, the lights, humidity, and room temperature,[5] are all components that contribute toward a collaborative "performance" among the multiple human and non-human parties.

Xu's installation enables the non-human beings, particularly the silkworms, to take over some degree of control and make their non-human agency visible. Within the several months in which the *Silkworm Book* installations were exhibited, the art assemblage underwent constant and gradual transformation. The silk created by the silkworms gradually thickened and covered the book, obscuring us from seeing the texts as the worms continued to wind their silk in all directions. On top of the white layer of silk, brownish stains were also increasingly seen which, as Koon describes it, presented "testimonial forces of [the silkworms'] existence and a haunting reminder of nature's fate that bring poignancy to the artwork" (Koon, 20). By obscuring human gaze and access to the book, the art installation forces viewers to focus on the silkworms and their alteration of the environment. Pheng Cheah argues that *Silkworm Book* aims to deconstruct anthropologism[6] and suggests that silkworms have a world, too:

> [I]t disrupts the teleological completion of the signification process by preventing our gaze from penetrating what is below the silk covering. And the means of this blocking of human access is precisely the world that the silkworm creates through its natural artifice. It is as though the silkworm is saying to the viewer: 'You have to enter my world and pass through it to get to yours. You can only achieve meaningfulness if my world gives you access to meaning. So you must acknowledge that I have a world.' (Cheah, 100)

Acknowledging the silkworm's world gives the material art assemblage a metaphoric significance in representing a multispecies condition in which non-human beings exert their agencies. With the silkworms spinning white silk to cover up all the words and obstructing us from reading the texts, the different versions of Silkworm Books, whether Mao's *Little Red Book* or Kafka's *Metamorphosis*, are all turned into the silkworms' world in which human constructs—such as the three "*wen*," *wenzhi* (words); *wenhua* (culture); *wenming* (civilization)—are all being obstructed by the non-human presence and actions. The worms and their activities have both physically and metaphorically overpowered the human world. Comparing that with the present environmental condition of our world, one could argue it reflects that in a time of environmental crisis, national and political agendas and everything anthropogenic, from human ideologies, culture to civilization, are all increasingly overshadowed by ecological and natural phenomena as represented in the installation by silkworms' natural practice of silk making.

While a multispecies reading of Xu's Silkworm installation enables us to see how non-human agencies operate in a way that overpower the human world, one could at the same time argue the *Silkworm Series* presents, after all, art artifacts that are created as a result of human design and interventions. They are still installations created for humans to gaze into non-human silkworms' lives. Are they therefore still being deprived of a world of their own? Framing it within an Anthropocene discourse, we are led to the question "are human intervention and control necessarily bad things?"

In the following section I wish to argue that the Silkworm Book is inviting multiple ecocritical approaches for interpretation: while placing it in a multispecies framework enables us to highlight the non-human agencies of silkworms and thus multispecies co-existence between human and non-human beings; an Anthropocene reading reminds us that human centrality is still a determining factor, both in the artwork and in our real world. Juxtaposing the two readings magnifies the tensions and interdependence between human and non-human agencies and pushes us to rethink the role of human beings in practicing environing toward the more-than-human world.

Hageman argues that a dialectical ideological critique is a necessary apparatus in ecocinema studies, since "all films are bathed in ideology," and "ideology always contains contradiction, internal and inherent to its structure" (Hageman, 64–5). Like films, artworks are ideological constructs. They too contain internal contradictions when placed under different frameworks for critical analyses. The discrepancies within the different approaches and analyses should be seen as "indices of the contradictions within the ideology that determines our current ability to think and represent ecology" (Hageman, 65). No clear position or answer is offered in Xu Bing's *Silkworm Series* under the presence and agential capacity of silkworms that remind people of multispecies co-existence in our world; while at the same time it urges us to contemplate upon the centrality of human agency, an eco-dialectical critique can also be made to open the issue up for further debates and discussion.

7.4 Anthropocene Metaphor and the Confucian Eco-Governance

Considering the growing magnitude and impacts of human activities on earth and atmosphere, Crutzen and Stoermer propose that humankind is finally moving beyond the Holocene, the post-glacial geological epoch that lasted for over ten thousand years. As they argue, it seems "more appropriate to emphasize the central role of mankind in geology and ecology by proposing to use the term 'Anthropocene' for the current geological epoch" (Crutzen and Stoermer 2000:17). In contrast with multispecies studies, Anthropocene studies highlight the centrality of humans in transforming the earth. In a way, the emergence of an Anthropocene discourse complicates the recent discussion and debates in posthumanist ecocritical studies, for it "stands in tension with those varieties of posthumanism that have, over the past three decades, sought to emphasize the central importance of nonhuman agents in social networks" (Heise 2016:203). According to Heise, Anthropocene's focus on human centrality has led to two opposing interpretations: it has been understood both pessimistically and optimistically. "For pessimists, the Anthropocene signals the enormous scope of negative human impacts on the environment; for optimists, it opens up the possibility of reimagining the nature of the future not as a return to the past or a realm apart from humans, but as a nature reshaped by humans" (ibid.). While Anthropocene and its Chinese translation *renlei shi* (人类世) have not yet become dominant keywords in Chinese ecocritical studies, a similar understanding of humans' relationship with the environment has been in consonance with Confucian tradition since ancient time, particularly in their common emphasis on the need of human dominance over the non-human world.

In *Mencius*—the Confucian classic published in 300 BC—Mengzi tells the legendary tales of several Chinese emperors in the section "Duke Wen of T'eng Book Two": "In the time of great floods during Emperor Yao's time, it was said that snakes and dragons emerged and intruded human lands, leaving humans with nowhere to settle. Emperor Yao appointed Yu to bring the waters under control. Yu dug furrows in the land and diverted the water into the sea. As a result, obstacles and dangers were rinsed away, snakes and dragons were driven into marshes, and ravaging birds and animals disappeared. People regained their farmland and settled down. However, soon after the capable emperors Yao and Shun's deaths, the disasters came back under the savage emperors' ways of rule. They left the fields unattended and built pleasure lakes and parks that caused the swarms of birds and animals to return. By the time of Tyrant Chou, the country was in complete chaos. Fortunately, the disastrous time ended with the arrival of good Duke Chou, who put an end to the tyrannical regime, conquered fifty nations, and drove away all the fierce beasts such as tigers, leopards, rhinoceros and elephants. People of the time were once again greatly delighted" (Mencius, 90–91).

Putting Sage rulers Yao, Shun, and Duke Chou in contrast to savage and cruel rulers like Tyrant Chou, Mengzi concludes that good human leaders bring about effective governance and thus harmony and stability to the world; while bad rulers turn animals predatory and the environment hostile to human beings. Confucian anthropocentrism proves its relevance in the contemporary world by offering us insights into the debate on humans' role in shaping the environment. From the words of Mengzi, the ancient Confucian Sage, we could see the Confucian stance toward human dominance: that is, humans both build and destroy the environment, subject to the intentions and virtues of rulers of the time.

Incorporating Confucian thoughts on environmental governance in our Anthropocene reading, Xu Bing's Silkworm Book presents a fitting metaphor of the ecosystem in miniature that leads us to contemplate humans' role in affecting the more-than-human world. All of Xu Bing's Silkworm installations involve a great deal of human intervention and management. Instead of being a static, unchanged art installation, Silkworm Book is designed to be an ongoing performance with human work throughout the exhibition period. Each day after the gallery closes, the silkworms are monitored, their droppings cleaned, and dead worms removed. Human intervention becomes necessary in order to ensure what viewers focus on is what the artist intends to present: not just a completed art piece created by a human artist, nor a demonstration of how nature operates without human intervention, but an ongoing performance of tensions created between humans and non-humans. Viewers are encouraged to observe what the silkworms have created and to experience the regulated process of the quiet tensions and peaceful interactions and wrestles between human beings and the silkworms throughout the period. The "artfulness" of the piece lies in the balance between what elements in the art installation are kept and what others are removed, in order to create a sense of interspecies harmony within an artwork constantly under transformation.

The multispecies world in the *Silkworm Book* resembles a miniature of the world in the Anthropocene era. From a Confucian perspective, the role of human artists to the world created within the art installation is comparable to the ancient rulers' role in shaping the environment of their time. Whether harmony and peace are attainable (in the artwork vs. in reality) depends on their intentions and virtues, which are translated into their care and attentiveness to the non-human environment under their influence and governance.

The Silkworm Book presents us a world that is increasingly defined by the dissolved boundaries between human/non-human; nature/culture; construction/destruction; and despite the non-human ability to impact the environment, human activities and interventions remain crucial in the worldmaking process. Because of such human intervention, the artwork transcends from a random and unnoticeable slice of reality (a "non-art") into an art assemblage highlighted by its materiality (human and non-human agencies) and relationality (human and non-human dynamics) that enable human viewers to make sense of the interspecies interactions and multispecies co-existence.

7.5 Concluding Notes: Rethinking Silk Road Culture and Civilization

Sung Ying-hsing, a Chinese scientist and public official of the Ming Dynasty, published the classic work *Chinese Technology in the Seventeenth Century* (also known as *Tiangong kaiwu*) in 1637, which documents the history of traditional Chinese technology ranging from agriculture, textiles, and mining to the manufacturing of weapons. Sung dedicated an extensive section of his publication on depicting the processes of producing clothing materials. As he suggests, "Nature has provided the materials for clothing" (Sung, 35). In this section, Sung documents in great detail how, in order to produce high-quality silk, the process begins with carefully raising silkworms. From bathing the silk-moth eggs, identifying the varieties of silkworms, and hatching and raising them, to preventing them from getting diseases, the book stresses the tremendous effort and time required for growing healthy silkworms (Figs. 7.6 and 7.7).

The careful breeding of silkworms and the skillful production of high-quality silk have long been considered a significant part of Chinese civilization and cultural history. Since the nineteenth century, the networks of trade routes between China and various Eurasian countries have been called the Silk Road. Silkworms and silk thus play an important role in defining the transnational image and early forms of globalization in Chinese history and civilization.

With the growing concerns for environmental well-being and animal welfare in recent years, these commercial practices of silk productions have, however, increasingly been deemed unethical. An animal rights organization points out that over 6600 silkworms have to be killed in order to produce one kilogram of silk, and billions of silkworms are killed each year in the world for silk productions.[7] With that in mind, it would seem necessary to re-examine whether the long Chinese tradition, as well as our understanding of Silk Road culture and civilization, is still in any way natural or ecological by today's standards. The changing ideologies

Fig. 7.6 Bathing silkworm eggs 浴蚕. *Chinese Technology in the Seventeenth Century*

and paradigmatic shifts in the age of Anthropocene mean that practices of environing require our constant self-reflections; looking ahead to the future; and adjusting our beliefs to lead more ecological ways of life.

In this chapter I have illustrated how Xu Bing's art installation, the *Silkworm Book*, offers a metaphor for the Anthropocene condition by offering a space for an eco-dialectical ideological discussion on human/non-human relationships in the present world. By using silkworms in his art installations, as well as showing considerable care and attentiveness to them, Xu is offering the silkworms a voice they never had before. While the *Silkworm Series* is not a harsh critique toward exploitation of

Fig. 7.7 Mature silkworms are placed on split-bamboo screens to spin cocoons 山箔. *Chinese Technology in the Seventeenth Century*

insect lives, it does leave us to ponder what Chinese culture, or more broadly, human knowledge and civilization are built upon, and at the expense of what?

Such ecocritical reflections of human knowledge, history, culture, and civilization are becoming more and more important in the Anthropocene age. Moreover, the aesthetic value of artworks like Xu Bing's *Silkworm Series* lies in their ability to provoke and challenge our predominating concepts of human centrality and dominance, as well as to ponder our relationships with non-human species, thus helping to build a future with better awareness of multispecies co-existence.

Notes

1. *Dear Climate*. http://www.dearclimate.net. Accessed 12 Dec 2018.
2. Kikkhah, Roya, "Damien Hirst condemned for killing 9000 butterflies in Tate show," *The Telegraph*, 14 Oct 2012.
   http://www.telegraph.co.uk/culture/culturenews/9606498/Damien-Hirst-condemned-for-killing-9000-butterflies-in-Tate-show.html. Accessed 12 Dec 2018.
3. The same can be said of Cai Guo-Qiang's *New York Earthworm Room* (1998). This chapter, however, focuses on Xu Bing's *Silkworm Series*, which the author visited in 2014.
4. From visual artists such as Yue Minjun and Fang Lijun who frequently depict themselves in paintings, to installation and performance artists like Zhang Huan, Sun Yuan, and Peng Yu who had in the past used their own bodies to create and perform personal art pieces, Chinese artists have been exploring issues of individual identity and human subjectivity, being in a society where individualism and collectivism are in constant conflicts.
5. As chief curator of the Xu Bing exhibition (2014) organized by the Asia Society Hong Kong Center, Yeewan Koon points out, "there were mitigating factors that affected this work, such as the environment of the exhibition space and the timing of the exhibition in May in Hong Kong that affected the lifespan of the silkworms" (Koon, 45).
6. Building upon Heidegger's work, Pheng Cheah uses the term "anthropologism" to refer to the philosophical conception with "man" as a basic philosophical category and asserts that only by proceeding from this can one develop a system of ideas about nature, society, and the world.
7. "The Silk Industry," *PETA-UK*. https://www.peta.org.uk/issues/animals-not-wear/silk/. Accessed 12 Dec 2018.

Bibliography

Broglio, Ron. 2011. "Meat Matters: Distance in Damien Hirst." In *Surface Encounters: Thinking with Animals and Art*. Minneapolis: University of Minnesota Press.

Chan, Sharon. 2014. "Silkworm Book: A Short Response." In *It Begins with Metamorphosis: Xu Bing*, ed. Yeewan Koon, 103–107. Hong Kong: Asia Society Hong Kong Centre.

Cheah, Pheng. 2014. "Entering the Silkworm's World: A Response to Xu Bing." In *It Begins with Metamorphosis: Xu Bing*, ed. Yeewan Koon, 98–100. Hong Kong: Asia Society Hong Kong Centre.

Crutzen, Paul J., and Eugene F. Stoermer. 2000. "The 'Anthropocene'." *IGBP Global Change Newsletter* 41: 17–18.

Elverskog, Johan. 2014. "Asian Studies + Anthropocene." *The Journal of Asian Studies* 73 (4): 963–974.

Hageman, Andrew. 2013. "Ecocinema and Ideology: Do Ecocritics Dream of a Clockwork Green?" In *Ecocinema Theory and Practice*, ed. Stephen Rust, Salma Monani, and Sean Cubitt, 63–86. New York/London: Routledge.

Haraway, Donna J. 2004. *The Haraway Reader*. New York: Routledge.

———. 2016. *Staying with the Trouble: Making Kin in the Chthulucene*. Durham/London: Duke.

Heise, Ursula K. 2008. *Sense of Place and Sense of Planet: The Environmental Imagination of the Global*. New York: Oxford University Press.

———. 2016. *Imagining Extinction: The Cultural Meanings of Endangered Species*. Chicago/London: The University of Chicago Press.

Iovino, Serenella, and Serpil Oppermann. 2014. "Introduction: Stories Come to Matter." In *Material Ecocriticism*, ed. Serenella Iovino and Serpil Oppermann, 1–20. Bloomington/Indianapolis: Indiana University Press.

Kirksey, Eben, and Stefan Helmreich. 2010. "The Emergence of Multispecies Ethnography." *Cultural Anthropology* 25 (4): 545–576.

Koon, Yeewan. 2014. *It Begins with Metamorphosis: Xu Bing*. Hong Kong: Asia Society Hong Kong Centre.

Manning, Erin. 2015. "Artfulness." In *The Nonhuman Turn*, ed. Richard Grusin, 45–79. Minneapolis: University of Minnesota Press.

Mencius. 2015. *Mencius*, Trans. David Hinton. Berkeley: Counterpoint.

Morton, Timothy. 2010. *The Ecological Thought*. Cambridge, MA/London: Harvard University Press.

Rose, Deborah Bird. 2011. *Wild Dog Dreaming: Love and Extinction*. Charlottesville: University of Virginia Press.

Silbergeld, Jerome. 2006. "Introduction." In *Persistence/Transformation: Text as Image in the Art of Xu Bing*, ed. Jerome Silbergeld and Dora C.Y. Ching, 19–24. Princeton: Princeton.

———. 2011. "The Space Between: Cross-Cultural Encounters in Contemporary Chinese Art." In *Xu Bing and Contemporary Chinese Art: Cultural and Philosophical Reflections*, ed. Hsingyuan Tsao and Roger T. Ames, 177–198. Albany: Suny Press.

Sung, Ying-hsing. 1966/1997. *Chinese Technology in the Seventeenth Century: T'ien-kung K'ai-wu*, Trans. E-Tu Zen Sun and Shiou-chuan Sun. Dover ed. Mineola: Dover Publications, Inc.

van Dooren, Thom. 2014. "Care." *Environmental Humanities* 5: 291–294.

van Dooren, Thom, Eben Kirksey, and Ursula Münster. 2016. "Multispecies Studies: Cultivating Arts of Attentiveness." *Environmental Humanities* 8 (1): 1–23.

Wang, Eugene Y. 2016. "Preface." In *The Zoomorphic Imagination in Chinese Art and Culture*, ed. Jerome Silbergeld and Eugene Y. Wang, vii–viii. Honolulu: University of Hawai'i Press.

Weintraub, Linda. 2012. "Eco Art Materials." In *To Life: Eco Art in Pursuit of a Sustainable Planet*, ed. Linda Weintraub, 43–50. Berkeley/Los Angeles: University of California Press.
Wiseman, Mary Bittner. 2011. "Subversive Strategies in Chinese Avant-Garde Art." In *Subversive Strategies in Contemporary Chinese Art*, ed. Mary Bittner Wiseman and Liu Yuedi, 3–20. Leiden/Boston: Brill.
Zhang, Zhaohui. 2005. *Where Heaven and Earth Meet: Xu Bing and Cai Guo-qiang*. Hong Kong: Timezone 8 Limited.

CHAPTER 8

Place, Animals, and Human Beings: The Case of Wang Jiuliang's *Beijing Besieged by Waste*

Haomin Gong

In the beginning of Wang Jiuliang's documentary film, *Beijing Besieged by Waste* (垃圾围城, 2011; BBBW hereafter), the director presents a rather grotesque scene in which a group of shepherds/scavengers, upon seeing a truck pulling backward into a landfill to unload its full load of waste, drive their sheep to the truck and let them feed on the waste. This scene, with all its filthiness, quite brutally disrupts our imagination of the lyrical beauty usually associated with sheepherding. It alerts us to the fact that animals, human beings, and waste are increasingly and inevitably linked with each other in our current ecological crisis, and their relationships can no longer be construed within a particular locality but have to be understood from a global or planetary perspective. In this chapter I explore specifically the significance of *place* and *misplacement* in the relationship among animals, human beings, and waste, an important issue in our articulation of *environing.*[1] In my view, what makes the scenes described above look so grotesque, among other things, is that things are *out of place*. In other words, things appear in the places where they do not belong, and this sense of *misplacement*, whether acknowledged directly or not, is, I will argue, cen-

H. Gong (✉)
Lingnan University, Hong Kong, People's Republic of China

C. Chang (ed.), *Chinese Environmental Humanities*,
Chinese Literature and Culture in the World,
https://doi.org/10.1007/978-3-030-18634-0_8

tral to the discourse of waste, as well as human and animal relationships to it, and by extension socio-cultural positionalities of wasted humans, trash animals, waste dumps, and other marginalized entities.

Critics have already mentioned the role that place plays in the construction of the discourse of waste. In her influential work, *Purity and Danger*, the British anthropologist Mary Douglas refers to the saying in the epigraph that is attributed to Lord Chesterfield, and points out the constructedness of purity and uncleanness. In her view, nothing is inherently dirty. Instead, they *become* dirty when they are put in a place to which they do not belong. For instance, "Shoes are not dirty in themselves, but it is dirty to place them on the dining-table; food is not dirty in itself, but it is dirty to leave cooking utensils in the bedroom, or food bespattered on clothing; similarly, bathroom equipment in the drawing room; clothing lying on chairs; out-door things in-doors; upstairs things downstairs; under-clothing appearing where over-clothing should be, and so on."[2] As she observes, Chesterfield's saying "implies two conditions: a set of ordered relations and a contravention of that order. Dirt then, is never a unique, isolated event. Where there is dirt there is system. Dirt is the by-product of a systematic ordering and classification of matter, in so far as ordering involves rejecting inappropriate elements."[3] In other words, dirt is only meaningful in a social system. Susan Strasser echoes Douglas's view by writing, "Trash is created by sorting."[4]

Here, I will focus on this process in which people classify things: a process in which people assign things to their due places in an order is an act by which people put not only the physical world but also their inner world. In other words, this is simultaneously a social and a *cultural* process. As Donald Worster remarks, "We are facing a global crisis today, not because of how ecosystems function but rather because of how our ethical systems function. Getting through the crisis requires understanding our impact on nature as precisely as possible, but even more, it requires understanding those ethical systems and using that understanding to reform them."[5] Ecological problems, in the final analysis, are rooted in our culture.

This chapter takes as a point of departure the *place* of waste, and investigates the ways in which places of waste, human beings, and animals are figured in Wang Jiuliang's eco-documentary *BBBW*, and the ecological implications that the discourse of waste has on a contemporary Chinese society marked, as it is, by raging urbanization and rampant commercialization. To be more specific, I focus on the concepts of *place*, *misplacement*, and *displacement* in the discourse of waste, and explore the ways in which they are inscribed in the narrative of the film and contribute to the construction of wasted nature, humans, and animals that are central focus of the film. I argue that waste, as an ecological problem and a social discourse, is a product of modern urbanization, commercialization, and consumerism, which not only defines the geographical place of waste but also constructs its social place. In contemporary China, waste signals not only the environmental crisis that accompanies the advent of consumer society but also the social and cultural problematics in China's developmentalism.

8.1 Place in *Beijing Besieged by Waste*

BBBW is the shared title both of a photographic work, which consists of a series of pictures of the waste dumps surrounding Beijing, and of the documentary film that is based on the photographic work, both produced by the Beijing-based freelance photographer and filmmaker, Wang Jiuliang. In the span of two years between 2008 and 2010, Wang traveled around 14,000 kilometers, visited and investigated about 500 waste dump sites, among numerous others, scattered in the outskirts of the Beijing city. Both the photographs and the film record, in a highly realistic and documentary manner, the startling consequence of irresponsible disposal of waste and the shocking processes by which domestic and construction-generated waste is crudely treated. They also expose some of the clandestine deeds of sanitation workers and government units in regard to waste processing and dumping. Overall, both works present powerfully the massive destruction of ecological system that the current waste-disposal system is bringing about, and the tremendous ecological, social, and cultural impacts this system has had on cities and the countryside alike, as well as on the people whose lives rely on waste, and those who seem to be remote from it but who in fact have no escape from it.

As the critic Li Nan has it, Wang's photographic work "not only exposes, beneath the glitzy veneer of the prosperity of Beijing, the lesser-known way of lives of the clans of scavengers and the collusive deeds [on waste disposal] of interest groups, as well as the global problem of environmental pollutions exemplified by this case. But more important, it demonstrates the photographer's admirable professionalism and the ability to expose problems under the condition of commercialism and consumerism, while he is a freelance independent photographer who has no official connections whatsoever."[6] Wang was awarded the Gold Medal of the Artist of the Year in the Fifth Lianzhou International Foto Festival in 2009 for this work, and named the Photographer of the Year by the Seyingwuji Web in 2010.

The problems of waste disposal in Beijing that Wang exposes are also happening in most of other cities in China on daily basis. Some places are even worse off in terms of extensiveness and magnitude of the pollution caused by waste. In a way, "cities besieged by waste" is an epitome of ecological problems that cities are faced with in the rapid urbanization in China, and it also conveys a strong symbolic meaning about limit and spaciality in the relationship between ecology and development.[7]

Place figures large in the film. First and foremost, besiegement, the central theme of the film, is a highly spatial image. It is presented mostly prominently at both the beginning and the end when the director marks waste dump sites that he has visited in bright yellow dots on a gloomy Google map (Fig. 8.1). Those hundreds of yellow dots, appearing one after another, virtually form a circle surrounding the city, like magic hoop constraining an expanding specter, visualizing the simmering tension between the urge for development and its environmental limits. This picture also exhibits a grotesque image of *environing*.

Second, the characteristic of the film is that every location presented, albeit all in the small precinct of Beijing area, is marked with particularly accurate degrees of its latitude and longitude. This marking, so rarely seen in other films, constitutes a crude sense of spatiality and reality, reminding the viewer, first and foremost, that those places in the film are real, existing

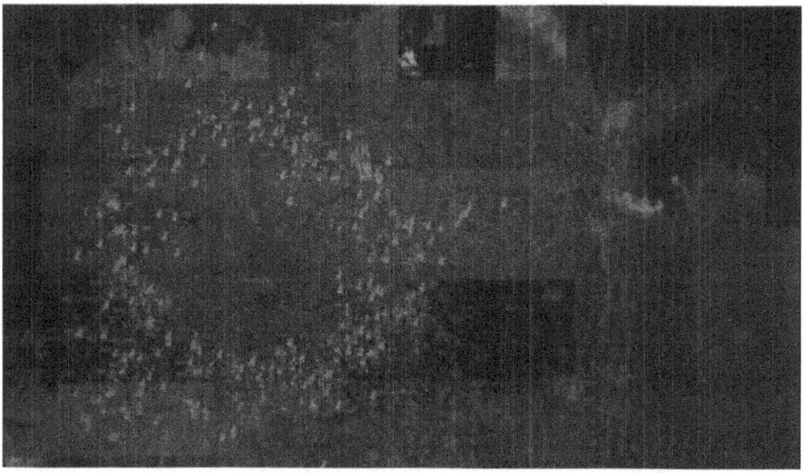

Fig. 8.1 Beijing City besieged by waste dump sites shown on a Google map

places on the earth, the single, only planet that all of us share. This meticulously registered sense of spatiality and reality, further, is an indispensable element that an eco-film should have for its political agenda in its intervention into the reality.

As some critics have pointed out, a thing *becomes* waste not because its physical functions disappear but more importantly because it has lost its social position. For instance, when discussing sociality of waste, the critic Wang Min'an remarks, "A thing, under any social condition, exists in a chain of signification. In other words, once it loses its own social function, or cannot find its own position in a social structure, it may find its place in waste dump.... A thing turns into waste when its signifying social chain is broken."[8] A special attention to the issue of place and misplacement in the discourse of waste urges us not only to see waste as an environmental issue but also to conceive it as a social and cultural issue, a social and cultural construction under the condition of urbanization, commercialization, and consumerism. In what follows, I will discuss the ways in which *place* of waste in urbanization and consumerism are inscribed in the narrative of the film, and human beings' and animal's positions in the process of this inscription.

8.2 Place of Waste and Urbanization

Waste, arguably, is first and foremost a product of urbanization. *BBBW* closely ties waste to the high-speed urbanization taking place in China. Admittedly, waste did not first appear with urbanization; human beings had produced waste long before this process began. However, as the French environmentalist Catherine de Silguy points out, waste used to go back to nature and enter into its eco-cycle in a natural way; it is precisely urbanization that blocked this cyclical process and made "useless" things waste. Perhaps it is more accurate, in a sociological sense, to call the "waste" in the pre-urbanization age things that have lost their functions, because by going back to the eco-cycle of nature, those "useless" things kept, or regained, their due positions in nature. The urban space, on the contrary, deprives them of their positions in it once they have lost their functions nor does it allow them to go back to nature in the way they used to return to nature.[9] As Wang Min'an observes, "To a great degree, waste is made by cities, created by the unfriendly hard shell of cities."[10]

However, the fact that cities do not allow functionless things to directly return to the eco-cycle, does not necessarily mean that those things immediately become waste. When human beings build cities, they also establish waste processing procedures that fit the urban lifestyle. They classify, collect, transport, recycle, compost, landfill, and burn waste, and these acts mark their efforts to put waste back into a man-made eco-cycle, taking the place of the natural eco-cycle that is becoming increasingly irrelevant with the rapid expansion of cities. In a sociological sense, these are also efforts that aim at re-conferring social positions to functionless things. Yet, these efforts, due to systemic, institutional, or practical factors, often fail to achieve what was intended, and thus deprive functionless things of their positions and render them into real waste. Thus perceived, waste—allow me to reiterate it—is not simply an ecological problem but also a social and cultural problem, or *problematic*, through which we can reflect on the ideologies embedded in our consciousness toward the environment.

In China, waste is particularly closely associated with its breakneck urbanization that has been going on for the past two decades. An examination of the position of waste in urbanization problematizes the historical impulse of developmentalism, a central theme that characterizes contemporary China, which, yet, is in a paradoxical relation with waste. In *BBBW*, besiegement by landfill sites and the expansion of urban boundaries form an unsolvable tension. On the one hand, rapid urbanization requires more space, and as a result, Beijing, as well as most other cities in China, is expanding its city limits in all directions outwards. On the other hand, with this expansion, the amount of waste produced is also growing at an accelerating speed and, thus, requires farther and larger sites to hold waste; but the existence of these landfills constrains the expansion of the city. The more rapid the city expands, the more waste it produces, and thus the more landfill dumps it needs, which in turn become obstacles to the expansion. Landfills and urban construction are, in fact, racing and fighting for the limited space that the city has.

There is a scene in the film in which construction materials are placed next to demolition and construction waste on the same site, where the latter is to be buried and the former are to be used for new construction almost simultaneously. This intriguing scene powerfully presents a historical irony that is going on pervasively in China: the impulse for development is so strong that construction and destruction are put on the same stage, struggling for the same space.

Or, we may also say, in the process of urbanization the developmentalist libido and development's excrement constitute an entangling paradox—the higher speed of development will bring out the larger amount of waste, which in turn constrains the space for development. As a result, some people choose to roughly bury waste, as the film exposes, without much professional, or responsible, treatment in order to make space for new construction. Symbolically, the phallic structures above the ground that exemplify masculine developmentalism seem to have accomplished a glorious conquest over the "abject" things that are named, in this case, waste.

The linearity of the teleological developmentalism exemplified here is, obviously, at odds with the cyclicity of the ecological view of development. Put in a different way, it is precisely by suppressing the cyclical view of eco-development that the linear view of developmentalism legitimizes its currency. But, intriguingly, those abject things called waste, which have been crudely removed out of sight, either being dumped in a remote area, or being buried without any treatment, nevertheless constantly remind us, like a specter, of their undispellable existence, and return to our lives, in different forms, such as polluted air, contaminated water, tainted vegetables, milk, and meat. As Julia Kristeva reminds us, the "abject," existing in a liminal space, disturbs our apparently clean order.[11]

Thus, the image that Beijing is besieged by waste reminds us the fact that cities and waste are mutually produced. In his discussion of the ways in which Chinese cities used to be defined by city walls and moats in pre-modern periods, while they are in nowadays defined, de facto, by waste dumping sites, Wang Min'an exclaims, "waste is structuring our cities."[12] Cities produce waste, but waste is defining cities. Arguably, waste is not only defining, in a geographical sense, the boundaries of cities—although these boundaries are too easy to be broken through—but also defining how cities are structured in terms of their treatments of waste. The ways in which a city deals with the waste it produces demonstrates its residents' values with regard to waste, and, interestingly, also reflects their values as to the issues that are apparently opposite to waste, such as purity, civilization, and so on. As Gay Hawkins demonstrates in her book, *The Ethics of Waste*, not only is waste produced and constructed by human beings but "what we want to get rid of also *makes* us who we are."[13]

Perhaps it is in this sense that we can make sense of the claim: "It is not that waste is besieging cities, but instead it is cities that are besieging/defining waste."[14] It is precisely our excessive urbanization that renders functionless things positionless and thus makes them waste. Along with this process of excessive urbanization, sustainable eco-development surrenders to the logic of linear modern development, and thus establishes the myth of developmentalism, or the "cult of progress" in Bauman's words.[15]

However, it becomes commonly acknowledged that in any definition, the defined cannot always be neatly included within the boundaries drawn in the discursive move of definition. The parts that fail to be included or spill over boundaries only remind us of the fact that no matter how dominant the social discourse that defines it, it may not neatly and entirely contain the object under definition. Thus, waste often finds its way back into the realm of cleanness, which, however, affirms its value by denigrating what is "valueless" into waste.

This cognitive shift from "waste besieging cities" to "cities besieging/defining waste," a shift, among other things, of subjective and objective positions, brings to our attention that we, urban residents or not, are not simply victims of this ecological problem but also perpetrators and accomplice. We are all participating, directly or indirectly, in the process of mindless urbanization, and contributing, in one way or another, to the social system that structures the current mechanism of developmentalism. As the director Wang Jiuliang poignantly asks, "Don't you think that the direction of our social development is wrong?"

8.3 Displacement of Commodities in a Consumer Society

In the end of the film, a few female scavengers tell the director, ashamed but also somewhat proudly, that their entire possessions have been dug up from waste: their homes are built with trashed construction materials; their clothes, and even some food, have been collected from waste. One of the scavengers points at the clothes she has on in front of the lens, and says that those have been found in trash cans in an "upscale residential compound" (*gaodang xiaoqu*). "Sometimes the bags of clothes they discarded were simply brand new," she remarks. Admittedly, such an irresponsible way of consumption is not a common practice in China today, but it is becoming increasingly noticeable. It quite explicitly reminds us that the rapid increase in the amount of waste that we produce every day may find its cause in our growing commercialization and consumer society.

Then, in what specific way do commercialization and consumerism shape the production of waste? In his investigation of waste in the current Chinese condition, Wang Min'an remarks that, "once its functionality is used up, a commodity would turn into the form of waste. In this sense, waste is what is left behind from the commodity—it is the corpse of the commodity."[16] Wang Min'an conceives waste as the opposite of commodity, as he continues to write: "Commodities are usually put on display, whereas waste is always hidden"[17]; "Modern stores are kingdoms of commodities, whereas landfill is a mirror image of this kingdom. Stores and landfills, two extremities of a modern city, echo each other across the borders of the city .... This juxtaposition of the kingdom of commodities and the kingdom of waste is a wonder of modern cities."[18] Here, waste is taken as a mirror of the urban commodity economy, reflecting, with all its "abjectness," the other side of the glitziness of the veneer of consumer society. This mirror-relation between commodity and waste is symbolized, or literalized, in a scene in which an airplane flies over the Wenyu River, a local river in Beijing known for many high-end apartments and facilities along its banks, and the director symbolically highlights this scene by capturing the image of the airplane reflected in waste-loaded water. Here, the director virtually puts airplane as a symbol of modern life in contrast with waste, which, on the one hand, is to be put away to legitimize the order of modernity, but, on the other, is precisely a product of modernity. Perhaps, as Wang Min'an remarks, we should "write a human history from the perspective of waste."[19]

Yet, if we only perceive waste as the remains of commodities after the latter's functions are used up—in other words, if we simply construe commodities and waste in a chronological sense—we may risk falling into a myth of functionality and physicality in understanding waste, and thus ignoring its social constructedness. The examples above have already shown that waste is not simply a form of existence *after* functions of commodities disappear. Strasser has reminded us that there is no direct connection between waste and the functions of an object—an object which has already lost its functions may not turn into waste if it is either fixed or put to another use; whereas another object may turn into waste even if it is still useful but has been discarded. Therefore, how an object *becomes* waste is a socio-cultural process.

The culture that determines the discourse of waste today is, arguably, the consumer culture that has swept the globe. If urbanization renders things placeless *after* they have lost their functions, consumerism then usually deprives commodities of their social places even *before* they have been "used up." Commodities prematurely falling into the realm of waste due mainly to our consumerist culture, again, suggests that waste has less to do with function than social place.

In *BBBW*, the relationship between places of waste and consumerism is exemplified in at least two aspects: (1) places of waste and nature in our consumerist social condition; and (2) places of human beings and animals in the discourse of waste. Waste is a product of consumer society because, first and foremost, capitalism, or to be more specific, capitalist mode of production and consumption, has structured human behaviors in a modern society. The production of commodities, which is fundamental to capitalism, has altered the relationship between human beings and their products, as Marx has observed. It will also inevitably alter the relationship between human beings and waste as well, it can be argued. Contemporary discussions of waste have testified to the latter argument: these discussions, whether from the perspective of functionality or sociality of waste—that is, whether they perceive waste as a negative image of the commodity or as a social construction—all situate the discourse of waste within the political economy of the commodity.

Production of commodities, among other things, replaces exchange value for use value and turns human labor into labor power, and, therefore, renders their labor power and their products commodities that can be freely exchanged in the market. People's relationship with things are therefore restructured. When we perceive things simply as commodities that can be traded freely in terms of common equivalent (money) in the market, a vision that distorts the relationship between concrete labor and the concrete products of this labor, then things inevitably lose their concrete, unique, and irreplaceable value. As a result, people also lose the sense of what Walter Benjamin calls the "aura" of things when they possess or discard them. This becomes a condition under which waste is massively produced. As Strasser remarks, those who keep and mend functionless things, who attempt to maintain social places for things used up, also develop special understandings and intimate relationship with things.[20] In other words, only when people re-establish or restore their concrete and unique, instead of abstract, relationship with things, can things maintain their social places for a longer term and avoid falling into waste. Yet, our commodity-centered society constantly destroys this intimate relationship, replacing use value with exchange value and creating a fetishism of commodity. Paradoxically enough, this commodity fetishism precipitates and accelerates the trashing of commodities.

Second, waste as a valueless thing is closely related to the issue of value in a commodity-based society. The value of a commodity is grounded on exchange value rather than on use value, as mentioned above. Gayatri

Spivak distinguishes two types of subjects—that which is grounded in consciousness and that which is grounded in labor power. As Ann Anagnost explains: "Spivak suggests that the concept-metaphor 'value' has different senses within these mutually exclusive conceptions of the subject. 'Truth,' 'Beauty,' and 'Goodness' are 'the three value spheres: cognitive, aesthetic, ethical in the realm of cultural value.' These three anchor enduring universal values on the one hand in contrast with that 'slight, contentless' thing on the other, a 'mediating ... differential which can never appear on its own' but which enables us to see the link between living labor and the commodity in exchange."[21]

In other words, the first kind of value signifies an inherent character, although this character is also situated within specific conditions, whereas the second type of value is what we call exchange value. A commodity has "value" because it is able to be exchanged, and its value is exhibited in its ability to be exchanged. Waste, it can be argued, is a thing that falls out of the signification of exchanges, and thus loses its place, and becomes valueless. In this regard, the valuelessness of waste, in a commodity-oriented society, means inability to exchange rather than being functionless. In other words, it means more a loss of its place in the magic space of the market than a loss of inherent value.

Third, as the film attempts to show, the "value" of the environment is unfortunately also tagged with a price and determined by our commodity-based society. The all-encompassing capitalism has inevitably incorporated nature into its mode of production—Neil Smith calls it the "capitalist production of nature."[22] Nature in a commercial society is perceived as an exchangeable resource, measured in terms of money, the only common equivalent, instead of deemed as an entity that possesses unique and unexchangeable values. As David Harvey points out, such a way of measurement not only reaffirms the fact that money is taken as the only universal measurement in current capitalist society but also reminds us that money exemplifies social power.[23] To measure the value of nature in terms of money and ignore the unique value of nature, inevitably exerts direct harm against nature. In a lamentable scene in the film, the director exhibits a bird's eye view of numerous quarries hidden in the dark of the mountains in suburbs of Beijing. The traces of destruction inflicted on the bodies of the mountains present, in this view, intriguingly bizarre shapes, visualizing an outcry of the atrocities that people have done to nature (Fig. 8.2).

Fig. 8.2 Grotesque image of quarries in suburbs of Beijing

Yet, when the "valueless" waste is also capitalized, the situation becomes even more deplorable. We learn from the film that the 13 million people living in Beijing produce about thirty thousand tons of waste every day, and a large portion of this waste goes to illegal landfills without much or any treatment. The reason, as one employee of a government-run waste treatment company says, is that they can hardly make their ends meet with the meager mandatory fees they collect from companies and residents. Even so, some companies and residents choose to "give" the waste they have produced for even lower fees to illegal landfills, who, with no state supervision, simply dump waste in remote open areas. This act, which is driven solely by economic interest, only reflects a distorted mindset on the part of people in regard to ecological issues. What is more disturbing is the sequence in which people herding sheep and cows drive their stock to the open landfill and let them feed themselves with whatever is edible or inedible there. In addition, toward the end of the film, some even "recycle" swill from restaurants in downtown Beijing, feeding their pigs with it, or making edible oil from it. Such "recycle economy" only displays how economic drives distort the human conscience.

Fourth, when the consumerist spirit dominates our society and, as a result, exchange value, with the help of popular media, gives its place to what Jean Baudrillard calls the "fashion value," which he explains thus:

> this sublime, de luxe wastage highlighted by the mass media merely replicates, on the cultural level, a much more fundamental and systematic wastage which, for its part, is integrated directly into economic processes, a functional, bureaucratic wastage produced by the production system as it produces material goods, a wastage built into them and, therefore, obligatorily consumed as one of the qualities and dimensions of objects of consumption: their fragility, their built-in obsolescence, their condemnation to transience. What is produced today is not produced for its use-value or its possible durability, but rather with an eye to its death, and the increase in the speed with which that death comes about is equalled only by the speed of price rises. This alone would be sufficient to throw into question the "rationalistic" postulates of the whole of economic science on utility, needs, etc. Now, we know that the order of production only survives by paying the price of this extermination, this perpetual calculated "suicide" of the mass of objects, and that this operation is based on technological "sabotage" or organized obsolescence under cover of fashion. Advertising achieves the marvellous feat of consuming a substantial budget with the sole aim not of adding to the use-value of objects, but of subtracting value from them, of detracting from their time-value by subordinating them to their fashion-value and to ever earlier replacement.[24]

What Baudrillard calls the fashion value, as can be seen, is deeply rooted in our consumer society's fashion-making, consumption-promoting mechanism, which has reshaped our culture and consciousness. It can also be traced back to the production of exchange value rather than use value that is fundamental to capitalism, only that Baudrillard's emphasis on the consumption end of the capitalist system brings to our awareness the acuteness of the production of waste in contemporary society.

Zygmund Bauman echoes Baudrillard by ascribing what he calls the "universal transience" of the commodity, so characteristic of what he calls "the liquid modern life," to human being's loss of the sense of the infinite. In infinity, Bauman argues, "everything falls into place."[25] But modernity makes that spiritual sense of infinity irrelevant, creates the "fear of life" that takes the place of the "fear of death," which can only be exorcised by "a 'calculating approach to life' that feeds on an unquenchable thirst for ever new possessions and the cult of 'progress'—by itself a meaningless idea, devoid of purpose."[26] What Baudrillard calls the detracting of "time-value" (which in a Marxian sense is closely associated with "use-value") in the quote above, resonates with what Bauman terms the "short-termism," which is signified in people's increasing impatience for the immediate satisfaction of desires. "The cure for the bane of 'short-termism in the search for pleasure," remarks Bauman, "is a shorter term yet"; but the speedy fulfillment of desire "also speed[s] up the fading of desires and their replacement with resentment and rejection."[27] Thus the massive production of waste.

Both Baudrillard and Bauman's observations are relevant footnotes to the aforementioned scene in which scavengers find in waste in high-end apartments bags of brand-new clothes. Those clothes do not become waste because their use value has disappeared but because they have prematurely lost their positions in the signifying chain and are, thus, replaced by more fashionable commodities, which are, in turn, destined to be replaced by even newer fashions soon. The speed of this replacement is accelerating and, as a result, the amount of waste is multiplying. Yet, this destined, hastening cycle of promotion-and-disappearance of fashions has already been inscribed in the very production of commodities, and our consumer society is rendering this production further and further away from use value and making time-honored value of objects increasingly irrelevant. A paradox in consumer society is that, with fashion value in place, "wasteful consumption has become a daily obligation"[28] and "affluence ultimately only [has] meaning in wastage."[29] Waste thus becomes a fundamental social displacement and inevitably trashes our lives. Liquid modernity, Bauman remarks, "is a civilization of excess, redundancy, waste and waste disposal."[30]

8.4 "Waste Becomes Politics": Human and Animal Positions in a Consumer Society

Waste, the "ability" to make trash, is regarded, in a consumer society, as a form of power, and consequently defines people's positions in society and culture. As Strasser remarks, the sorting, or categorization or classification of objects is ultimately "an issue of class."[31] Similar to the situation in contemporary United States, which usually boasts of its material abundance, in China "disposability" is also increasingly "promoted for its ability to make people feel rich."[32] To some degree, waste "orients the whole system"[33] and shapes our society and culture.

BBBW seems to allot scanty presentations of those who are "able" to produce a large amount of waste. Their existence is instead indicated in the things that represent them, such as the tremendous volume of waste they have made, the remains of demolition in massive magnitude, as well as the "legal" enforcement for forced demolition backed by bloodsucking real estate developers and their conspirators in local governments. These people may not be the focus of the film, but the "ability" or the "power" they represent is in fact the political source of trash-making.

Instead, the film presents more the group of "unable" "wasted humans" migrating to Beijing from all over the country. Their positions in this city resembles the waste that they rely on—they are a non-existing existence and an invisible group in city life. As Bauman describes, "[t]he stage is set for the meeting of human rejects with the rejects of consumer feasts; indeed, they seem to have been made for each other."[34]

Wasted humans, according to Bauman's observation, is a product of the order-building of modernization, economic progress, and globalization: "each order casts some parts of the extant population as 'out of place', 'unfit' or 'undesirable'"; economic progress "cannot proceed without degrading and devaluing the previously effective modes of 'making a living' and therefore cannot but deprive their practitioners of their livelihood"; and globalization is "the most prolific and least controlled, 'production line' of human waste or wasted humans."[35] Indeed, the new order-building in post-socialist China's social restructuring, particularly the process of the large scale privatization and commercialization that dismantled the socialist system since the 1990s, has rendered a large swath of the population unfit and out of place. Economic progress with huge geographic and demographic disparity has further created a large pool of redundant population from "less developed" areas and older and less educated people. Globalization may seem a lesser player in the production of the wasted humans in the narrative of *BBBW*, but it is a central force in the production of both physical waste and human waste in Wang Jiuliang's second documentary, *Plastic China* (塑料王國, 2016).

In the beginning of the film several trash workers are standing among other city residents in Tiananmen Square, watching the flag-raising ceremony. This scene seems to remind us that these trash workers are no different from others in terms of national identity, and should enjoy an equal political status. However, when the ceremony ends, they quietly scatter from this center of the city, and of China, and are submerged into the sea of people, and retreat to the marginal and invisible place that they believe they belong to.

Yet, compared to those who live by and in landfills, these trash workers can be said to live a normal life, because, as registered residents, they have legal social status and are visible beings, whereas the former are linked with the city solely by waste. Those scavengers seem to have come closer to the center of Chinese society, physically and symbolically, when they left their hometowns and came to Beijing, the capital city, to seek a more decent way of life; yet, this closeness, paradoxically, only brings them a sense of alienation and deprivation, as they are away from the place with which they have close connections yet are not accepted by their new "homes." They are uprooted from their localness, only to be replaced in an alien land. That is why they are unwilling to face the camera, lest their townspeople may get to know what exactly their lives in Beijing are like.

Apparently, the scavengers in the film live on the margins of the society. In the current social hierarchy with urban centers on the one end and landfills on the other, which is paralleled by the one, which often times assumed by consumers, with commodities on the one end and waste on the other, scavengers, obviously, occupy the very end of the latter, living an abject life. However, it is precisely their "abjectness" that supplements and constructs the life on the other end. It is their "dirty" work in remote, underground areas that makes the "decent" life in urban centers possible; yet, it is also by denigrating their wasted lives that what we call "normal" lives assert their normality.

In China, it is thousands of such scavengers who have been undertaking the job of classifying and recycling waste at the very end of the life cycle of objects. However, it does not have to be this way. As Silguy observes, if we can recycle and compost "waste" in the stage of its production, that is, at the moment when individuals throw away "waste," then we will not only produce much less "waste" but also restore the harmony between people's environmental consciousness and the environment.[36] Yet, the fact is that in China, this process is conducted in the very last stage of landfill, and it creates a "wasted" class of scavengers that is deemed to match the status of the object it deals with. As the film shows, they build their homes in landfills, places that are excluded from others' sights, but these people "depend on landfills, on which they build their dreams." Even so, they are constantly threatened by the police and other state powers, and are driven away from their landfill homes in the name of cleanliness—a move that is reminiscent of the "purification" process in modernization, but, in the Chinese condition, is also a political game used for a legitimization of the accomplice of the state and commercial powers. The wasted humans thus become even more invisible.

The lives of wasted animals, as the grotesque scenes of feeding sheep and cows at the beginning and the end of the film powerfully exhibit, are equally abject in a consumer society. Those animals are cast to the position of waste as the result of, first and foremost, the order-building of a modern consumer society, in which they are perceived simply as commodities. They are raised, or "produced," for consumption by atomized, unknown consumers with no organic or intimate connections with them. It, therefore, makes no difference to those "free" consumers whether these animals are raised in a landfill dump or on an animal farm, where the animals are no less ill-treated, just as it does not really matter to iPhone fans, for example, that their devices are produced in sweatshop factories in China, as long as their devices satisfy their ever-changing desires. The severance of intimate ties between humans and their animal "products" in a commodity-based society, contributes to the production of wasted animals. These animals are placed, as their owners are, among the waste they rely on, even though they will eventually come back to serve as food or other products for those who have participated, whether they are aware of it or not, in trashing the former's lives, a coming-back that is as inevitable as it is vengeful.

Yet people sometimes tend to romanticize these wasted animals, which results in a further misplacement. For instance, an interesting sequence in the film shows a young couple taking their wedding photos with professional photographers by the bank of the Wenyu River mentioned earlier. The river is severely contaminated upstream, but its lower stream is ironically studded with expensive apartment buildings and even golf fields. It so happens that at that moment a herd of sheep, which possibly have just come from a landfill, is passing by and the bride, not aware of this, asks to hold one for her pictures. This scene is ironic in that purity and abjectness converge here, and it constitutes a tension that points to our misplaced society, culture, and consciousness.

Toward the end of the film, lives of scavengers are depicted in a slightly sentimental manner that is in contrast with the overall detached tone of the film. Two such episodes deserve our attention: first, a sequence in which a scavenger's *family* life is highlighted in pictures documenting their daily life of work, family meals, fun time, and so on. Accompanied by soft and melodic background music, this sequence is the only moment with some degree of emotional warmth in the entire film. This sequence shows how this group of people's desire for a normal life and their efforts to create one within their limited capacities are not different from everyone else's. This representation, loaded with warmth and hope, however, only brings to mind the cruel fact that these people are simply stuck in their current places and do not see much hope for change or mobility.

Second, in the depth of a massive landfill, an elderly scavenger has lived in a "house" built with waste materials that he'd collected over the years. It is obvious that the elderly man has invested tremendous efforts in building this house of his and takes it as his only home. But some day, he died all alone, and nobody noticed it until days later. This is a typical "waste" life, as the entire life of the elderly man has been locked with waste and the meaning of this life is equally wasted. His body is surrounded by waste, "corpses" of commodities, symbolizing an ironic "match" of wasted humans and wasted objects. But the question still remains: they are wasted by whom?

These depictions do not mean that the director means to romanticize those people living at the bottom of the society. We also see exposed in the film that they mishandle waste and contaminate water and air; raise sheep and cows on landfills; feed their pigs with scrap food; water their vegetables with contaminated water, and so on. They are not simply victims of waste but also perpetrators of environmental problems. To make the matter worse, some of them even conduct illegal business such as making edible oil from swill and selling it back to restaurants.

The director admits that when he took candid pictures of those sequences, he fell into a moral dilemma: on the one hand he knew that they were doing something illegal and immoral to the environment; but on the other he also understood that if he were to expose them, they would fall into further trouble and their fragile life would be more threatened. He does not give us an answer to this dilemma, but urges us to reflect on it. As he remarks, contemporary discourse of waste is a mirror, "reflecting the insatiable desires of human beings. When numerous resources and energies become massive waste because of our unnecessary demands, and when we are facing the scene of waste besieging and devouring our cities, are we aware that we have to re-examine our consumerist culture immediately!"[37]

8.5 Conclusion

Ecocinema, as Rust and Monani point out, is a reaction to the "dominant, consumerist modus operandi."[38] *BBBW* is an eco-documentary about place, displacement, and misplacement, and this sense of displacement and misplacement is most poignantly conveyed in the tension between what is perceived as clean with what is deemed as trashy, and in the irony that the "trashy" is constructed by, supplements, and deconstructs the "clean."

As an eco-documentary, *BBBW*, arguably, also finds its own place in our society by engaging its audience into self-reflection and intervening into real-life environmental problems. As the director says, "we want to disseminate our ideas through our work and words; we want to effect changes to the situation we are faced with through our work and all that we've done."[39] This film made a huge splash among its audience and relevant institutions after its exhibition on various occasions. Although at first it encountered some resistance from some institutions, who wanted to deal with such a public issue in a "conventional" way of concealing them or suppressing them, preventing them from further dissemination, the municipal government finally took action and addressed most of the problems included in the film. In this sense, the film's intervention into public discourse was effective.

Finally, in terms of place, *BBBW* also reminds us that waste as an ecology issue is not confined to a specific place—although the film focuses specifically on Beijing, it will, in a variety of forms, impact people in other places in our interpenetrating ecosphere, an issue that is further and more poignantly addressed in his next eco-documentary *Plastic China*. But there is an interesting scene in this film that implies this global perspective: a waste transporting truck is pulling back and put in focus, where the slogan of the 2008 Beijing Olympics, "One World, One Dream," is printed on its body. Indeed, as the filmmaker intriguingly shows in this scene, ecological issues are never restricted to any specific locality, particularly more so as our lives are increasingly globalized, and eco-documentaries such as *BBBW* constantly remind us of this fact.

Notes

1. I refer to Chia-ju Chang's definition of "environing" in her "Introduction" in this volume—"consideration of modes, aesthetics, ethics and politics of environmental inclusion and exclusion." My idea of "place and misplacement" resonates with Chang's definition in that waste as a misplaced object is always already involved in the politics of environmental inclusion and exclusion—waste is discarded but also surrounds us, marking the boundary between us and the other. It bears the socio-cultural significance of marginality and supplementarity.
2. Douglas, *Purity and Danger*, 37.
3. Ibid., 36.
4. Strasser, *Waste and Want*, 5.

5. Worster, *The Wealth of Nature*, 27.
6. Jack, "2010 Seyingwuji niandu shi da sheyingshi jiexiao."
7. For instance, "Jingji ban xiaoshi" (Economy in 30 Minutes), a signature program in the Economic Channel of the China Central TV (CCTV) consecutively broadcast special reports on the subject "Pojie laji weicheng" (Resolving besiegement of cities by waste) in the end of 2014 and beginning of 2015. These reports covered dire situations of waste disposal in Wuhan, Three Gorges Dam, Danjiangkou, Leshan, Haikou, and cities in the Southeast coastal area.
8. Wang, "Lun laji," 214.
9. See Silguy, *Histoire des hommes et de leurs ordures.*
10. Wang, 212.
11. See Kristeva, *Powers of Horror*, passim.
12. Wang, 221.
13. Hawkins, *The Ethics of Waste*, 2.
14. See Chen and Tao, "Bushi laji baowei weicheng, ershi chengshi baowei laji."
15. Bauman, 100.
16. Wang, 213.
17. Ibid., 217.
18. Ibid., 215–16.
19. Wang, 217.
20. Strasser, 10–11.
21. Anagnost, "The Corporeal Politics of Quality (*suzhi*)," 204.
22. Smith, *Uneven Development*, 33–65.
23. Harvey, *Justice, Nature, and the Geography of Difference*, 150–51.
24. Baudrillard, *The Consumer Society*, 47.
25. Bauman, 95.
26. Ibid., 99–100.
27. Ibid., 110–11.
28. Baudrillard, 48.
29. Ibid., 45.
30. Bauman, 97.
31. Strasser, 9.
32. Ibid.
33. Baudrillard, 46.
34. Bauman, 59.
35. Ibid., 5–6.
36. See Silguy.
37. Zhou Shiping. "Yong liangxin 'guancha' laji," 26.
38. Rust and Monani, "Introduction," in *Ecocinema Theory and Practice*, 3.
39. Zhou, 22–23.

Bibliography

Anagnost, Ann. 2004. "The Corporeal Politics of Quality (suzhi)." *Public Culture* 16 (2): 189–208.

Baudrillard, Jean. 1998. *The Consumer Society: Myths and Structures*. London/ Thousand Oaks: Sage.

Bauman, Zygmunt. 2004. *Wasted Lives: Modernity and Its Outcasts*. Cambridge: Polity.

Chang, Chia-ju. 2017. "Global Animal Capital and Animal Garbage: Documentary Redemption and Hope." *Journal of Chinese Cinemas* 11 (1): 96–114.

Chen, Xuelian 陈雪莲, and Tao Hong 陶虹. 2011. "Bushi laji baowei weicheng, ershi chengshi baowei laji" 不是垃圾包围城市, 而是城市包围垃圾 ("It Is Not that the Waste Besieges Cities, but It Is Cities that Besiege the Waste"). *Guoji xianqu daobao*, June 8.

de Silguy, Catherine. 1996. *Histoire des hommes et de leurs ordures: du Moyen-âge à nos jours*. Paris: Le Cherche midi éditeur.

Douglas, Mary. 1966. *Purity and Danger: An Analysis of Concepts of Pollution and Taboo*. London: Routledge.

Harvey, David. 1996. *Justice, Nature, and the Geography of Difference*. Cambridge, MA: Blackwell Publishers.

Hawkins, Gay. 2006. *The Ethics of Waste: How We Relate to Rubbish*. Lanham: Rowman & Littlefield Publishers.

Jack. 2010. "2010 Seyingwuji niandu shi da sheyingshi jiexiao" 2010 色影无忌十大摄影师揭晓 ("2010 Top ten photographers of Seyingwuji"). December 14. http://ww.xitek.com/html/expert/award2010/news/201012/14-56050_9.html. Accessed 20 Dec 2018.

Kristeva, Julia. 1982. *Powers of Horror: An Essay on Abjection*. Trans. Leon S. Roudiez. New York: Columbia Univeristy Press.

Nagy, Kelsi, and Phillip David Johnson II, eds. 2013. *Trash Animals: How We Live with Nature's Filthy, Feral, Invasive, and Unwanted Species*. Minneapolis: University of Minnesota Press.

Rust, Stephen, and Salma Monani. 2013. "Introduction: Cuts to Dissolves—Defining and Situating Ecocinema Studies." In *Ecocinema Theory and Practice*, ed. Stephen Rust, Salma Monani, and Sean Cubitt, 1–13. New York: Routledge.

Smith, Neil. 1984. *Uneven Development: Nature, Capital and the Production of Space*. Oxford: Blackwell.

Strasser, Susan. 1999. *Waste and Want: A Social History of Trash*. New York: Metropolitan Books.

Wang, Min'an 汪民安. 2010. "Lun aji" 论垃圾 ("On Waste"). *Meishuguan* 1: 210–223.

Worster, Donald. 1993. *The Wealth of Nature: Environmental History and the Ecological Imagination*. New York: Oxford UP.

Zhou, Shiping 周仕凭. 2010. "Yong liangxin 'guancha' laji: fang ziyou sheyingshi Wang Jiuliang" 用良心 "观察" 垃圾: 访自由摄影师王久良 ("Observe the Waste with Conscience: An Interview with the Freelance Photographer Wang Jiuliang"). *Huanjing jiaoyu* (7): 21–26.

CHAPTER 9

Land, Technological Triumphalism and Planetary Limits: Revisiting Human-Land Affinity

Xinmin Liu

With the onset of Anthropocene, we humans have come to exert substantial geological and meteorological impact on earth. Ironically, the power of such agency has also made us eerily aware of the cost and loss of the man-led conquests and achievements; we now have to reckon with the fact that the feat of human progress can cause backlash and negate what human agency has hitherto harnessed and reclaimed on earth. At the heart of the many environmental crises we face today lies the grave human error of the *selective enlightenment* (Nixon, 14). We have, in the name of civilizational progress, opened a "can of worms" that have not only aided the infectious spread of technological abuse but worn down our wisdom in guarding against the partial and excessive use of instrumental reason. While regretting over our past oblivion of the planetary limits, we now face the ultimate impasse our modern progress has brought us to: our planet is rashly depleted in its resources and overly strained with its restorative balance due

X. Liu (✉)
School of Languages, Cultures, and Race, Washington State University, Pullman, WA, USA

C. Chang (ed.), *Chinese Environmental Humanities*, Chinese Literature and Culture in the World, https://doi.org/10.1007/978-3-030-18634-0_9

to unbridled human conquest of the natural world; we cannot but wonder as to how much longer and in what ways we can keep the engine of modern development running?! Given the urgency of this crisis, we need to confront those unsettling yet deep-seated concerns: have we pursued a design of civilizational progress that is bound to overdraft and underpreserve the natural assets of the earth's biodiversity? Is this biotic richness, harnessed to cater exclusively to human demands, ever able to keep up with the ceaseless drive for market-driven pursuit of profits? In the end, we will have to question ourselves: are we going to co-exist with land, water, climate and other earth-bounded forms of life and resource so that our offspring will have a decent chance to survive together with theirs in the future?

If we answer in the positive, we then have to accept the fact that, in terms of human-land relationship, our view of the nonhuman world has become mostly incorrect and uncannily dysfunctional; we must admit that nonhuman nature is now changing in unknowable ways, and that the logic of human reason governed by means of statistically stable variations has been shaken to its root. We ought to bring ourselves to confess that the triumphs of human technological prowess over the nonhuman lifeworld have been profoundly tarnished because some of them were decidedly motivated and accomplished by the insatiable human desire for power and wealth. Hence it comes as no surprise that some critics of Anthropocene are already calling for a more accurate and deserving term, such as "Econocene," in place of Anthropocene.[1] While disagreeing with Econocene as a replacement term, I heartily embrace its critical thrust as it aims to launch an intense probe into the flawed workings of the modern capitalistic economy driven by a reckless thirst for profits at the expense of planetary livelihood.

I take issue with the detrimental impact of Anthropocene in the Chinese context as an environmental humanist because I am keen on knowing more about the correlation between human existence and the planetary limits and remain committed to advocating the idea of "human-land affinity." The issues I have explored in my research on land-based human endeavor bear on an area intersecting aesthetics, ethics and ecology, and at the center of my inquiry are these questions: why should we renew our awareness of human-land affinity in an age buoyed by global economic "miracles?" How can we regain a sense of the *interdependence* between humans and the nonhuman lifeworld by way of human-land affinity, once revered by our agrarian traditions, and strive to restore and rebuild a livable habitat for both? It is a rather compelling fact that, since the onset of China's Reform Era (1980s–present), there have been aggravating trends of bringing the physical environs (both wild and reclaimed) under heedless human control in the interests of modern development, that is, to engage in massive industrial construction and unchecked urban housing construction mostly for commercial gains. This has resulted in a skidding imbalance between man-made landscapes and wilderness, and an ever-worsening conflict between the humans and the nonhuman species in battling for livable habitats. Take, for example, the proverbial saying that hangs on the lips of many technocrats-turned-entrepreneurs: "we can't let the river water flow by in vain!" (*Buneng rang shui baibai de liuzou!*)[2] In reality, their mantra can be applied to any natural resources at hand in order to satisfy the desire of these profit-craving technocrats—coal, petroleum, natural gas, minerals, you name it. Putting it plainly, what they mean is that when a resource is *not* presently utilized it is deemed an income foregone. It follows that the society at large must follow their lead in acting on the "cash-in imperative" *here and now* and give them the consent to "cordon off" lands, rivers and lakes for whatever money-making projects they set their minds on and allow them to consume the resources therein in order to bring in revenue as freely as they will and as fast as they can.

Accustomed to seeing the biophysical world merely as inanimate objects at human disposal, these technocrats have latched on ambitious strategies and designed extravagant projects to reclaim, rearrange and reconstruct land, water and other natural settings solely by way of seeking utmost profit. And being lured by easy material gains has clouded their vision to the risks of pushing the consumption of water, timber, minerals and other natural resources way past an irreparable limit. It is evident that such an overly utilitarian mindset has blunted their moral sensibility, causing them to overstep the baseline of their professional ethics and compromise the integrity of their professional knowledge about ethical limits of resource utility. As a result, contrary to being the educated elites of the Chinese nation, they have failed to live up to the public's confidence and trust in them as the leaders of political reform and social progress, thus not just tarnishing their image as role models for the entire nation but also causing the general public to feel disoriented and vulnerable toward the inexorable ups and downs in carrying out the reform while still keeping faith in the common good to be attained through it.

9.1 Misconceiving Tech-Induced Modernity

There is clearly a loss of moral bearings on the part of these technocrats. How did it happen? It is not as if China's educated elites have stumbled onto some unchartered waters in dealing with the mission of science and technology in promoting social wellbeing. After all China's crossing into the modern era around mid-1800s was expressly motivated by ethical renewal as well as technological upgrade in its wholesale borrowing of Western values, of which the most favored ideas were democratic thinking and scientific knowhow—widely dubbed as **Mr. D** for Democracy and **Mr. S** for Science. Even in its infancy, Chinese modernity had posited ethical goals while the leading intellectuals were trying hard to embrace the Western concept of "evolution" much in the spirit of "change-as-improvement" to guide the cause of social progress.[3] The progress to be garnered through this kind of change for social betterment was to be actualized on the level of the society and community to the end of the common good. As an example, emphasis was laid by Yan Fu (1854–1921), Ma Jianzhong (1845–1900) and Qian Xun (1854–1927) on the need to embed technological knowhow adopted from the West in China's frantic strife for social reform and national strength. These pioneering individuals had been trained as diplomats in Western Europe and earned the trust of the learned gentry of the late Qing era, having worked at one time or another as the personal secretaries for influential Qing officials such as Li Hongzhang (1823–1901) and Zhang Zhidong (1837–1909). Yet as they shifted their focus to translating texts of evolution theories (in the case of Yan) or compiling encyclopedias of scientific knowledge from the West (as with Ma and Qian), they stood firm by their faith in Confucian humanism nurtured by their classical upbringing and by way of aspiring after ideals of Confucian ethics. Despite the assertive rhetoric of their anti-traditionalist attitudes, when they sided with the Foreign Affairs Movement,[4] they never hesitated to rally around eminent social reformists like Kang Youwei (1858–1927) and Liang Qichao (1873–1929) who, as staunch humanists, applauded science and technology as ingredients of a "Commonwealth of Science." Even as they compiled thousands of entries of scientific knowledge in the hope of making them accessible to the commoners, they seldom wandered off the course of fulfilling a role subordinate to the overall cause of social wellbeing.[5]

However, this holistic way of envisioning social progress ended up being outpaced and overpowered by the hasty pursuits of the newly founded Chinese state to avenge her feeble and humiliated past at the hands of foreign imperialist powers and be swiftly restored to her former glory and splendor. This state-led mandate persuaded many elite intellectuals to embrace a linear path to national supremacy on the strengths of wealth and power, which has posed itself as the irreversible path to social betterment. Before long it nurtured a full bloom of a kind of teleological thinking that mimicked the rise to technological dominance of first the Soviet Russia, by way of graduated but centralized economy in the 1950s, and then Euro-America, by way of market-driven venture capitalism since 1990s. Overlooking the irreparable gap between the two cold-war ideological camps, the Chinese state has patched up and put into practice a brand of socialism that ineptly and callously mixed ingredients of market-driven entrepreneurship with mechanisms of its ideological control. The PRC's nation-building project since 1949 has been fraught with blunders in which many in the science community caved in to a wide spread of instrumental misuse of resources spurred by statist agendas adorned in idealist guise. The decades of "battling against the Heaven" (*yu tian fendou*) have resulted in campaigns of landscaping on massive scale, such as "removing the hills and filling up the seas" (*yishan tianhai*), South-North Water Diversion Project (*Nanshui beidiao*), that ironically paralleled the cruel and injurious political campaigns under Maoist socialist revolution; by "bringing the mountains and waters under control" (*zhishan zhishui*), a great many geological oddities across the land have been "reformed" as if they had taken on the malicious persona of the nation's common political enemy.

The outcome is flatly predictable: what started off as some naïve idea to rearrange the natural world so as to improve water distribution or reclaim more acreage of farmlands have ended up triggering an onslaught of environmental disasters: pristine forests have been mowed down, arable lands have been contaminated, water shortage has grown worse and regional climates have turned rogue. Akin to the environmental loss, equally detrimental erosion has occurred by way of the fabric of social conscience: many in the ranks of technocrats have veered off the path of moral sensibility and ethical obligation, and, by willingly suspending their moral judgment, they have become accomplices in rash acts of mangling the lands, waterways and mountains or have been blindly involved in despoiling the various resources found in the natural world. For all their scientific acumen and technological wisdom, they could not help lending a hand to the State's campaign-style drives in refashioning nature, turning blind to their ethical obligations and uncaring of humanistic values and ecological concerns.

My approach to critiquing the developmental hubris of the past decades has led me to focus on the harrowing impact of man-made landscapes that have made profound inroads into our physical surroundings, but, more importantly, into the ways in which we are compelled to voice our social conscience or disclose our failure to do so. It has made me aware how technological advance at once inspires our confidence and impairs our judgment, and in what ways the pursuit of technological primacy at whatever the cost can be as risky and perilous as it is instrumental and efficient when it comes to the utility of land-based resources. Putting it plainly, we all have to determine ***how much is too much*** when judging the extent to which we seek growth by means of science and technology. I thus believe that it is imperative to revive our dimming awareness that planetary life forms and human beings are infinitely interconnected by virtue of the land. One way I deem most viable is to reawaken our sensibility to the affinity between landscapes and the humans as an antidote to reign in reckless urban and industrial growth. To foster such a state of mind, we need to firmly limit and undo the impact of de-territorialized (i.e., being torn from our visceral ties to the land) life styles enticed by market-driven consumer culture. We seek to bring alive our ethical impulse for landscapes, both natural and man-made, and our first step ought to be advocacy of the human affect as a safeguard when the wellbeing of the land and its ecosystem comes under threats posed by heedless pursuits by those humans caught in the boom-or-bust fever. To stay active and vigilant despite our toxicized bodies and impaired sensibilities, we need to enhance the sense of being embedded in our immediate habitats, particularly of cohabiting them with other residential species, by virtue of our shared corporeal being. In doing all this, we should seek to not just help preserve the intersubjective bonding via social and cultural memories of the land-nurtured heritages but affirm and ensure how human-land interdependence ***can effectively steer the rudderless chase of power and wealth.***

To resist the greed-driven abuse of land-based resources, I have drawn on the critical insights of new materialism and taken cue from its theoretical probe of the human neglect of the materiality of ecosystems. I do so in the hope of tracing the human triumphalist mindset and its oversized ambition to its early root—the utilitarian values of Western modernity. When Alison Byerly addresses the complexity of modern scenic travel, she questions the essence of humans' domination over nonhuman nature:

> Does it (the wilderness) have an inherent value, or is its value dependent on the use to which it is put? At what point does the art of appreciation become an act of appropriation in which its intrinsic qualities are sacrificed to the agenda of its audience?[6]

In the same vein, leading critics of the utilitarian mode of disparaging the Earth's materiality have taken the question even further; Jane Bennett, author of *Vibrant Matter: A Political Ecology of Things*, chastises the abusive control and wasteful use of earthly resources by modern technologies. She offers the most pointed insight in critiquing modernity by way of such terms as *resource*, *commodity* and *instrumentality*, and the aim of her approach is "to present human and nonhuman *actants* on a less vertical plane than is common ... and to bracket the question of the human."[7] In a patently cognitive move, she shifts the focus of her scrutiny from the human subject alone to an interplay between the human and nonhuman. By averting the former as the dominant center, she now relies on interacting with the "material agency or effectivity of nonhuman or not-quite-human things." In order to rid ourselves of the "earth-destroying fantasies of conquest and consumption" as the cause of triumphalist hubris, she believes, we need to beware that "[t]hese material powers, which can aid or destroy, enrich or disable, ennoble or degrade us, in any case call for our attentiveness, or even 'respect.'"[8] She draws particular attention to the importance of "induc(ing) in human bodies an aesthetic-affective openness to material vitality" as one of the primary goals to motivate humans to "move selves from the endorsement of ethical principles to the actual practice of ethical behaviors."[9]

Bennett's approach is bold, provocative and even subversive. Nevertheless, when she articulates the role of the affect, I have also felt unhinged at times by her tendency to leave out altogether the human factor in this material-human interplay. In a section named "Impersonal Affect," for instance, on one hand, she avows her allegiance with the camp of the affect-oriented theorists who deem it politically imperative to embrace the field of "micropolitics" (such as religious practices, media rituals, neuroscientific experiments, etc.) and underscore the need to form and reform ethical sensibilities and social relations; on the other, she then takes an abrupt turn in her train of thoughts by declaring that "I branch out to an 'affect' not specific to human bodies. I want now to focus **less** on the enhancement to human relational capacities resulting from affective catalysts and **more** on the catalyst itself as it exists in nonhuman bodies."[10] When she shifts her discursive gears like that, it tends to throw the readers off in a disorienting move and weakens the strength of her otherwise persuasive argument.

What I believe matters more in this type of "micropolitics" in the Chinese context is precisely the infinite bond between the humans and nonhuman world—the human-land affinity especially—and the role of "re-enchantment" in reviving and enhancing human relational sensibilities amidst the nonhuman world. In other words, rather than exploring the catalyst proper as it emanates from material things, we need to focus more on the dynamic of the intermesh as it occurs between humans and things because there has not been such a clear-cut conceptual divide between the human subject and the nonhuman object due to residual impact of traditional Chinese philosophy. In the Chinese concept of subjectivity, the emphasis is laid more on *the imminent synthesis* as it is being actualized through the interconnection between humans and things, and less on the antithesis that brings into sharp focus the discordant relationship between the subject and object.[11]

In the light of this, I applaud the vibrant impact of "impersonal affect," owing not so much to the catalytic force in nonhuman bodies in and of itself as to the triggering and enhancing effect on the human sensibilities and the ensuing revival of their ethical concerns in the midst of dealing with nonhuman materiality. In concrete, I want to explore the layered ambiguity of presenting landscapes by "disinterested (human) observers" to tease out the apparently factual and objective posture in their one-sided claims and bring to light its deep-seated exploitative traits toward the material world. Contrary to their profess disinterest, I stress that observers of material nature like the technocrats tend to imagine the natural resources in a way that appears impersonal and serviceable, but in the long run invariably dominating and despoiling. I thus argue that an unreflective view of technological primacy in the guise of scientific neutrality can easily mislead to humans' triumphalist hubris, which, in its turn, gives way to excessive pursuit of wealth and power. This mindset goes to disclose a kind of selective and short-term enlightenment launched by the modern West; it causes humans to be oblivious to the fact that humanity as a whole must depend wholly on livable habitats on earth for their future generations to carry on as humans. It has thereby hardened into a costly blind sight to the fact that there exist exhaustible limits of the Earth's ecosystem, and that this fact is a categorical No Outlet warning to the linear path of civilizational progress. We need not look far to find convincing evidence from many literary, filmic and cultural works of our times of ecological endangerment. I now dwell on a couple of such cases to drive my point home.

9.2 Unmasking the "Wizardry" of *Cash Changeover*

Let us first focus on the case of such frantic consumption of the planetary resources with dire disregard to the disruptive risks in damming China's major waterways, and we will review a documentary film, *The Silent Nu River* (*Chenmo de Nu jiang*, 2003) by Hu Jie and an investigative TV News Probe named *Damming the Nu River: From Start to Halt* (*Nu Jiang jieliu: Cong kaifa dao tingjian*, 2016) by China's National TV Channel (CNTV).[12] I use the notion of human-land affinity here as a way to delve into the mindset of technocrats and like-minded local leaders stationed along the Nu River banks to see how they have grappled with technological primacy in the context of river-damming. The Nu River is widely known as the last free-flowing major waterway in southwest China, and those familiar with the topographical features of this region are instantly reminded of the fact that it runs parallel to two other major rivers in this region which form the renowned Tri-River Corridor (*Sanjiang bingliu*).[13] Lying at the foot of the Himalayas and fed by the melt ice-capped mountain ranges, the three river giants exist as a time-honored life-sustaining ecosystem to the whole of southwest China. The two documentary footages, *The Silent Nu River* and *Damming Nu River: From Start to Halt* cover a time span in which the controversy of damming the Nu River waxed and waned, depicting a roller-coaster course of decision-making and -unmaking, yet eventually offered us a compelling testimony to how a hard-won solution to excessive pursuit of growth came about against the odds of a global mega-damming drive. The two journalistic works jointly testify that excessive damming has proven disastrous to the local ecosystems and unsettling to the livelihood of local multi-ethnic inhabitants. More importantly, they present a strong case in which technological primacy as a normal way of thinking has been diffused or dispelled, thanks to the persistent efforts by environmental activists across China to evoke a sense of human-land affinity in their campaigns to resist the 13-tier hydraulic damming project slated to be built along the Nu River in 2010s.

In the heyday of China's economic boom (circa 1995–2005), the demand for energy across the densely populated urban and especially coastal regions of China resulted in a "hydraulic turn" in energy production. Sensing their opportunity for a fast-track growth had arrived, the local leaders and technocrats who had long envisioned hydraulic projects to be built on the Nu River now seized the moment and laid out a grand scheme of multiple, cascading power stations to be constructed. They had also been inspired by the success of mega-damming projects being built on the *Jinsha River* and the *Lancang River* nearby at the time that had set the trend to "go big or go home" in hydraulic engineering; as their key to booming success in growth and profits, they were chomping at their bits for a technological upgrade to bring fast lucrative profits in return. How did these technocrats—many of them are ethnic natives—change from cherishing the value of the long-lasting local ethnic heritages, such as the towering snowy ridges, roaring rivers, combined with diverse and colorful ethnic customs, to going along with demolishing them hastily in order to realize their dream of reaping huge return of profit? The key to understanding this cherish-to-demolish change lies in how they dissected the human-land affinity and implemented a virtual yet potent formula known as "cash changeover" (*bianxian*). In practice, the formula went like this: firstly, obtain a statistically tabulated sum total of the local resources, such as the lands, the mineral deposits underground, the timber to be harvested from the forests, and the energy to be extracted from the flowing rivers; next divide the sum total by five times to obtain a one-fifth figure as "expendable cash," and then sell the actual resources such as timber, mineral and other resources to the volume of one-fifth worth as if it were loans made available from an investment bank; lastly spend the cash on the planned lucrative projects or contracts they have the authority to approve and undertake.[14]

Deplorable as it might sound, this formula was a typical brainchild of the revenue-craving technocrats who sought "creatively" to justify their excessive utilization of local resources so that they could undertake the largest possible number of industrial enterprises and commercial properties and, in return, reap the maximum revenue before their tenure as local officials expired. As was often the case during the 1990s and early 2000s, their feverish can-do spirit, emulating the success of China's coastal boomtowns, was in part driven by "the quota incentive" enforced by the State's top-down policies, and in part by these individuals' own greed for personal gains when tempted and misguided by the illicit gains via embezzlement and racketeering. Once caught between the two vices, it was inevitable that they resorted to the tactic of cash changeover, which had been hitherto proven as a statistically conceived "wizardry" of fast-track success.

However, while it is explicable to blame their surrender to the lure of monetary profits on the frailty of human nature, it is more intolerable to realize that these technocrats have principally forfeited their professional and ethical accountability. When they turn a blind eye to how woefully the cash changeover formula distorts and subverts scientific knowledge, they have violated an essential ground rule of their professional ethic as scientists, that is, to be honest and faithful to the fact that what essentially constitutes the particular ecosystem of the local environ is its integral symbiosis, for which all kinds and species of life are the inalienable parts of a whole. Parts, if not their entirety, of such knowledge and the keen sensibility derived from it should be foundational to their earlier training in scientific learning. The American environmentalist Aldo Leopold says,

> The land is one organism. Its parts, like our own parts, compete with each other and co-operate with each other. The competitions are as much a part of the inner workings as the co-operations. You can regulate them—cautiously—but not abolish them. ...
>
> A thing is right when it tends to preserve the integrity, stability and beauty of the biotic community. It is wrong when it tends otherwise.[15]

This spells out the bottom line of hydraulic engineering, or of any earth sciences for that matter. The biophysical assets of the Nu River basin have thrived over centuries to foster such a rare sanctuary of geological deposits, plants, insects and animals owing largely to the fact that these life forms are interconnected and complementary to one another as a whole. Moreover, they are totally immersed in the local climate that has nurtured them into a symbiotic entity. No formula can be more unscientific and unethical to treat them as if they were machine parts to be individually unhinged and replaced at will.

But as the cash changeover formula has been numerically conceived, the local policy-makers are misled to commit the error of "forfeiting" the fundamental law of symbiotic unity governing the natural world and opting for slicing up those readily cashable assets from the integral biota. They allowed themselves to "cash in" on these assets as if they were itemized human belongings, inanimate and isolated from each other. To their mind, adopting a policy like this would take a mere act of crunching the numbers, hence, it is only logical for them to "carve up" a part—one-fifth—of the total value, and do whatever they deemed fit with it. They were as a result totally blind sighted to the detectable risks of interfering with the cohesive stability of the local ecosystem, which happens to sustain the lifeline of myriad species of residential life here, humans included.

Furthermore, on the ethical level, their entrepreneurial drive is fundamentally flawed due to a lack of concern for the life-supporting needs of local ethnic groups, such as the *Nu*, the *Lisu*, the *Dulong* and other ethnic minorities, whose dependence on the ecosystem already hangs in delicate balance.[16] It is often the case that, where the material resources, such as water, air, soil and forests, are found in abundance, there surely exist diverse ethnic minorities whose life styles blend in seamlessly with their ecological settings. Damming, mining or lumbering in such regions would pose an instant threat of ruining their farmlands and pastures, tip the balance of their biotic wellbeing, and result in uprooting these ethnic minorities from their ancestral homelands and disrupting their time-honored livelihoods. But these technocrats-turned-entrepreneurs, bent on profiting by way of technological supremacy, look callously away, ignoring the vital fact that there have existed lifelines shared by the humans and the nonhuman world for centuries in this region.

I now present a telling instance of the cash changeover formula being implemented on the Nu River and the imminent loss and ruin to be triggered. The case is found in the documentary film, *The Silent Nu River*, in which Director Hu Jie's camera moves up close and personal to focus on a local leader's candid reaction to the cash changeover. Presently, a middle-aged man, a Nu native, is heard praising the magical wonder bestowed on them by the foothills of the Himalayas. Then, deliberately, facing the camera, the Nu leader begins loquaciously to "inform" us how he and his Nu community have come around to a full grasp of the prospects of lucrative "returns" by turning their river into a money-maker. By now we know that the Nu cadre has been introduced to the cash changeover formula, and it is no surprise that he glowingly conveys his new "vision" by pointing to the water behind him and declares: "beneath all that is where the money is hidden, and we are ready to scoop it all out!"[17] He is, of course, referring to the imagined profits to be gained by means of the grand scheme of 13 cascading hydraulic stations to be built on the Nu River; even though he gives no verbal clue as to where and how he learned about the cash changeover formula, it is evident that without it none of these power stations would have made any sense to him in the first place.

For a long while, I had been perplexed by the ease and eloquence with which the Nu leader "flaunted" his vision of the damming project as if it had already been completed and were churning out piles of cash. What genuinely unnerved me was the fact that he, while still in awe of the towering snow-capped mountains behind him and the timeless roaring of the river currents in his earshot, could so rudely abandon the interdependency that binds the sublime landscapes with their ample resources and the livelihood of the hitherto self-sufficient human and nonhuman species residing locally. There is no doubt that his life up to this point has been immersed in this bond with the multifarious fauna and flora that have called this region home since ancient times. So why this abrupt, zealous flaunting of their ancient wisdom? This close-up image of an "enlightened" local leader captured by Director Hu has disclosed the hidden perils of the instrumental reason—he has bought into it despite himself. His words serve to illustrate how the local cadres had been coerced into a mindset driven by the cash changeover and have been inadvertently plugged into the global circuit of venture capitalism. With a sleight of hand, the Nu cadre is so enthralled by the instrumental reason that he abruptly suspends his ethnic identity, severing his affective ties from the environing community that bears witness to the time-honored endurance and survival of his Nu ancestors.

9.3 Mending Human-Land Affinity

Needless to say, such a demolition of the nonhuman life forms entails costs, that is, tipping the symbiotic balance of the biotic environ toward the brink of collapse and in the long run bringing human beings to face their own perils. To be expected, though, they begin initially with acute endangerment for the land, land-based life forms, then the livelihood of the land-dependent ethnic cultures and, more permanently, the symbiosis blending them in a holistic way. Nonetheless the perils remain "unseen" and "unreal" to the local technocrats seeking investors and developers so long as all elements—air, vegetation, soil, aquifer and other planetary elements—are regarded as portable and insular rather than being interconnected within the ecosystem and functioning through its harmonious whole. However, it takes only one mind embedded in the local conditions, a local Tibetan named Ah Luo, to see through the guise of the entrepreneurial hype and disclose what severe ruins and disasters would descend upon the local inhabitants once the cash changeover policy is put into practice. In a calm voice, Ah Luo presently speaks to Director Hu on camera, saying that if the policy to dam the Nu River is enforced, the local ethnic communities would have to confront a much dreaded reality: their way of life would be profoundly altered, starting with their most familiar daily routines: they would have no decent pastures to herd their cattle, no body-cleansing hot-springs nearby to take a dip, and no sandy river banks for young lovers to hold song-and-dance rituals. In brief, he warns, their modes of existence would be drastically lost through no fault of their own. He says:

> When we are relocated in a strange land, we will be totally disconnected from what our ancestors have handed down to us. I would have no courage to open my lips when my children ask me what the Nu River looked like, and I could only answer by pointing to photos of the Nu in the album. … Would there be any sense of pride left in me after being uprooted from my most prideful belongings?![18]

Ironically, in the above instances, what gets some tongue-loose and others tongue-tied are the opposite sides of the same scenario if the local technocrats were to carry out their "miracle" formula. They are the result of the local technocrats' willing complicity with and conformity to a market-driven world view: it is the self-serving gains measured in monetary value. Their professional expertise, when decoupled from its ethical conduct, only serves to fuel their resolve to go on a "demolish-and-consume" spree as its overriding objective. Devoid of his ethical guidance, the Nu cadre becomes complicit with the mad-dash-for-cash trend, turning himself into an easy prey of the unchecked pursuit of revenue and material affluence to the exclusion of his ethnic heritage and "lived" wisdom of ecological wellbeing. Equally problematic is the télos underpinning their blind faith in the logic of modern development—a single-track, linear path of progress configured in quantifiable data for raising GDP, and little else. What it entails is the necessity for fast-track economic growth, which has unevenly mapped out the pace, scale and timing of economic growth for different regions of China. As a result, a lopsided pattern has emerged according to which, in terms of resource utility, those regions with little or no access to land-based resources are entitled to receiving priority choice of fiscal investment and technological knowhow simply by virtue of being *urban* and *coastal*. In stark contrast, those regions with plentiful and direct access to natural resources are more likely to miss out their opportunity for timely investment to fund development because they are *rural* and *out-of-the-way*. To make the situation worse, these regions must yield their natural rights to resource utility to other more developed regions and have their own homelands stripped and depleted for the sake of other undeserving ones. Such is the lopsided nature undergirding the state's developmental mindset!

Unfortunately, the local technocrats have failed to detect the falsified logic behind the cash changeover; therefore, they have failed to challenge the unjust and inequitable distribution of benefits for this region. They have also been resigned to the fact that, when aligned along this linear and single-track strategy, those at the lower stages of development (such as China's southwestern rural and border regions) are fated to give up their lived habitats and biodiversity so that those at higher stages of development (urban centers like Beijing, Shanghai and Shenzhen) can go pursuing their goal to further heights, thus widening rather than closing the gap of economic disparity. What makes their practice even more inexcusable is their being oblivious to the belief that no alternative model for societal betterment is feasible simply because it is not as "profitable" and "high-achieving" as the market-driven economy. To me this is indeed the root cause of their moral neglect and their failure in ethical intervention due to an unthinking faith in technological triumphalism.

What gives us redeeming light of hope through this study is to observe how the two documentaries come to resonate with each other in attesting to the reawakening of ethical accountability toward the nonhuman world in China. Shared by both the freelance filmmaker and the official TV journalists, a renewed moral sensibility has urged them to interact with alternative viewpoints in better grasping the integral wholeness of the local biotic communities. They do so by revealing the correlation between *ethical obligation and the technological knowhow* as the key to resolving a decade-long dispute over how to balance between ecological wellbeing and economic growth along the Nu River Basin. In March 2016, the Governor of Yunnan Province declared a moratorium on all the hydraulic damming projects, large or small, along the Nu River.[19] Given a boost in ethical concerns, the journalists of the News Probe conducted on-the-ground interviews with local technicians at the small hydraulic stations built along the *Dulong* River, a key tributary of the Nu River. Thanks to their effort to be up close and personal in their investigation, we are offered a vantage point for viewing some astonishing facts of how the local climate and terrain together wielded an unseen chain reaction after the monsoon rainfall set off torrential flooding down the Dulong that year, and, by proximity, the detriments of damming the rivers in this region as well. A case in point is their visit to a hydraulic power station on the Dulong during which the honest and sober-minded station chief revealed, while speaking on camera to the journalists, that their generator's turbines could not be operated during the rainy season due, surprisingly, to insufficient intake of river water to run the power generators. He went on to explain that the bizarre situation was caused by the floating trash of deadwood, plants and rock residues in the river that almost blocked the water intake opening to the turbines, causing them to slow to a paralyzing stall. It was an eye-opening moment when he learned in the end that the floating mass of trash had been in its turn caused by the loss of topsoil and whatever had grown on the mountain slopes along the Dulong upstream. What an uncanny and punitive reminder of the forgotten basics of hydraulic engineering!

Undoubtedly, for the local technicians it turned out to be a lesson about what NOT to utilize in a terrain like the Dulong when planning for damming projects, and indeed a lesson learned in a timely and profound manner! It echoes what Ah Luo, the Tibetan herdsman interviewed by Director Hu earlier, voiced in his warning that the topsoil of the river banks was relatively thin and already overburdened by vast and densely grown forests and other plants. Thus, the technicians and the viewing public are faced with an astounding but genuine fact that there exist correlating innerworkings that coalesce and conserve a balanced state of growth and wellbeing of the entire river basin. Disruptive and excessive human efforts to divert and slow down the rapid currents would only upset the biotic equilibrium and lead to unintended outbreaks of deterioration and disaster in the livelihood of the human and the nonhuman alike. Rather than knowingly carve up the land-based and water-dependent resources numerically, it is far more important for local technocrats to make it known how precarious the symbiotic nature of the human-land relationship is and what limits and restraints they should place on their own tendency to go beyond sustainable limits in resource utility. It is their obligation to make such an ethical choice; it is a choice that begins with a self-motivated vigilance against any sense of hubris derived from technological prowess, it is a choice to make a public stand against the "myth" that their profession is totally immersed in objective truths best suited for making rational judgment. Likewise, they are obligated to be duty bound not to allow intrusive technologies to deplete and ruin the verdant landscapes of these frontier regions and lead to uprooting the local ethnic minorities from the land of their ancestry. To do all this, they need to revive the original ethical spirit of seeking technological progress typical of the early phase of China's modernization: becoming a "seeker" of the overall planetary wellbeing.

In their turn, the ethical choice that the minority communities need to make is to preserve and sustain their ancestors' ***reverential obligation*** to the landscapes around them, which helps them reenchant their affective ties with the ethos of peace and harmony born and bred of these biotic environments; they can rely on their generations' bond with the biophysical environ as the source of an ethical deterrent, empowering themselves to stand up to excessive and abusive utility of land-based resources for the sake of monetary gains. Instead of resigning to the fate of forced relocation to make way for future damming projects, they can seek their own development schemes based on sustainable integration of their types of life needs with the natural habitats; they need to be proactive in averting any type of violence and disruption of the symbiotic inner workings. In sum, I believe that this is precisely where critical intervention should be engaged to help advocate a two-way transference of values between ethical reasoning and technological knowhow.

Notes

1. There has been discussion at symposiums and conventions to debate and authorize the adoption of this geological term for over a decade now—it is generally acknowledged that the term "Anthropocene" was coined in 2000 by the astrospheric chemist Paul Crutzen and ecologist Eugene Stoermer. The consensus among the science and humanity circles had been expecting for the term to be formally approved at the 35th International Geological Congress held in August 22–27, 2016, in Cape Town, South Africa. The term "Econocene" was first coined by the Berkeley scholar Richard B. Norgaard, "The Econocene and the California Delta," 1–5. Additional notation is in order here due to the timing of occurring events centering round the term "Anthropocene."
2. This mantra was touted by many local leaders as the unchallengeable pretext to design and implement their own lucrative projects that involve large-scale alteration of local landscapes and huge consumption of local natural resources. A good example of that is the case of the damming projects along the Nu River in Yunnan Province in southwest China. Mr. Li Haishu, the Director of the local Forestry Bureau defends his belief with such a mantra on camera when he is interviewed. The case will be discussed below.

3. Yan Fu is arguably the first Chinese writer to work on an integrated model of science and ethical wellbeing in his translation of both Thomas Huxley and Herbert Spencer (ca. 1895), which "transvaluates" Spencer's social Darwinian notion so as to render it amenable to social progress. For further details, read Chapter 2 of my book, *Signposts of Self-Realization: Evolution, Ethics and Sociality in Modern Chinese Literature*, 14–27.
4. Foreign Affairs Movement, also known as the Self-Strengthening Movement, lasted from 1861 to 1895. It started off with a court reform to push for modernization of arsenal and ship industries, but gradually expanded its impact on politics, diplomacy and education in late Qing China. For its impact on social thought, readers can refer to Ted Huters' *Bringing the World Home: Appropriating the West in Late Qing and Early Republican China.*
5. Ma Jianzhong and Qian Xun were contemporaries of Yan Fu, and like him, both traveled to Europe in the late 1890s to study hard sciences, and then were employed as diplomats of the Qing government on account of their classic upbringing and their personal experiences of mastering the basic knowledge of Western sciences. Both became actively involved in the Foreign Affairs Movement and later became leading Encyclopedists who compiled volumes of classified terms and ideas of Western scientific know-how in 1901 and 1902 respectively.
6. Alison Byerly, "The Uses of Landscape: The *Picturesque* Aesthetics and the National Park System," 52–68.
7. Bennett, *Vibrant Matter: A Political Ecology of Things*, ix.
8. Ibid., ix.
9. Ibid., x–xi.
10. Ibid., xii. Emphasis is mine.
11. For a detailed analysis of such essential difference that exists between Chinese and Western subjectivity, read the first section "Some Uncommon Assumptions" in David Hall and Roger Ames, *Thinking through Confucius*, 11–25.
12. Hu and Wang, *The Silent Nu River*. Accessed on 1 June 2017 at https://www.youtube.com/watch?v=0Jet0AdWK4
13. The other two are the Jinsha River which eventually flows east as the Yangtze River and the Lancang River which runs southeast to become the Mekong River. The Tri-River Corridor constitutes a hugely crucial ecosystem not only for China's southwest but for many southeastern nations such as Myanmar, the Laos, Cambodia and Vietnam.
14. A number of local leaders adopted this policy of *bianxian* because it had allegedly been promoted by certain Japanese economists, and they embraced it as an already proven and reliable tactic for local economic planning. A case in point for using such a *bianxian* idea is explained in note

2 above, in which Mr. Li Haishu, Director of the Forestry Bureau of the Nu Jiang Autonomous Region in Yunnan Province defended his adoption of it as legitimate practice of a theory proposed by a Japanese economist.
15. Leopold, *A Sand County Almanac*, 190.
16. These ethnic minorities, the *Nu*, the *Lisu*, the *Dulong*, are among the many living on the banks of the Nu River.
17. Hu, *The Silent Nu*.
18. Ibid.
19. News Probe, *Damming the Nu River*, the concluding installment, CNTV, 2016. It is important to note that this concluding footage attributes the changed stand as a follow-up act by the Yunnan governor after a recent visit by Xi Jinping, the CCP leader. It is understandable for the TV journalists to abide by political protocols, but I hasten to add that it should not overshadow the crucial role played by the News Probe journalists who had put their feet down on the ground and conducted a series of interviews along one tributary of the Nu River long before the visit of the CCP chief.

Bibliography

Bennett, Jane. 2010. *Vibrant Matter: A Political Ecology of Things*. Durham: Duke University Press.

Byerly, Alison. 1996. "The Uses of Landscape: The *Picturesque* Aesthetics and the National Park System." In *Ecocriticism Reader: Landmarks in Literary Ecology*, ed. Cheryll Glotfelty and Harold Fromm. Athens: The University of Georgia Press.

Hu Jie 胡杰, and Wang Yongchen 汪永晨. 2003. *Chenmo de nujiang* 沉默的怒江 (*The Silent Nu River*). Caifeng dianying youxian gongsi (Caifeng Film Corporation Limited).

Huters, Theodore. 2005. *Bringing the World Home: Appropriating the West in Late Qing and Early Republican China*. Honolulu: University of Hawai'i Press.

Leopold, Aldo. 1970. *A Sand County Almanac*. New York: Ballantine Books.

Liu, Xinmin. 2014. *Signposts of Self-Realization: Evolution, Ethics and Sociality in Modern Chinese Literature*. Boston: Brill, Netherlands Publishing.

Nixon, Rob. 2011. *Slow Violence and the Environmentalism of the Poor*. Cambridge, MA: Harvard UP.

Norgaard, Richard B. 2003. "The Econocene and the California Delta." *San Francisco Estuary & Watershed Science: A Broad Perspective*. October Issue. http://escholarship.org/uc/item/4h98t2m0. Accessed 17 May 2016.

TV News Probe. 2016. *Damming the Nu River: From Start to Halt*. 怒江截流: 从开发到停 (*Damming Nu River: From Start to Halt*). Beijing: China National TV Official Channel.

CHAPTER 10

Ecomedia Events in China: From Yellow Eco-Peril to Media Materialism

Ralph Litzinger and Fan Yang

In the opening of Bong Joon-ho's 2013 sci-fi film, *Snowpiercer*, we are told that technological capitalism destroyed the planet twice. It first destabilized the earth's climate by making it dangerously warmer. Then, a corporation in the global north, against the objections of environmentalists in the global south, decided to release the chemical CW-7 to cool down the planet. Instead, it freezes and the earth becomes uninhabitable. However, we soon discover that a small fragment of human life persists on a single long train. The train is divided into closed compartments, each guarded by heavily armed security forces. As the train encircles the frozen orb of the Earth, the last remnants of humanity are divided into the wealthy and the poor, unmistakably mirroring the gross inequalities and ravages of global capitalism today.

The drama quickly heightens. A man named Curtis organizes a rebellion, only to learn in the end that the entire uprising has been masterminded in order to get rid of excess adult bodies that are disrupting the

R. Litzinger (✉)
Cultural Anthropology, Duke University, Durham, NC, USA

F. Yang
Media and Communication Studies, University of Maryland, Baltimore County, Baltimore, MD, USA

C. Chang (ed.), *Chinese Environmental Humanities*, Chinese Literature and Culture in the World,
https://doi.org/10.1007/978-3-030-18634-0_10

closed and fragile ecosystem on board. However, Namgoong, a Korean engineer who once designed the train's security gates, uses an industrial waste by-product called Kronon that he has been collecting to blow through a door to the outside. The explosion triggers a massive avalanche, which, in turn, destroys the train and almost everyone in it. At the end of the film, the only two survivors, Namgoong's daughter Yoni and a mixed-race boy named Timmy, walk away from the wreckage into a world of snow and ice, watched over by a lone polar bear moving cautiously along a mountain ridge.

Adapted from a best-selling graphic novel in France (written by Jean-Marc Rochette and called *Le Transperceniege*), *Snowpiercer* easily lends itself to a reading through the lens of commodity fetishism, particularly in a perpetually accelerating digital age. Only by demystifying the digital commodity—an epistemological "blowing up," as it were—can one begin to engage the object's deep history and ultimately destroy the system of exploitation that produces it.[1] Yet the ending of *Snowpiercer* is instructive in other ways as well. The spectacle of explosion it stages to dismantle the violent and dehumanizing remnants of the industrial, fossil-fuel and chemical dependent capitalist system is more strikingly a story about climate-based racial injustice and non-White multi-species post-apocalyptic futures. For the destruction of train leaves behind two people of color, the sole survivors set against the backdrop of an almost blinding non-human whiteness, in the form of ice, snow, and a lone polar bear. We are invited to ponder a cautionary tale about the entanglements between the human and the non-human, and, at the same time, to bear witness to a vision of a possible multi-species future in which white people cease to exist.

We open this chapter with these ruminations on *Snowpiercer* to advance a wider discussion about the relationship between China, seen by many as a threat to both the ecological health of the planet and ecomedia and digital capitalism research in the environmental humanities. In particular we draw on new work in the emerging field of media materialism studies. Although *Snowpiercer* is not about China specifically, we argue that it problematizes a persistent Western-centric bias in both the environmental humanities and in the literature on media materialism. As we enter the second decade of the twenty-first century, perhaps more so than any Western nations, China has generated a growing visual archive of industrially produced catastrophes, caused by dramatic chemical and industrial explosions. These are not cinematic fantasies, though they are densely mediated, most often by digital technologies and social media sharing platforms. While they do not directly address the ambiguous possibilities of multi-species non-white European futures we see in the end of *Snowpiercer*, they have been just as violent and deadly, and are often deeply racialized. This visual archive of images has opened up spaces for debate among scholars, journalists, and bloggers about China's often ecologically deleterious role in global economies of production, consumption, and resource extraction, in short, its role in the future of the planet itself.

To unpack these issues, we explore two fairly recent chemical explosions, one which took place in Kunshan (2014) outside of Shanghai and the other in a newly developed residential area in the city of Tianjin (2015). These explosions flooded the Chinese and global news media with gruesome images of humans and non-human debris, scattered across a haunting landscape of dust, death, and destruction. As we reflected on these explosions, and how knowledge of them was enabled by digital platforms and social media disseminations, it seemed to us that the catastrophe fantasy of the *Snowpiercer* film, which itself ends on an explosion made possible by our dependency on chemicals to fuel dreams of endless capitalist growth, appears to be all too alive in the People's Republic of China. In Western media journalist reporting, for example, these chemical catastrophes are routinely depicted as somehow solely China's problem.[2] They point to the failure of non-democratic government regulation. Or to the corruption that persists between capital and an authoritarian state. We rarely encounter discussions about what these explosions tell us about China's role in the global economy as a key manufacturing and recycling site for global consumer goods, including media gadgets.[3]

We call this genre of representation "Yellow Eco-peril."[4] Invariably and perhaps predictably in a geo-political landscape that will not let the Cold War end, it depicts China as a polluting and polluted Other seen as slowly destroying the planet. The excessive and prolific circulation of images and almost obsessive media coverage in recent years documenting the "airpocalypse" smog of Beijing is another representative example. We argue that these images, like the smog itself, reveal as much as they obscure. Captured, shared on social media, circulated and commented upon again and again, they are both ideological distortions and discursive productions of a racialized Eco-Otherness. For us, they problematically disconnect China's air pollution from the very systems of globalized production and consumption—the deep earth mining, the factory work, the digital consumption practices—that have propelled and intensified both China's stupendous development and its ecological challenges. We aim in this chapter to offer a counter-politics to this discourse of Yellow Eco-peril.

We approach these real-world explosive scenarios as *ecomedia events*. This concept, we argue, offers a possible new perspective for addressing how to think life, death and destruction in ecological catastrophes, but also issues of environmental justice in China, across Asia, and globally. Spectacular and ephemeral, ecomedia events are instantaneously mediated and circulated occurrences that enact the ecological footprints of the Anthropocene. They are moments in which the material processes of digital production link the "old" forms of resource extraction with our new lives of electronic gadgetry and media tool dependency. For us, China, and the Asia Pacific more broadly, constitutes what Alain Badiou calls an "evental site," a locality where certain elements previously excluded from dominant representation can be brought into view during specific occurrences.[5] Ecomedia events are precisely these occurrences that defy the dominant representational lens of Yellow Eco-peril with regard to China by "blasting open," as it were, the unequal and uneven experiences of climate change under the new and changing configurations of "digital capitalism."[6]

10.1 Ecomedia Events

We begin in a world of explosive dust.

At 7:37 am on Saturday, August 2, 2014, a massive aluminum dust explosion occurred at the Zhongrong Metal Products polishing factory in the city of Kunshan, just outside of Shanghai. At least 64 migrant workers, working unmasked and in an improperly ventilated room, were polishing the interior wheel and brake frames for GM cars. Another 187 people in and around the factory site were hospitalized with serious injuries. The entire building was gone in a moment. All of the destruction, the mangled bodies and body parts, the mourning relatives, those who rushed to the scene to give blood for the few survivors, all of this was caught on mobile phone cameras and then shared almost immediately through social media, until the government called for a stop to online image sharing and commentary.

The labor activist community in China went to work almost immediately, but mostly online, publishing reports, distributing and signing petitions, trying to get people to the scene. These early reports, prepared within days, argued that there was a total lack of regard for workers' safety at the Zhongrong workshop.[7] One enterprising reporter found out that daily aluminum dust concentrations were so high that by noon on the day of the explosion work surfaces were covered in a layer of dust "as thick as a coin," and that each worker could sweep up a paint bucket full of dust. These reports were widely circulated on social media sites.[8] They argued angrily that workers were given useless low-quality masks and gloves once a week, and that long-term exposure to aluminum dust left many workers suffering from "Black Lung disease" (*chenfei bing* 尘肺病), or pneumoconiosis. Another report stated how workers exhausted themselves by taking on repeated overtime shifts, greatly exceeding the statutory limit set in the labor law. The explosion occurred during a statutory non-working time; workers were supposed to be resting, surrounded by families and friends. And yet 261 people signed in to work that day, seemingly attracted by the promise of double-time wages on the weekend. During earlier interviews done at this factory, researchers were alarmed to discover that the Zhongrong migrants worked from 7 am to 10 or 11 pm daily, with no weekends off and just one or two days rest each month.

Both of us witnessed the Kunshan explosion on social media. Like many people in China and around the world who followed this deadly dust explosion and its grisly aftermath of death and destruction, neither of us imagined that something even more horrific would occur just a year later. On August 12, 2015, a great fireball shot into the night sky in the large and prosperous port city of Tianjin, in northeastern China. A privately run warehouse complex containing dangerous chemicals, mostly used in the mining and refinement of gold, exploded, claiming over 160 lives and injuring hundreds of others. Most of the residents living in the area had no idea they had bought apartments within kilometers of warehouses that stored some of the most deadly and explosive chemicals in the world. These chemicals are not only used to extract gold and other minerals from deep within the strata of the earth, they are also used to charge and fire the very media technologies that were used to document, in the very moment of their occurrence, such destructive industrial and chemical explosions.[9]

These tragic industrial explosions are not, of course, accidents. As one labor activist in China put it in a private conversation on WeChat, "to call them 'accidents' is an insult to the dead." Nor can they be written off as aberrations in China's quest for economic development, as official government statements often claim. There are many scholars and activists who want to see the total dismantling of the capitalist system and an end to the global economy's continuing modes of deep earth resource extraction. While we are sympathetic to these arguments, we want to highlight how the material and chemical processes of capitalist production and consumption are linked to "older" forms of resource extraction (long histories of mining, for example) and to our new digital lives, our fascination with slick polished aluminum surfaces, our willing submission to corporate planned obsolescence of consumer electronics, and our complex forms of media saturation. To reiterate our point above, the migrant workers, citizens, journalists and expats who captured the explosions and their aftermath shared the images almost instantaneously using the very smart phones, mostly produced in China, that were made possible by the very same processes of mineral extraction that caused these deadly explosions in the first place.

These instantaneously mediated events are intimately connected to the extractive processes and material resource needs of China's industrialization, and to its rise as a global economic power. They are also about the production and circulation of technologies that have produced new ways of living with and through digital media—the Selfie, Instagram, Facebook, Weixin, and Weibo. Not only are they forming an emergent archive of China's purportedly imminent eco-apocalypse, they also point to long histories of resource extraction that have brought us to our current moment of climate change crisis. More importantly in our view, what links these events together is their capacity to provoke an understanding of "our entanglement with media not just on a sociocultural but also on a biological level."[10] Like Kronon in *Snowpiercer*, the dust and debris from the Kunshan and Tianjin explosions can be mobilized to rethink the relationship between media, event, and the environment. Media, in these instances, are no longer merely tools used by humans to document and visualize ecologically implicated events. Rather, they engage humans in co-constitutive processes of *mediation* that break down distinctions between nature and culture, the human and the non-human, the social and the geologic.

10.2 Media as Mediation

Our conception of *ecomedia events* is indebted to a range of recent scholarship rethinking standard film studies approaches that see media as an inert object for textual criticism, or which approach eco-critical work in the humanities simply in terms of the ethics of representation. Mediation is a key theme in much of this work. In his book *EcoMedia*, for example, Sean Cubitt has argued for mediation "as a core concept for understanding the ways in which the physical, technological and political worlds evolve inextricably together."[11] Likewise, for Kember and Zylinska, writing in their book, *Life After New Media*, "mediation becomes a key trope for understanding and articulating our being in, and becoming with, the technological world."[12] It is through the reframing of media as mediation, what Cubitt describes in his more recent book, *Finite Media*, as "the material processes connecting human and nonhuman events,"[13] that we have in part developed the concept of the ecomedia event to make sense of the imbrication of the human and the non-human, the physical, technological and the political, and to challenge the discourse of Yellow Eco-Peril in analyses of China's ecological challenges.

As visualizations of industrial catastrophes, the events of Kunshan and Tianjin are distinct from the "ceremonial" media events discussed by Daniel Dayan and Elihu Katz.[14] They might better be understood as what Andreas Hepp and Nick Couldry call "media events in a global age"—that is, "transcultural phenomena" that are "produced not only by the mass media (television, radio) but also by the Internet and other digital media,"[15] involving multiple forms of mediated communication that transcend national boundaries. These events also differ from grand-scale disasters that occur "in nature." Yet just as the rising frequency of natural catastrophes cannot be dissociated from human actions, the explosiveness of ecomedia events is enabled simultaneously by the affective surplus generated by lost human lives and the instantaneous transmission via communication networks. In this sense, ecomedia events share with those "live" events of ritual order—what Dayan and Katz refer to as "high holidays of mass communication"[16]—a sense of "interruptions" of everyday routines.[17] Much like Kronon's stopping of the train in *Snowpiercer*, they provide occasions to contest the "representationalism" typifying the Yellow Eco-peril discourse.

For our purposes, as we indicated at the outset of this section, we understand representationalism as the reductive understanding and use of media as mere instruments of representation. As Karen Barad puts it, "representationalism marks a failure to take account of the practices through which representations are produced."[18] This failure characterizes much of Western journalistic coverage of China's environmental crisis. An iconic image of these accounts of Yellow Eco-peril, for instance, is Chinese people wearing facemasks in order to survive Beijing's "airmageddon."[19] Captured by the journalistic lens, images of this kind often visualize China in a manner akin to a post-apocalyptic Hollywood film. Emphasized are the terror (i.e. air quality) and the technicality of solutions (i.e. masks), not unlike the narrative that opens *Snowpiercer* about the release of CM-7 as a geo-engineering intervention to treat global warming. What is left out of these representations of a China slowly devouring itself is any mention of the material production of global capitalism, and China's key role in this process as a destination for outsourced manufacturing and as a location for imported waste recycling, at least until the implementation of the "National Sword" policy in January 2018 banning the latter. A polluted China, in other words, appears as a distant *object* to be represented, rather than as a *subject* shaping the material conditions that make this representation possible.

The problem of representationalism is arguably dramatized as the differences between the passengers in the front and back of the train in *Snowpiercer*. "The first world" in the film, as Aaron Bady argues, seemingly "works less to provide its citizens with pleasure than to shape their desire by constructing others through their pain, lack, and death."[20] The lack of clean air in China, in this sense, serves to remind journalistic audiences in the developed West of their "fortune" of inhabiting the part of the earth in which air pollution is not (or no longer) part of everyday reality. As David S. Roh, Betsy Huang, and Greta A. Niu point out, "the rhetoric of 'pollution'" shared by media representations of major Asian economies, including China, often implies that "even as Asia has finally reached modernity, it does so irresponsibly, without regard for the supposed lessons learned by the West during its periods of rapid industrialization over the course of the twentieth century."[21] Yellow Eco-peril discourses of this kind thus function ideologically, much like the hallucinating drug of Kronon, to assuage the latent anxiety among denizens in the West toward environmental crisis by presenting it as more relevant to their lesser-blessed Asian Others. They also work to perpetuate the longstanding Western view that Asian countries such as China are "behaving like recalcitrant children, refusing to abide by its tutelage, insisting on growing up too quickly, and thus warranting its constant surveillance."[22]

If *Snowpiercer* presents a poetics and politics devoid of place, of specific global situations and struggles,[23] the industrial and chemical explosions at Kunshan and Tianjin are catastrophic occurrences in *real places* and in *real time*. Taking place on Chinese soil, these are events created in part by the migrant workers, citizens, journalists and expats who captured them, using the very smart phones assembled in China. As such, they defy the representationalist framing of Western journalism precisely because during these events, “China” and “environment” can no longer be understood as preformed objects that are in turn “mediated.” Instead, images of China herein emerge as “stabilizations of the media flow,” temporarily fixing the “originary process” of mediation.[24] This is a process that involves not only the people who produce and use the digital tools, in China and beyond—what Jack Qiu calls manufacturing and manufactured “iSlaves”[25]—but also those participating in the generation of energy that powers these tools in places like China.

To further develop these ideas, we turn in the next section to the emerging literature on media materialism as a critical framework that allows us to better discern how ecomedia events disrupt the representationalism of Yellow Eco-peril. What gives shape to ecomedia events, we argue, is no longer the various representational apparatuses that sustain them but rather the materiality of media. To reiterate our point again, this materiality provides a methodological intervention that differs from conventional humanities approaches of close reading and textual analysis, as previously demonstrated in works on ecomedia in China, such as Sheldon Lu and Jiayan Mi's pioneering *Chinese Ecocinema*. In Lu's Introduction to the volume, for instance, he suggests that as “cinema with an ecological consciousness,” Chinese ecocinema “articulates the relationship of human beings to the physical environment, earth, nature, and animals from a biocentric, non-anthropocentric point of view.”[26] While we believe this articulation remains important to unpack, as may be seen in our reading of *Snowpiercer*, we also wish to explore the cultural and political implications of ecomedia events not just as preformed media products (like films) that partake in the “re-imagination of locale, place, and space”[27] but as explosive moments that bring into visibility the deep entanglements between human bodies and non-human matters.

10.3 Media Materialism: Time, Body, and Matter

Media materialism is, in our usage, a reorientation in the study of media that attends to the materiality of media technologies. As Jussi Parikka has put it, it “refers to the necessity to analyze media technologies as something that are irreducible to what we think of them or even how we use them.”[28] Media materialism draws attention to the dynamic involvement of the human and the non-human in the production, consumption, and recycling of media devices. It also encourages a rethinking of humans and media as co-constitutive of each other, as “entangled processes of becoming.”[29]

While its roots may be traced to the works by German media theorists like Friedrich Kittler, media materialism can be said to share a set of concerns with the constellation of theoretical interventions now grouped under the name "new materialisms." As Diana Coole and Samantha Frost state in their introduction to one of the first edited collections on the subject, new materialisms defy binary and oppositional modes of thought, such as the distinctions between the organic and the inorganic, the inanimate and the lively, the human and the non-human. Instead, they push us "to recognize that phenomena are caught in a multitude of interlocking systems and forces and to consider anew the location and nature of capacities for agency."[30]

New materialists therefore turn to nascent developments in natural sciences to rethink the vitality of matter while mobilizing breakthroughs in biology to reconsider the conditions of life and death. They also revisit the long-standing tradition of historical materialism to call for a "critical and non-dogmatic reengagement with political economy."[31] In other words, new materialists are concerned as much with embodiment as with de-centering the human, at once challenging the nature-culture divide while promoting a more rigorous investigation into the relationships between power, corporeality, and the agentic forces of substantive and non-substantive matter. In Jane Bennett's words, they call for "a willingness to theorize events ... as encounters between ontologically diverse actants, some human, some not, though all thoroughly material."[32]

The recent "ecological" turn in media and communication studies reflects precisely such an effort to radically remap the materiality of media. This is exemplified in works like John Durham Peters's *Marvelous Clouds*, Jussi Parikka's *A Geology of Media*, and Sean Cubitt's *Finite Media*. Peters, for one, reverses the conventional claim that "media are environments" to argue that "environments," be it fire, water, sky, or earth, "are also media."[33] For Parikka, an emphasis on geology "leads us to track the importance of the nonorganic in constructing media before they become media: the literal deep times and deep places of media in mines and rare earth minerals."[34] Cubitt likewise links media to energy and matter to argue that "eco-politics must begin in the assertion that our environments are not only capable of communication, but are constantly communicating."[35] All three situate the study of media technologies and digital culture within broader discussions of political economy, proposing a radical reimagination of politics through the materiality of media in light of the risk and reality of ecological crises.

Despite the attention paid to decolonization and indigenous populations in works such as Cubitt's, media materialism has not resonated with postcolonial studies scholars concerned with the persisting uneven power relations between the West and the rest. Works such as Brian Larkins's *Signal and Noise*, Ravi Sundaram's *Pirate Modernity*, and Weihong Bao's *Firey Cinema*, to be sure, have engaged media materiality in non-Western settings that challenge the Euro-American-centric paradigm of media studies. But media materialism, like new materialism, has tended to disavow the perpetuation of racial and other inequalities in a global context. "Interpreting and describing our entanglements with non-human, materialist forces," as Diana Leung argues, "are not enough to account for, much less dislodge attachments to, social categories and representational arrangements."[36] This need to address racial difference when attending to the non-human is the lesson we find in the ending of *Snowpiercer*, when two persons of color survive the train crash, only to find themselves at the behest of non-human whiteness. The film is seemingly cautioning us not to fetishize materiality as "a transcendental signified that merely replaces language or culture as an organizing principle."[37] Instead, it is necessary to more carefully unpack how non-human forces are entangled with racialized "discursive or psychic formations"[38] that undermine the global vision of eco-politics.

To sum up our argument to this point, we have analyzed Yellow Eco-peril as one such discursive formation that positions China as an Other—a major polluter destroying the planet—which effaces the material connections between economic globalization and "Chinese" pollution. We have argued that the ecomedia events that take place in China are prime opportunities to establish these otherwise obscured material linkages. It is to this end that we find media materialism a particularly useful analytical framework. We now want to show how it can shed light on three closely intertwined dimensions of ecomedia events: time, body, and matter.

First and foremost, media materialism draws attention to the *temporal*—and not just representational—aspect of an ecomedia event's making. To be sure, the instantaneous capturing and sharing of the Kunshan and Tianjin explosions easily invoke the trope of speed often associated with digital media. But what makes these events distinct from other picture-worthy spectacles prone to virality is their ability to bring into view what Parikka terms an "alternative deep-time strata of our media culture."[39] Rather than blindly celebrating the instantaneous transmission capacity of internet-equipped smart phones that make these events visible, media materialism asks us to consider the geological conditions of possibility for such transmission. After all, both the chemicals that caused the explosions and the metals within the digital gadgets that were used to document them rely on extractions of minerals from deep within the strata of the earth. Moreover, such explosive events often cut into the sped-up cycles of planned obsolescence championed by global tech giants like Apple, whose iPhone 6 production in China, for example, was halted by governmental investigations into dust control in all factories after the Kunshan incident. Examining these events from this time-based perspective, then, allow us to uncover the shared origins between the objects *and* instruments of representation of ecomedia events deep in the geological history of the planet.

Media materialism's attention to geological deep time also opens itself out to a more critical engagement with two other interlinked dimensions of ecomedia events: *body* and *matter*. The explosions are intense catastrophic events marked by a sense of temporal presence. Yet these moments of "quick deaths," once grasped through the timeframe of geology, also evoke what Rob Nixon calls "slow violence."[40] Despite their spectacular qualities, these events result from an accumulation process akin to what Nicolas Shapiro describes as the "chemical sublime." Contrasting with Enlightenment ideas of the sublime that celebrate the triumph of human reason over nature, Shapiro's notion of the chemical sublime "elevates minor enfeebling encounters into events that stir ethical consideration and potential intervention."[41] This is precisely what we observed during the Kunshan explosion, when the same dust that caused the explosion also enters the lungs of miners and causes illnesses such as "Black Lung disease." And yet, stunningly, almost all of the mainstream media coverage, in both China and in the West, seemed to miss the connections between the dust that blows up a factory and the dust that enters the lungs of China's miners, factory and construction workers, the occupations most susceptible to Black Lung. Both the aluminum dust in an improperly ventilated factory in Kunshan and the dust that enters the lungs of a coal or gold miner, or someone blasting concrete, accumulates into yet another kind of ecomedia event, revealing not just the connection between dust in a mine underground and dust floating in the air of a polishing factory but the deep entanglements between body and matter.[42] It was only when former miners and labor activists in China began to make these connections visible through petitions and online reporting, which was quickly censored, did we begin to see the emergence of an ecomedia event that, again in Shapiro's words, both connected the Kunshan explosion to Black

Lung and stirred deeper ethical and political considerations about the deadly potentialities of dust.

Though "Black Lung" has been recognized in China since the early 1930s, it was only recognized as an "occupational disease" among China's mining communities in the post-1949 period and then noted as a national health emergency beginning in the 1980s. The China Coal Miner Pneumoconiosis Treatment Foundation, a foundation overseen by the State Administration of Work Safety Supervision, created in the 1980s, reported in June of 2015 some 720,000 patients. These figures are contested. According to the studies conducted by social organizations such as "Love Save Pneumoconiosis" (*da'ai qingchen* 大爱清尘), which provides care and raises funds for sick miners, there are likely more than six million people in China suffering from the illness, and probably even more given the long gestation period of the disease and the fact that there is great stigma attached to a worker sick with Black Lung. Whatever the numbers, Black Lung constitutes something like an astonishing 90 percent of occupational disease population in China.[43] Worse yet, there is no known cure. Depending on the stage of the disease, a patient's life can be prolonged by anywhere from months to decades with adequate care and medication, though many workers, especially those without labor contracts or working in mines outside of the state enterprise system, never receive the care they need because their former employers simply deny any responsibility. Black Lung is so personally and socially debilitating in China because those workers not protected by legal forms of compensation know their bodies as disposable, their labor power no longer needed.

With Black Lung, we are into the dark world of Rob Nixon's slow violence, as the dust and debris of mining and polishing in factories slowly eats away at the lungs and makes breathing almost unbearable. Despite the prevalence of the disease among miners, many of the treatments and medications for Black Lung are not covered by public health insurance. For instance, the most effective drugs are inhaled medications and lung washing (*xifei* 矽肺), which are still not covered by public insurance, despite protests and petition efforts on the parts of miners, their families, and social organizations. Without health insurance, it is nearly impossible for patients to gain access to these kinds of treatments, as the cost of one lung "washing" or lavage treatment averages around 10,000 RMB, about 3–5 months of salary for an average miner. The situation is further exacerbated by the fact that many miners lack an understanding of and confidence in the existing public health insurance systems, but it is also a consequence of the mining boss's often intentional failure to insure, protect against and compensate for their employees' work hazards.

These linkages between mining, pneumoconiosis, and China's long-standing labor struggles are poignantly illustrated in photographer-turned-filmmaker Zhao Liang's 2015 film *Behemoth* (*Bei xi moshou* 悲兮魔兽). With a title inspired by the Old Testament and a voiceover narrative drawing on Dante's *Divine Comedy*, Zhao's cinematography takes us through a lyrical yet heart-wrenching journey—from the open cast mines on the steppes of Inner Mongolia to the miners' residencies, from the hospital rooms where workers undergo lung washing treatments to the graveyard of those who died from pneumoconiosis, from the factories that turn raw metals into building material to Ordos, a "ghost city" filled with empty high-rises. In a brief but perhaps not unimportant scene shot in a miner's dormitory, a young worker is seen carefully plugging his mobile phone into an adapter to charge the battery before retiring in his bunk bed.

Moments like this encapsulate the kinds of contradictions we seek to bring to the forefront in our engagement with media materialism. In *Behemoth*, the electricity that powers the communication technology comes from none other than the coalmine at which the worker labors. As the dust and debris of mining devours the workers' bodies, rendering them "the same damaged matter" as "natural landscapes,"[44] Zhao's narration, that "this is who we are—we are that monster—the monster's minions" at the end of the film, takes on a new level of meaning. While the critique of China's rampant developmentalism and unfettered urbanization is unmistakable, what becomes salient here is the ways in which new digital lives have come to implicate *all of us* reliant on information communication technology, or ICT, to stay connected.

To be sure, a cinematic text like *Behemoth* is well positioned to represent ecological crises in a manner similar to anthropological inquiries into the social life of things and objects. But ecomedia events like the Kunshan explosion implore us to attend more critically to the "matter-energy"[45] of dust and debris. This is because such an explosion directly enacts the agentive force of an inert matter like aluminum dust. As a form of "inconspicuous waste,"[46] dust does not in itself contribute to the composition of a media object like the mobile phone, whose social relations and meanings can be delineated. Yet its making is inseparable from the production process of media (and other) commodities. "Dust," therefore, "takes us—and our thinking—to different places and opens up multiple agendas."[47] One such agenda is precisely to investigate "a materiality that is itself heterogeneous, itself a differential of intensities, itself *a* life."[48] This is a kind of "vital materialism"[49] that urges us to connect dust with the workers who "make their livings" in the production process that generates dust as a by-product, and therefore "know and experience" it "in their senses and their cells."[50] The explosion, in other words, can be seen as an instance in which non-human matter has found a way to communicate its presence. It has staged a hyper-visible theatrical moment that condenses the extended temporality of Black Lung disease as a form of unseen "cellular dramas of mutation."[51]

As for the Tianjin incident, the proximity of the warehouse to residential areas means that many living in the vicinity are in a prime location to document the multi-part explosion before evacuating their homes. In many of these amateur videos now available on YouTube, the mobile-phone-users-turned-documentarians can be heard screaming and cursing repeatedly while urging their fellow spectators to "run" for safety.[52] One survivor Daniel Van Duran, an American living in a building nearby who had captured the blasts on his cell phone camera before fleeing the scene, for example, found his home devastated upon return.[53] Viewing the footage that emerged from the explosion, one can not only sense the handheld devices reverberating with the spectacles captured on screen but also feel the awe in the inadvertent voiceovers of the videographers. In one collage of videos that garnered over two million views on YouTube, a few of the clips end abruptly by cutting to blank screens, leading several viewers to wonder if the people who shot these videos had died on the spot. This is a concern shared by viewers on the Chinese forum Zhihu, one of whom felt that people were "taking videos with their lives."[54]

Our use of media materialism thus enables us to highlight that neither the camera phones nor the media event they have captured can be reduced to "a discrete entity that can be said to have one-way 'effects' on other entities."[55] Instead, what the event enacts is "an environment or a field of forces" that links the media artifacts—be it texts or technologies—with "the acts and processes of producing and temporarily stabilizing the world into media, agents, relations, and networks."[56] While the debris from the explosion is rarely calculated as a raw material in ICT production, it has manifested itself in spectacular forms, "invading" the private homes and bodies of the camera phone users. The chemicals (such as ammonium nitrate and potassium nitrate) that caused the explosion may not find their way into the final product of ICT. Yet their role as explosives in coal mining and other mineral extraction helps to generate the electricity that powers the mobile phones and computers distributing the online videos. When the spectators make use of their mobile phones to document, visualize, and disseminate imageries of such an industrial catastrophe, they also generate large quantities of digital data, which requires physical storage facilities in various parts of the world. The ecomedia event, then, is best understood as a transnationally circulated "accident in geologies of media culture" that illuminates media as "multitemporal planetary environments in which planetary pollution becomes perceptible and sometimes also experienced."[57] It allows us to understand media as a dynamic milieu wherein human bodies are simultaneously agentive in carrying out actions that have environmental consequences *and* vulnerable to such non-human forces as the accumulation of dust.

During these ecomedia events, dust, chemicals, and debris, become "a narrative that enters us."[58] They become a part of who we are, especially as consumers of media tools like iPhones and computer technologies of all kinds. In this sense, media materialism urges us to understand dust and debris as agents that forge what Nicolas Mirzoeff calls a "countervisuality"—a visual practice countering hegemonic representations of "Anthropocene visuality."[59] Just as Anthropocene visuality blinds us to the destructive effects of the Western-led, neoliberal faith in market forces, we have suggested that Yellow Eco-peril systemically disconnects China from the material processes of global (media) production, consumption, and, until recently, e-waste recycling. A turn to a countervisuality is therefore consistent with "a decolonial politics that claims the right to see what there is to be seen and name it as such: a planetary destabilization of the conditions supportive of life."[60] Rendered as a force, dust and debris as vibrant matters offer a glimpse into the workings of digital global capitalism. More importantly, in our view, they invite new political questions about life and death in digital-age China, and globally.

10.4 Chinese Ecomedia Events in Global Contexts

The Tianjin and Kunshan explosions and the slow violence of deep earth mining and Black Lung disease are all events that took place in China. However, we have argued that once examined through some of the analytical tools provided by new work in media materialism, they allow us to see how intricately entangled they are in the logics and processes of global digital capitalism. Nonetheless, the challenge remains whether a non-human matter like dust, as a kind of evidence of disease, or something that explodes and kills when not properly handled, will create new spaces of hope and a new politics of digital activism that defy the racialized representationalism of Yellow Eco-peril.

This ultimately gets at the larger political question of China's place in debates about the future of the planet. To borrow from the authors of *Screen Ecologies*, the Asia-Pacific region (of which China is part) forces us to "go beyond the conventional backyard of media studies of Europe and the United States."[61] After all, this is a region that is "undergoing rapid urbanization, has the globe's largest percentage of megacities, and is the site of much ICT manufacturing and consumption."[62] Countries within the Asia-Pacific are also subject to greater risks of natural disasters due in part to the uneven effects of climate change, as attested by the increased frequency of typhoons in countries like the Philippines. In this sense, our contemporary digital lives in the First World may also pose tremendous danger for the future wellbeing of many people living in the global south, who in comparison contribute much less to climate change in the present than those in the so-called developed world.

The Asia-Pacific, as Rob Wilson argues, is very much an imaginary that informs the "Korean global film" of *Snowpiercer*.[63] In addition to the metaphoric power of Kronon that inspired our conception of ecomedia events, we are also taken by the film's depiction of the miserable and seemingly depleted people packed into the back of the train as unruly subjects. These are people with desires to better understand the conditions of their oppression—why are we here? how does the system work?—and to take control of the engine that keeps the train going. They are not unlike the activist volunteers in China who constantly defy state surveillance and harassment from mine bosses and local government officials alike to advocate for the miners' wellbeing, often using online digital platforms as well. In Muchuan county, Sichuan province, for example, since the late 1980s, tens of thousands of peasant workers have worked the lead and zinc mines in Leshan City. In 2003, large-scale closure and the rapid fire selling of private mines left many workers without jobs, and without the possibility for adequate health exams. Many of those who returned home started to show symptoms of lung disease. In 2007, under the leadership of a former miner turned activist named He Bing, 60 workers with Black Lung in Muchuan county decided to pressure the government to hold these private mining companies accountable. In organizing the activist community, they elected a workers' committee led by worker representatives, who forwarded their demands through letters and appeals to the government. During their resistance, they raised technological literacy through group-based education. They also mobilized social media and public news, attracting public attention and donations from all over the country.

The localized resistance movement in Muchuan extended to eight adjacent counties and small cities. The government eventually listened, and created a new program called "the rescue methods for Black Lung workers in Leshan City." This new fund, now several years in the works, primarily focuses on issue of poverty, workers' medical expense, life compensation and children's education. Much of this, of course, came too late, since during the five-year struggle, 15 members passed away. Additionally, activists like He Bing do not like how the government equates the prevalence of Black Lung disease to poverty, which they feel ignores the insufficient industrial regulations and social infrastructures that produce Black Lung in the first place. Yet, the story of collective resistance in Leshan City has had a far-reaching impact. Following its lead, in Jiulong County, a Tibetan Autonomous County, for example, 146 pneumoconiosis workers pressured the local government and private mining companies to eventually take responsibility for the massive outbreak of pneumoconiosis in the area.

With their bodies having "become environmental"[64] by way of exposure to dust, the Leshan miners-turned-activists have also shown that "communication is the material form in which not only speech but action occurs."[65] Digital media tools—deeply implicated in histories of resource extraction, in the making of precarious lives, and surrounded by the ever-present specter of death—can indeed open up possibilities for different forms of activism in China. But more importantly, the activists behind the Leshan victory demonstrate that when "the distinction between environment and humanity … is breaking down in favor of a distinction between market and environment," people who have become part of the environment by way of their exclusion from the market would then be "in a position to build a completely unprecedented commons."[66]

It is precisely this kind of communicative possibility that media materialism can help us envision with regard to ecomedia events that are ostensibly "Chinese." "The political event occurs," Cubitt reminds us, "when order is confronted with what it has excluded, yet over which it has exercised command or asserted stewardship."[67] We have analyzed Yellow Eco-peril as a discursive regime that depicts China's eco-crisis in such a way as to exclude the participation of Chinese miners, workers, and other subalterns in the global economy that creates it. Ecomedia events in

China, in this sense, are political in that they bring into visibility the material processes of digital capitalism that are otherwise obscured. Grasped through the framework of media materialism, they prompt us to connect the lives and activism of Chinese miners and workers to other sites around the world where similar struggles have taken place.

In the mountains of Peru and Ecuador, at Standing Rock, in the Congo, and so many other places, we continue to engage in the political battle over what Naomi Klein among others has called extractivism—"a nonreciprocal dominance based relationship with the earth, one purely of the taking" and "is the opposite of stewardship, which involves taking but also taking care that regeneration and future life continue."[68] Our use of media materialism, and our concept of the ecomedia event, makes us uneasy with this dichotomy, since extractivism is not in any simple way the opposite of Klein's somewhat romantic notion of earthly stewardship. Extractivism is what fuels and fires and energizes life today, especially our digital lives, and the materials of media culture upon which we depend. It links the deep geological history of the sub-strata of the earth to the atmospherics of everyday life. In our discussion of a few select, yet widely covered and discussed ecomedia events in China, we have tried to show how new theoretical thinking in the field of media materialism allows us to draw connections in novel and more critical ways between these seemingly disparate ecological and cultural spheres. Establishing these linkages, we argue, are crucial to situate ecomedia events in China within a wider global context, a country which is under increased pressure to address the deleterious effects of the production, consumption, and recycling of media technologies.[69]

10.5 Conclusion

In *Snowpiercer*, the industrial waste Kronon works simultaneously as a hallucinating drug and an explosive, one that ultimately brings an end to the world of industrial capitalism and its grisly aftermath of brutality and oppression on the train. In our thinking about the Kunshan and Tianjin explosions and about the activism that surrounds miners living the slow violence of Black Lung, the film and its ending kept haunting us. We needed to think not just about the materiality of Kronon itself—a by-product of waste that contains within itself a potent vitality—but also about the ambiguous question of hope after the train is spectacularly exploded. *Snowpiercer* leaves us with a world devoid of white Europeans, in which two people of color survive, alone with a polar bear, who may devour them, or care for them. In Bong's cinematic imaginary, the post-Anthropocene future is radically uncertain, wildly contingent, completely up for grabs.

China, of course, is not post-Anthropocene. It is very much of the Anthropocene, entangled in myriad debates about the present and about planetary futures. Scholars are writing about e-waste and the toxic discarded debris of digital consumerism.[70] Increasingly we encounter stories and images of bicycle graveyards, calling into question the easy disposability of the shared economy.[71] Questions about whether China can wean itself off of coal, the geo-politics of building oil and natural gas pipelines across national borders in Myanmar and Central Asia, and citizen reporting on the deadly environmental effects of chemical and industrial run animate all discussions about China's ecological futures.[72] Given the 280 million plus rural migrants who have toiled in China's factories, built its global cities, and worked the mines, China must be part of any global discussion about precarity, about whose lives matter, and whose are expendable. Thinking through work on media materiality, representationalism, and non-human vitality, we have highlighted the role of digital devices in documenting and circulating images of industrial explosions. We have argued that the dangerous dust that forms in factories and in the lungs of miners involved in coal production, mineral extraction, and deep earth mining are inextricably part of the processes that produce our digital lives. In writing against the ideological distortions of what we have called Yellow Eco-peril—that see China and its developmental model as always already destroying the planet, bringing forth, through its insatiable desire for economic growth, a planetary ecological catastrophe—we have tried to connect our digital present to the material conditions—always global, and always linked to supply chains—that underpin them. We want to take seriously scholars and activists who have called for an end to digital fetishism.[73] We want us "users" to better understand the larger supply chains that make the digital possible, to link gold mining in Guizhou to coal mining across the north of China to mineral extraction in Tibet and Xinjiang to cobalt mining in the Congo to lithium production in Argentina to Foxconn manufacturing plants across China, and to consumers around the world who make use of all this "invisible" stuff every time they turn on their computers, their smart phones, snap a selfie, record a protest, document an industrial explosion.

Finally, we also want to call attention to the fact that our own tools of research and ethnographic storytelling—the mobile phones, the digital cameras and images, the social media platforms, the batteries, cables and the clouds—are implicated in the deadly dust that enters the everyday lives of miners and industrial factory workers, in China and elsewhere. In *Learning to Die in the Anthropocene*, Roy Scranton writes, "The study of the humanities is nothing less than the patient nurturing of the roots and heirloom varietals of human symbolic life. This nurturing is a practice not strictly of curation, as many seem to think today, but of active attention, cultivation, making and remaking."[74] This collaborative chapter is an attempt to enact this mode of nurturing. We want to cultivate a reflexive sensitivity toward the uneven politics of life and death as they are intertwined within new, and ever-changing configurations of global capitalism. By carefully placing the ecomedia events we have examined within wider, globalizing worlds of capitalist production and consumption, and writing against views that recycle Yellow Peril discourses in our Anthropocene present, we think it is time to make the dust and the debris of mining, the spectacular destruction and death from aluminum explosions, and the slow dying from Black Lung disease and smog part of any debate about digital media in the environmental humanities. The relevance of digital media—its global production, circulation, and consumption—to China's so-called eco-apocalypse is too intimate to dismiss.

Notes

1. Commodity fetishism and the possibilities and limits of revolutionary resistance have indeed been the two dominant tropes in Leftist readings of the film. See, for example, Aaron Bady, "A Snowpiercer Thinkpiece, Not to Be Taken Too Seriously, But For Very Serious Reasons"; Gerry Canavan, "'If the Engine Ever Stops, We'd All Die': Snowpiercer and Necrofuturism."
2. A representative example of this view is found in Elizabeth Economy, *The River Runs Black: The Environmental Challenge to China's Future.* See also "China's Environmental Crisis," https://www.cfr.org/backgrounder/chinas-environmental-crisis, which features a lead image, shared widely throughout the journalistic world, of two older men pulling a cart of coal in the dead of winter in Shanxi Province, with ominous coal towers spewing pollution into air thick with smog.
3. The "Made in China" development model has been thoroughly documented in policy, scholarly and journalistic circles. Only in recent years have we seen critical work on global waste imports to China. Since the

1980s China has been the world's largest importer of waste. In 2012, for example, up to 56 percent of global exported plastic waste ended up in China. Imported plastic waste alone reached a peak of almost nine million tons in 2012. Proactively seeking to rectify this situation, in July 2017, the Chinese government announced a new policy banning 24 types of waste under four categories: certain types of mining slag, household waste plastics, unsorted waste paper and waste textiles. The policy went into force on January 1, 2018. This has sent shockwaves through the global trade in waste recycling. For an early report, see http://www.greenpeace.org/eastasia/press/releases/toxics/2017/Chinas-ban-on-imports-of-24-types-of-waste-is-a-wake-up-call-to-the-world%2D%2D-Greenpeace/. The best filmic treatment of waste in China is found in Wang Jiuliang's 2016 masterpiece, "Plastic China." See https://www.plasticchina.org. One of the first studies on recycling and the global circulation of waste in China is found in Adam Minter, *Junkyard Planet: Travels in the Billion-Dollar Trash Trade*. For an excellent, more recent ethnographic discussion on rural pollution, including e-waste in China, see Anna Lora-Wainwright's groundbreaking, *Resigned Activism: Living with Pollution in Rural China*.
4. We see Yellow Eco-peril as part of the long-standing racist narratives regarding China (and Asia more broadly) primarily produced and propagated by the West. More recently, thanks to China's economic rise, these narratives have proliferated in diverging ways in a global context, leading Franck Billé to pluralize the term as "Yellow Perils." See Franck Billé, "Introduction," in *Yellow Perils*, 5.
5. Alain Badiou, *Being and Event*.
6. Dan Schiller, *Digital Capitalism*.
7. See, for example, "The Tragedy of the Village that Built Kunshan Zhongrong." *China Labour Bulletin*, Aug. 20, 2014. Survivors of the Kunshan Disaster Face Grim Future, *China Labour Bulletin*, Aug. 13, 2014.
8. A plethora of Chinese-language reports appeared on social media platforms in China, but were quickly removed by the government.
9. Drone footage of the immediate aftermath of the explosion can be found on many You Tube sites. For two examples, uploaded by New China TV on August 14, 2015, see: https://www.youtube.com/watch?v=NBjbqnAt1BI and https://www.youtube.com/watch?v=C5GsQXPOcPA
10. Sarah Kember and Joanna Zylinska, *Life after New Media*, xviii.
11. *Eco Media*, 5.
12. *Life after New Media*, xv.
13. Sean Cubitt, *Finite Media*, 3.
14. Daniel Dayan and Elihu Katz, *Media Events*, 7.

15. Andreas Hepp and Nick Couldry, "Introduction: Media Events in Globalized Media Cultures," in *Media Events in a Global Age*, 11.
16. Dayan and Katz, *Media Events*, 1.
17. Ibid., 5.
18. Karen Michelle Barad, *Meeting the Universe Halfway*, 53.
19. See the images included in, for example, Oliver Wainwright, "Inside Beijing's Airpocalypse—a City Made 'almost Uninhabitable' by Pollution."
20. "A Snowpiercer Thinkpiece, Not to Be Taken Too Seriously, But For Very Serious Reasons."
21. Roh, David S., Betsy Huang, and Greta A. Niu, "Desiring Machines, Repellant Subjects," in *Techno-Orientalism*, 226.
22. Ibid., 226.
23. Rob Wilson, "Snowpiercer as Anthropoetics: Killer Capitalism, the Anthropocene, Korean Global Film."
24. Kember and Zylinska, *Life after New Media*, 21.
25. Jack Linchuan Qiu, *Goodbye ISlave: A Manifesto for Digital Abolition.*
26. Sheldon H. Lu, "Introduction: Cinema, Ecology, Modernity," in *Chinese Ecocinema*, 2.
27. Ibid., 1.
28. *A Geology of Media*, 1.
29. Kember and Zylinska, *Life after New Media*, xvi.
30. Diana Coole and Samantha Frost, "Introducing the New Materialisms," in *New Materialisms*, 9.
31. Ibid., 7.
32. *Vibrant Matter*, xiii–xiv.
33. *The Marvelous Clouds*, 3.
34. *A Geology of Media*, 5.
35. Cubitt, *Finite Media*, 177.
36. Diana Leong, "The Mattering of Black Lives: Octavia Butler's Hyperempathy and the Promise of the New Materialisms," 11.
37. Ibid., 10.
38. Ibid., 11.
39. Parikka, *A Geology of Media*, 42.
40. *Slow Violence and the Environmentalism of the Poor.*
41. "Attuning to the Chemosphere: Domestic Formaldehyde, Bodily Reasoning, and the Chemical Sublime," 369.
42. This section of the chapter draws on a range of work on Black Lung. In English, see, "Deadly Dust: The Silicosis Epidemic among Guangdong Jewelry Workers and the Defects of China's Occupational Illnesses Prevention and Compensation System," China Labour Bulletin CLB Research Series: No. 1 December 2005 and "The Hard Road: Seeking justice for victims of pneumoconiosis in China," CLB Research Series,

April 2010. On this history of silicosis and pneumoconiosis in the Appalachia and the United States since the Depression, see David Rosner and Gerald Markowitz, *Deadly Dust: Silicosis and the On-Going Struggle to Protect Worker's Health*. The best study in the UK context is found in Arthur McIvor and Ronald Johnston, *Miner's Lung: A History of Dust Disease in British Coal Mining*.
43. "Da'ai qingchen," 2014.
44. Christian Sorace, "Paradise Under Construction": http://www.chinoiresie.info/paradise-under-construction/. Accessed July 4, 2018.
45. Bennett, *Vibrant Matter*, 57.
46. Cubitt, *Finite Media*, 119.
47. Parikka, *A Geology of Media*, 87.
48. Bennett, *Vibrant Matter*, 57.
49. Ibid., 57.
50. Cubitt, *Finite Media*, 119.
51. Nixon, *Slow Violence and the Environmentalism of the Poor*, 6.
52. See, for example, *2015 TIANJIN PORT CHINA HUGE EXPLOSION, HD Every Angle Synced*.
53. Shanshan Dong, "Tour American's Home Destroyed by China Blast," NBC News, August 20, 2015, https://www.nbcnews.com/news/china/tianjin-blasts-tour-american-survivor-dan-van-durens-wrecked-home-n412986. Accessed July 4, 2018.
54. "How Are the People Who Took Videos of the Tianjin Explosion? (Tianjin Baozha Paishe Shipin de Ren Dou Zenmeyang Le?)," *Zhihu*.
55. Kember and Zylinska, *Life after New Media: Mediation as a Vital Process*, 77.
56. Ibid., 77.
57. Jussi Parikka, "Deep Times of Planetary Trouble," 285.
58. Parikka, *A Geology of Media*, 102.
59. "Visualizing the Anthropocene," 226.
60. Ibid., 230.
61. Larissa Hjorth et al., *Screen Ecologies: Art, Media, and the Environment in the Asia-Pacific Region*, 5.
62. Hjorth et al., 18.
63. Wilson, "Snowpiercer as Anthropoetics: Killer Capitalism, the Anthropocene, Korean Global Film."
64. Cubitt, *Finite Media*, 180.
65. Ibid., 181.
66. Ibid., 180.
67. Ibid., 178.
68. *This Changes Everything: Capitalism* vs. *The Climate*, 169.
69. Richard Maxwell and Toby Miller, *Greening the Media*.

70. Lora-Wainwright, *Resigned Activism*.
71. Thomas Bird, "China's bike sharing bubble has burst—drone photography captures its casualties rusting in 'bicycle graveyards.'"
72. Bryan Tilt, *The Struggle for Sustainability in Rural China: Environmental Values and Civil Society*.
73. There is a quite vast and diverse literature on digital fetishism. For one widely cited and circulated example, see Wu Ming, "Fetishism of Digital Commodities and the Hidden Exploitation: The Case of Amazon and Apple." http://www.metamute.org/community/your-posts/fetishism-digital-commodities-and-hidden-exploitation-cases-amazon-and-apple, most recently accessed on July 20, 2018.
74. *Learning to Die in the Anthropocene*, 99.

Bibliography

2015 TIANJIN PORT CHINA HUGE EXPLOSION, HD Every Angle Synced. 2015. https://www.youtube.com/watch?v=dgurTdK0PTA. Accessed 4 July 2018.

Badiou, Alain. 2013. *Being and Event*. Trans. Oliver Feltham. New York: Bloomsbury Academic.

Bady, Aaron. 2014. "A Snowpiercer Thinkpiece, Not to Be Taken Too Seriously, But for Very Serious Reasons." *The New Inquiry*, July 29. http://thenewinquiry.com/blogs/zunguzungu/a-snowpiercer-thinkpiece-not-to-be-taken-too-seriously-but-for-very-serious-reasons-or-the-worst-revenge-is-a-living-will/. Accessed 4 July 2018.

Barad, Karen Michelle. 2007. *Meeting the Universe Halfway: Quantum Physics and the Entanglement of Matter and Meaning*. Durham: Duke University Press.

Bennett, Jane. 2010. *Vibrant Matter: A Political Ecology of Things*. Durham: Duke University Press.

Billé, Franck. 2018. "Introduction." In *Yellow Perils: China Narratives in the Contemporary World*, ed. Sören Urbansky and Franck Billé, 1–34. Honolulu: University of Hawaii Press.

Canavan, Gerry. 2014. "'If the Engine Ever Stops, We'd All Die': Snowpiercer and Necrofuturism." *Paradoxa* 26. http://epublications.marquette.edu/cgi/viewcontent.cgi?article=1298&context=english_fac. Accessed 4 July 2018.

Coole, Diana, and Samantha Frost. 2010. "Introducing the New Materialisms." In *New Materialisms: Ontology, Agency, and Politics*, ed. Diana Coole and Samantha Frost, 1–43. Durham: Duke University Press.

Cubitt, Sean. 2005. *Eco Media*. Amsterdam/New York: Rodopi.

———. 2017. *Finite Media: Environmental Implications of Digital Technologies*. Durham: Duke University Press.

"Da'ai qingchen [Charity Fund of Love Save Pneumoconiosis]." Da'ai qingchen. 2014. http://www.daaiqingchen.org/list.php?fid=49. Accessed 4 July 2018.

Dayan, Daniel, and Elihu Katz. 1992. *Media Events: The Live Broadcasting of History*. Cambridge: Harvard University Press.

Dong, Shanshan. 2015. "Tour American's Home Destroyed by China Blast." *NBC News*, August 20. https://www.nbcnews.com/news/china/tianjin-blasts-tour-american-survivor-dan-van-durens-wrecked-home-n412986. Accessed 4 July 2018.

Hepp, Andreas, and Nick Couldry. 2010. "Introduction: Media Events in Globalized Media Cultures." In *Media Events in a Global Age*, ed. Nick Couldry, Andreas Hepp, and Friedrich Krotz. London/New York: Routledge.Comedia.

Hjorth, Larissa, Sarah Pink, Kristen Sharp, and Linda Williams. 2016. *Screen Ecologies: Art, Media, and the Environment in the Asia-Pacific Region*. Cambridge: MIT Press. http://www.jstor.org/stable/j.ctt1c2crsq. Accessed 4 July 2018.

"How Are the People Who Took Videos of the Tianjin Explosion? (Tianjin Baozha Paishe Shipin de Ren Dou Zenmeyang Le?)." *Zhihu*. https://www.zhihu.com/question/34523697. Accessed 4 July 2018.

Kember, Sarah, and Joanna Zylinska. 2012. *Life After New Media: Mediation as a Vital Process*. Cambridge: MIT Press.

Klein, Naomi. 2015. *This Changes Everything: Capitalism vs. The Climate*. Reprint edition. New York: Simon & Schuster.

Leong, Diana. 2016. "The Mattering of Black Lives: Octavia Butler's Hyperempathy and the Promise of the New Materialisms." *Catalyst: Feminism, Theory, Technoscience* 2 (2): 1–35.

Lu, Sheldon H. 2010. "Introduction: Cinema, Ecology, Modernity." In *Chinese Ecocinema: In the Age of Environmental Challenge*, ed. Sheldon H. Lu and Jiayan Mi, 1–14. Hong Kong: Hong Kong University Press.

Maxwell, Richard, and Toby Miller. 2012. *Greening the Media*. 1st ed. New York: Oxford University Press.

Mirzoeff, Nicholas. 2014. "Visualizing the Anthropocene." *Public Culture* 26 (2): 213–232. https://doi.org/10.1215/08992363-2392039. Accessed 4 July 2018.

Nixon, Rob. 2011. *Slow Violence and the Environmentalism of the Poor*. Cambridge: Harvard University Press.

Parikka, Jussi. 2015. *A Geology of Media*. Minneapolis: University of Minnesota Press.

———. 2016. "Deep Times of Planetary Trouble." *Cultural Politics* 12 (3): 279–292. https://doi.org/10.1215/17432197-3648846. Accessed 4 July 2018.

Peters, John Durham. 2015. *The Marvelous Clouds: Toward a Philosophy of Elemental Media*. Chicago: The University of Chicago Press.

Qiu, Jack Linchuan. 2017. *Goodbye ISlave: A Manifesto for Digital Abolition*. Urbana/Chicago: University of Illinois Press.

Roh, David S., Betsy Huang, and Greta A. Niu. 2015. "Desiring Machines, Repellant Subjects." In *Techno-Orientalism: Imagining Asia in Speculative Fiction, History, and Media*, ed. David S. Roh, Betsy Huang, and Greta A. Niu, 221–226. New Brunswick: Rutgers University Press.

Schiller, Dan. 2000. *Digital Capitalism: Networking the Global Market System*. Cambridge: MIT Press.

Scranton, Roy. 2015. *Learning to Die in the Anthropocene: Reflections on the End of a Civilization*, City Lights Open Media. San Francisco: City Lights Publishers.

Shapiro, Nicholas. 2015. "Attuning to the Chemosphere: Domestic Formaldehyde, Bodily Reasoning, and the Chemical Sublime." *Cultural Anthropology* 30 (3): 368–393. https://doi.org/10.14506/ca30.3.02. Accessed 4 July 2018.

Wainwright, Oliver. 2014. "Inside Beijing's Airpocalypse—A City Made 'Almost Uninhabitable' by Pollution." *The Guardian*, December 16. https://www.theguardian.com/cities/2014/dec/16/beijing-airpocalypse-city-almost-uninhabitable-pollution-china. Accessed 4 July 2018.

Wilson, Rob. 2015. "Snowpiercer as Anthropoetics: Killer Capitalism, the Anthropocene, Korean Global Film." Presented at the 2015 Yonsei Summer Workshop on "Cinema, Spectacle and Visuality in the Korean Context," Seoul, July 17.

SECTION III

Sustainability, Organic Community, and Buddhist Multispecies Ethics

CHAPTER 11

The Paradox of China's Sustainability

Christopher K. Tong

While the term "sustainability" is used today across political, economic, scientific, and cultural discourses, there is no single definition that satisfies the diverse intents and purposes. In the narrow sense, "sustainability" refers to the ability of an ecosystem to remain diverse and productive over an indefinite period of time; in the broad sense, it involves the management of human impact on the environment, as societies transform ecosystems and extract resources to meet their present and future needs. The modern concept of "sustainable development" is especially relevant to the latter case. One of the sources for this concept is *Our Common Future*, commonly called the Brundtland Report, published in 1987 by the United Nations' World Commission on Environment and Development.[1] Chaired by Gro Harlem Brundtland, Prime Minister of Norway at the time, the commission consisted of members representing 21 industrialized and developing countries from five continents. Among the members were William Ruckelshaus, the first head of the Environmental Protection Agency (EPA), and Ma Shijun, an ecologist from the People's Republic of China (PRC). Balancing environmentalist concerns and developmentalist agendas, the report describes the intent of "sustainable development" as follows:

C. K. Tong (✉)
Asian Studies, University of Maryland, Baltimore County, Baltimore, MD, USA

C. Chang (ed.), *Chinese Environmental Humanities*,
Chinese Literature and Culture in the World,
https://doi.org/10.1007/978-3-030-18634-0_11

> Humanity has the ability to make development sustainable to ensure that it meets the needs of the present without compromising the ability of future generations to meet their own needs. … [S]ustainable development is not a fixed state of harmony, but rather a process of change in which the exploitation of resources, the direction of investments, the orientation of technological development, and institutional change are made consistent with future as well as present needs.[2]

The Brundtland Report constituted an international contribution to the longstanding discourse of environmental constraints to human growth that had spanned from Malthus to peak oil, and to works widely read in the years prior such as *The Population Bomb* and *The Limits to Growth*.[3] As sustainability became a ubiquitous term, scholars debated its definitions, discursive functions, and social content, returning to a set of perennial questions that Timothy Luke summarizes as: sustainability of what, for whom, for how long, at what scale, and under what conditions?[4] In recent years, alternative frameworks seeking to move beyond the sustainability discourse have gained momentum. Often cited as articulating the post-growth and degrowth positions is Tim Jackson's *Prosperity Without Growth*, which questions whether progress, as measured in economic terms, should continue to be a legitimate goal for industrialized countries given the worsening disparity between rich and poor.[5] Meanwhile, an example of scholarship discussing the postsustainability and unsustainability positions is given by John Barry who advocates mitigating cases of "actually existing unsustainability" here and now before making any claims on future sustainability.[6] Sustainability, "one of the most contested words in the political vocabulary" as one scholar puts it,[7] is perhaps showing signs of wear and tear in Western discourses.

Meanwhile, sustainability is more relevant than ever in China, but with a difference. Translated as *kechixu fazhan*, "sustainable development" has become a key concept in academic discussions and government policies in the PRC.[8] Most recently, "ecological civilization" (*shengtai wenming*) has emerged as an umbrella term for the type of sustainable development that China's political leadership envisions for the country. Incorporated into the PRC Constitution at the 2018 session of the 13th National People's Congress, "ecological civilization" is the newest among five guiding principles for China's development in the post-Mao era.[9] However, despite its resonances with Western understandings of environmental protection and sustainability, "ecological civilization" serves ideological purposes as well: it signifies a broad set of strategies by Chinese authorities to address environmental issues, while offering rhetorical flexibility with which to interpret, adapt, and implement state policies.

China's aspirations for sustainable development are at odds with its status as one of the world's worst polluters. Ironically, "sustainable development" entered Chinese discourses in the late 1980s and early 1990s at a time when slogans such as "pollute first, manage later" (*xian wuran, hou zhili*) spoke to China's rapid economic expansion. Public outcry over environmental health problems has grown in recent years, as reports spotlight the contamination of food supplies with agrochemicals, "cancer villages" along polluted rivers, and the effect of smog on the life expectancy of city dwellers. Of these, Chai Jing's 2015 documentary film *Under the Dome* (*Qiongding zhi xia*) is notable, given its favorable reception by the Chinese public and subsequent censorship by the authorities.[10] China is currently the world's largest producer of energy-related carbon dioxide emissions, having surpassed the United States in 2008.[11] Relying on coal for more than two-thirds of its electricity needs, China accounts for nearly half of the world's coal consumption.[12] These figures are likely to change in the next decades, as the growth of China's gross domestic product (GDP) slows and as services play a larger role in an economy traditionally driven by energy-intensive manufacturing.[13] In 2014, the PRC government announced goals to peak its carbon dioxide emissions around 2030 and expand zero-emission sources to account for 20 percent of its total energy consumption,[14] but it remains to be seen how changes to energy policies will be carried out at the local, provincial, and national levels. The enforcement of environmental regulations has been complex and uneven in China, as they are difficult to reconcile with economic development.[15] That the Ministry of Natural Resources (*Ziran ziyuan bu*), the Ministry of Ecology and Environment (*Shengtai huanjing bu*), and the Ministry of Agriculture and Rural Affairs (*Nongye nongcun bu*) were among the new ministries created in the restructuring of the State Council in 2018 suggests that China's approach to environmental management is multifaceted.

With its claims to be one of the oldest civilizations in world history and a rising superpower in the twenty-first century, China is a unique case study in the interplay between economic development and environmental protection. On the one hand, the geopolitical and sociocultural entity that we now call "China"[16] has managed to sustain its population for several millennia. On the other, China has an extensive history of modifying, exploiting, and even damaging its terrains and waterways. To address the question of China's sustainability, we should first recognize the paradoxical nature of Chinese interactions with the environment. The goal of this chapter is to provide an overview of China's history, culture, and environment in relation to the concept of sustainability, broadly understood. Rather than a historical or empirical study, it is intended to be an analysis of the arguments and rhetorical strategies that researchers use to describe Chinese attitudes and behaviors toward the environment. This chapter offers a critical survey of key positions on Chinese environmental history, literature, and thought to illuminate what scholars perceive to be the paradox of China's sustainability.

11.1 "Three Thousand Years of Unsustainable Growth"

China's prolific textual tradition, along with its archaeological record and material culture, has allowed scholars to tell stories about the Chinese civilization from a spectrum of vantage points. In the field of Chinese environmental history, a question has been raised: if a civilization thrives by modifying its environment—and has done so for several millennia—does it make sense to call this civilization a "sustainable" one? Consider the following passage in Mark Elvin's *The Retreat of the Elephants: An Environmental History of China*:

> Through more than three thousand years, the Chinese refashioned China. They cleared the forests and the original vegetation cover, terraced its hill-slopes, and partitioned its valley floors into fields. They diked, dammed, and diverted its rivers and lakes. They hunted or domesticated its animals and birds; or else destroyed their habitats as a by-product of the pursuit of economic improvements. By late-imperial times there was little that could be called "natural" left untouched by this process of exploitation and adaptation.[17]

This concise account encapsulates the story that Elvin wants to tell about China's environment: as the population that came to be known as the "Chinese" expanded from their ancestral homeland across the eastern Eurasian landmass, they modified the territories known as "China" and exploited their ecosystems in destructive ways. To indicate the temporal and spatial scale of this enterprise, Elvin observes that the natural habitats of Asian elephants in the region have reduced to a few ecological niches in the Southwest. The "retreat" of the elephants—the metaphor after which

Elvin's book is titled—illustrates the millennia-long expansion of human settlements over the lands collectively called "China."[18] Elsewhere, Elvin describes this process as China's "three thousand years of unsustainable growth."[19]

Elvin's sweeping thesis that the Chinese refashioned China for several millennia is at once powerful and provocative. Powerful, because it sheds light on how the growth of a human population altered ecosystems, terrains, and bodies of water, destroying the habitats of nonhuman species in the process. Provocative, because it leaves certain questions about "Chineseness" open or unanswered.[20] For example, what transformations did the population that came to be called "Chinese" experience over these three thousand years? And what intersection of ethnic, cultural, and political identities does the term "Chinese" now imply? How did China's majority interact with other peoples living on the same lands and elsewhere? Since it is unavoidable that humankind extracts nutrients, fuel, fiber, and building materials from its surroundings to survive, what aspects of human impact on the environment are particular to the Chinese civilization? Given that the Chinese agricultural system has provided for a growing population for a remarkably long period of time—albeit at the cost of biodiversity—how does one make sense of the contradiction between China's environmental degradation and its "sustained" growth? These questions invite further consideration of Elvin's thesis in dialogue with critical race studies, global environmental history, and environmental humanities, as I will discuss below.

When Elvin argues that the Chinese refashioned China for more than three thousand years, which "Chinese" does he have in mind? "Chinese" is neither a static nor a self-evident term: its meanings are contingent upon ethnic self-identifications, cultural allegiances, and political realities. The "Chinese people" (*Zhonghua minzu*), as it is understood in the modern sense, is a collective of ethnic, cultural, and political identities forged in the rise of Chinese nationalism. Its conceptual flexibility is further enabled by the fact that the Chinese term *minzu* refers to a spectrum of ethnicities, nations, and bioracial categories that range from communities with shared ancestry to groups fashioned by political authorities for administrative purposes. The narrative of the Chinese people as a single, unified population that has existed from ancient times to the present is a product of nationalist thinking in the early twentieth century and its subsequent reformulations.[21] As a modern nation-state, "China" did not come into being until the founding of the Republic of China in 1912. The PRC, established in 1949, calls itself a multiethnic state consisting of the Han ethnic majority (*Hanzu*) and 55 officially recognized ethnic minorities (*shaoshu minzu*). With more than 90 percent of PRC citizens identified as Han, this nomenclature breaks down the monolithic nature of the Chinese people to some extent, but leaves the Han majority as largely self-evident and, again, monolithic.

If we interpret Elvin's thesis to mean that "Han Chinese" refashioned China for more than three thousand years, we encounter the following issues with regard to Han as a category of identity. In *Critical Han Studies: The History, Representation, and Identity of China's Majority*, Thomas S. Mullaney argues that when the scale and composition of the Han ethnicity are taken into account, it "appears less like a coherent category of identity and more like an umbrella term encompassing a plurality of diverse cultures, languages, and ethnicities."[22] Furthermore, the category of Han is so intertwined with the category of Chinese that they often appear to be interchangeable or commensurate in academic discourses. For instance, Hanban, the agency that oversees the Confucius Institute worldwide, translates *Hanyu*, or the purported "Han language," simply as "Chinese."[23] As Mullaney observes, naturalizing the equivalence between Han and Chinese identities contributes to the majority's "powerful and hegemonic neutrality," which is not unlike how whiteness tends to be perceived in American society.[24] For Mullaney, whereas non-Han ethnicities are subject to the state's classification and scrutiny, there is "a virtual silence over the formation of Han identity as something apart from the overall discussion of 'Chinese nationalism' more generally."[25] As a corrective to nationalistic interpretations of China's history, it is important to remember the internal complexity of the Han ethnicity and the diversity of the Chinese as a whole. In the context of China's millennia-long environmental transformations, we should also ask how other peoples may have thrived or suffered in this process.[26]

Various peoples have lived on the lands that we now call "China." Historically, one of the features that distinguished ancient Chinese culture from non-Chinese ones—at least from the Chinese perspective—was agriculture: grains were especially significant as a cultural marker.[27] Chinese-style agriculture began with the growing of millet in the northern loess highlands and wet-rice cultivation in the Yangzi River Valley and along the southeastern coast. Agriculture, as Robert Marks argues, was one of the main driving forces of China's millennia-long transformation of the environment. The Chinese empire enforced policies based on "Chinese assumptions about land and the highest uses to which it could be put," which typically valued farmland above all.[28] This preference for farmland resonates with what James C. Scott calls "state landscaping"[29] or a state's systematic efforts to alter landscapes and waterways, settle populations for the cultivation of land, and encourage taxable subsistence activities. As Scott explains in *Against the Grain: A Deep History of the Earliest States*, the making of early states depended largely on their ability to maintain and expand agricultural holdings that generated surplus in the form of taxable cereal grains. Scott's argument that "grain states were restricted to a narrow ecological niche that favored intensive agriculture"[30] appears to support Marks' account of China's historical preferences in land use.

As the Chinese empire expanded across the eastern Eurasian landmass, the concomitant spread of Chinese-style agriculture transformed a variety of terrains, including those inhabited by non-Chinese peoples, into cultivated land. During the Han dynasty, for example, settlers relocated from the North China plain first to the Gansu corridor west of the Yellow River and later northwest into the Tarim Basin, converting steppes into farms through military-agricultural colonies (*tuntian*).[31] Over time, Chinese preference for intensive agriculture would redefine the environments of the core regions and take priority over other subsistence activities in the frontier regions such as hunting-gathering, nomadic pastoralism, agro-pastoralism, and shifting agriculture. From forests to grasslands, from hillsides to islands, from the heartland to the frontier, China's majority refashioned diverse landscapes into settlements and croplands according to the ideals and norms of Chinese agriculture. Generally speaking, the family farm was the predominant unit of organization, while military-agricultural colonies consolidated contested frontiers. As techniques of waterway management and land reclamation developed, Chinese communities converted swamps, lakes, riverbeds, and coastal flats into additional land for settlement and cultivation. These patterns of environmental change provided for China's population but at the cost of increasingly labor-intensive agriculture and severe environmental degradation.[32]

At the same time, China was not unique in modifying its environments to extract nutrients, fuel, fiber, and building materials for its population. Prior to the nineteenth century, environmental degradation in China may not have been as dramatic as we often imagine it to be. As Kenneth Pomeranz suggests, the environmental history of China tends to be distorted in part by the projection of nineteenth- and twentieth-century ecological disasters and population problems back to earlier periods.[33]

When we compare China with other regions of the world, we may see similarities in their respective environmental conditions. Take forest cover in the region we call "Europe," for example. Deforestation was an apparent issue for both China and Western Europe by the nineteenth century, but, interestingly, key regions in China may have been better off—or at least not worse—than their European counterparts.[34] By 1750, approximately 25 percent of China's overall land surface was forested, and that figure would sink to 5–10 percent by 1950.[35] As a comparison, it is estimated that about 16 percent of France was forested by 1789, while other parts of Europe such as Britain, Italy, Spain, Portugal, and the Low Countries had forests on 5–10 percent of their land surface by 1850.[36] As Pomeranz notes, "China's problems with forest cover and fuel supply were … serious, but probably not as bad as we often think, and—surprisingly—not clearly worse than those of western Europe."[37] In domains such as forestry and agriculture, China may have even outperformed Europe. Pomeranz goes as far as to claim that, in the eighteenth century, Europe was "catching up with China and Japan in both best and average practices… rather than blazing new paths,"[38] since these East Asian countries used arable land efficiently to support their growing populations. While both China and Europe experienced ecological constraints during this period, the respective regions would later resolve their limitations in drastically different ways.

By the early nineteenth century, the conditions for what Pomeranz calls the "Great Divergence" between China and Europe were in place.[39] Whereas China continued to mitigate its ecological vulnerabilities through labor-intensive agriculture and domestic trade, Europe found sources of ecological relief on its continent and overseas. In the case of England, the availability of coal coupled with low transportation costs laid the groundwork for an industrial breakthrough, while forests in Eastern Europe functioned as an ecological "buffer" for the core regions. Meanwhile, the New World acted as Europe's key trading partner and supplied mined fertilizers, such as guano and nitrates, to replenish European fields. China had not resolved its ecological vulnerabilities as Europe did around the time of the Industrial Revolution, for Europe benefited in ecological terms from the colonization of the New World, the destruction of its indigenous populations, and the exploitation of enslaved peoples.[40] Acknowledging that China is "comparable in size, population, and internal diversity to Europe as a whole,"[41] Pomeranz argues that neither region had a clear ecological advantage over the other in the middle of the eighteenth century, before the time of the Great Divergence. An advantage of putting China's environmental practices in a global context is avoiding the implication that China was somehow "exceptional" in its environmental degradation. In fact, countries in premodern East Asia relied largely on ecosystems within their respective borders to support their populations.[42] While this reasoning is not intended to justify China's imperialistic practices and its over-consumption of resources through international trade, it does balance China-specific studies with comparative ones.

A more comparative approach to environmental history recontextualizes Elvin's thesis of China's "three thousand years of unsustainable growth." As Elvin himself admits, his analysis of China's millennia-long growth results in a contradiction:

> The styles of economic exploitation of the environment adopted, at least after the ecologically self-conscious restraint of the first period *were not, over the very long run, sustainable in a steady form*, especially given the apparently uncontrollable tendency of the population not only to grow, but to grow at an ever-increasing rate. *The Chinese economy survived these millennia, and in fact did remarkably well for a time*, quite possibly leading the world in the middle centuries of the middle ages, because it mastered new technological skills such as hydraulically sophisticated irrigated rice-farming, and because it was continually expanding into fresh resource areas such as Tairwan [sic] in the later seventeenth century, the far south-west in the early nineteenth century, and Manchuria beyond the Liaor [sic] river valley in the early twentieth century.[43]

Although Elvin claims that the Chinese economy is not "sustainable" over the long run, he concedes later in the paragraph that it has not only survived for several millennia but also led the world in terms of agricultural innovations and practices for a period of time. Does Elvin not refute his thesis with his own analysis? Indeed, Elvin calls his argument a "paradox in appearance."[44]

What is crucial is grasping the unarticulated reasoning underpinning Elvin's paradoxical thesis.[45] A common view among historians, and Elvin agrees, is that China relied largely on ecosystems within its borders to sustain its population for a remarkably long time. However, sustainability concerns not only time but also space. The paradox of China's sustainability lies not in China's long-term population growth per se, but rather in its transformation of a range of environments into Chinese-style settlements and croplands that were highly efficient but relatively homogeneous. This reasoning is corroborated by Marks' account of the millennia-long processes by which the Chinese empire expanded from the Han heartland across the eastern Eurasian landmass, altering various territories according to Chinese ideals and practices of land use. What has degraded China's environment—and some may not view this as "degradation"—is the conversion of diverse terrains and bodies of water into settlements and croplands in the first place. If there is an unsaid truth to Elvin's paradox, it is the fact that Chinese-style land use has undermined China's biological and ecological diversity in the long run.

The land that we call "China" is a populous one: it has sustained between one-sixth and one-third of the world's human population at any given time. While the agricultural systems that supported this population proved to be efficient, transforming the environment into Chinese-style settlements and croplands undermined China's biodiversity.[46] As the Chinese empire expanded across the eastern Eurasian landmass, it transformed more and more environments according to Chinese ideals of land use—a process that may be interpreted as what Chia-ju Chang calls in this volume "environing" or "encircling territory" but over the course of centuries and millennia. After the establishment of Chinese-style farmlands, however, the systems in place could be remarkably sustainable and efficient. Over time, China was able to provide for more people with less farmland per capita: even though the total acreage of cultivated land increased between 2 CE and 1848 CE, the ratio of cultivated land per capita decreased on average.[47] Chinese agricultural norms and practices also played a role. Rice paddy fields increase in fertility and workability over time in contrast to conventional dry-land farming that tends to deplete the soil of nutrients.[48] As Marks mentions, polders created out of swamps and marshes at least a millennium ago still produce rice today, while farmland created two millennia ago continues to be farmed.[49] Perhaps, the best-known examples of sustainable farming from China are symbiotic ecosystems such as fish ponds with fruit trees, mulberry trees, or sugarcanes planted on the embankment (*guoji yutang*, *sangji yutang*, *zheji yutang*). These anthropogenic ecosystems integrate agriculture, pisciculture, and sericulture in ways that promote the recycling of energy. Nonetheless, these relatively sustainable systems of food production come at the cost of the diverse ecosystems that existed beforehand. As China's population grew, environmental degradation became increasingly apparent in the nineteenth and twentieth centuries, and Maoist campaigns, such as the Great Leap Forward, further devastated China's landscapes.[50] In the post-Mao era of economic reform, especially the first decade of the new millennium, China saw the majority of its population move and settle in urban areas. Land and property development, often enabled by administrative divisions at local levels, has become an aggregate force of environmental change across the country. As China's environment is further transformed in the twenty-first century, the paradox of China's sustainability will continue to be a relevant problem.

11.2 "Green Paradoxes"

East Asian cultures have informed scholars from a range of disciplinary backgrounds on modern environmental thinking. Environmental philosophers, such as J. Baird Callicott and Donald Swearer, have long cited the philosophical and religious traditions of East Asia, especially Buddhism, as sources of comparison and inspiration.[51] Since the late 1990s, Harvard University's Center for the Study of World Religions has published a series of edited volumes on the intersection between religion and ecology, three of which are devoted to Confucianism, Daoism, and Buddhism, respectively.[52] Meanwhile, scholars of East Asian thought such as Cheng Chung-ying, Tu Wei-ming, Roger T. Ames, and Graham Parkes have explored the varying possibilities of reading the philosophical and religious traditions of China from the perspective of environmentalism.[53] Affirming the validity of such an approach, Tu observes an "ecological turn" in the New Confucian Humanism of recent decades, while James Miller argues for Daoism's relevance to the discourse of sustainability.[54] Although other scholars express reservations toward labeling Daoism and Buddhism, in particular, as inherently "environmentalist,"[55] it is clear that East Asian thought has contributed to the development of environmental ethics as a field and the broader discourse of environmentalism in the United States and elsewhere.

In *The Retreat of the Elephants*, Elvin claims to have diagnosed another paradox, this time, in premodern Chinese thinking about Nature. For Elvin, the sustainability of the Chinese civilization is paradoxical, not only because it transformed its environments in exploitive ways for several millennia but also because its textual traditions ironically praised Nature as an ideal to be respected. In particular, Elvin and others argue that the behaviors of Chinese individuals, communities, and governments do not match their stated attitudes toward the environment.[56] In his assessment of various genres of premodern Chinese writing, such as philosophical and religious texts, poetry anthologies, farming manuals, local gazettes, and government reports, Elvin concludes:

> A paradox thus lay at the heart of Chinese attitudes to the landscape. On the one hand it was seen, not as an image or reflection of some transcendent being, but as a part of the supreme numinous power itself. Wisdom required that one put oneself into its rhythms and be conscious of one's inability to reshape it. On the other hand the landscape was in fact tamed, transformed, and exploited to a degree that had few parallels in the premodern world. … This paradox shows that the relationship of a representation to a reality—so far as we can understand the first and reconstruct the second—may be complex. Even as is the case here, at least superficially contradictory. Behind all such studies of perceptions of the environment there lurks a *trap*: that of assuming that people's actual behavior was on the whole necessarily in accord with the ideas and feelings expressed in our sources.[57]

In the sources that he has assembled, Elvin finds a nontrivial contradiction between the stated attitudes of the Chinese and their actual behaviors toward the environment. More specifically, Elvin's use of the word "trap" appears to be his warning against any naïve reading of these Chinese sources. To be sure, the sense of discrepancy between attitudes and behaviors—not to say, hypocrisy—is neither new to humanity nor unique to the Chinese. The question is: what would one learn by studying this discrepancy as the essence of a culture or cultures?

Karen Thornber tackles this difficult question in *Ecoambiguity: Environmental Crises and East Asian Literatures*. First, Thornber defines *ecoambiguity* as "the complex, contradictory interactions between people and environments with a significant nonhuman presence,"[58] but its meaning is perhaps best conveyed by an anecdote about her experiences on family trips:

> At some point the paradox hit. My family's love of nature meant that we spent gallons of fuel driving or flying to get to nature, and considerable time disrupting it. No matter how careful we were, didn't our mere presence harm the landscapes to which we became so attached? But had we stayed home, would we not simply have disturbed other ecosystems?[59]

Her rhetorical questions anticipate answers in the affirmative. Thornber rightly criticizes certain privileged modes of interacting with the environment, in which she conscientiously includes herself as participating. Elsewhere in her book, she uses "green paradox," "green hypocrisy," and "loving nature to death" to describe more destructive versions of this phenomenon.[60] For Thornber, human actions are necessarily detrimental to the environment, despite efforts to mitigate or reverse those effects: such a stance resonates with schools of environmental thought such as Deep Ecology. However, Thornber claims that this phenomenon is especially pronounced in East Asian societies:

> East Asian peoples long have had a heightened consciousness of human-induced damage to environments, but despite the commitment of individuals, organizations, and governments to repairing extant damage and limiting further harm, overall environmental degradation in the region shows every sign of persisting.[61]

Given the traditional emphasis on harmony between humankind and Nature in East Asian cultures, Thornber believes that it is important to highlight the discrepancy between stated attitudes of environmental appreciation and actual behaviors leading to environmental degradation:

> To be sure, ecoambiguity appears more prevalent in literature from East Asia than in other textual corpuses. And its irony is certainly deeper, considering the region's long cultural history celebrating the intimate ties between humans and nature even as its peoples severely damaged environments.[62]

The implication is that East Asians, more than other peoples, have failed to live up to the ideals and values in their cultures. As a result, the critical energies reserved for investigating human contributions to environmental degradation are redirected toward the subset of East Asia in particular.

While broad criticisms of Western culture, such as those made by Lynn White Jr. and John Passmore in the 1960s and 1970s, respectively, call attention to how environmental problems may be rooted in Western values and habits,[63] generalizations about any set of cultures or peoples are seldom attempted in academia today. The issue is even more delicate when such arguments are made by scholars in a comparative or intercultural context. At the same time, Thornber's assertion about East Asian cultures, as provocative as it may sound to readers who interpret it as approaching essentialism, reiterates a widely held view in Asian studies. Rhoads Murphey, for instance, offers the following assessment in "Asian Perspectives of and Behavior toward the Natural Environment," published in the edited volume *Landscapes and Communities on the Pacific Rim* in 2000:

> All Asian cultures ... have long been noted for their admiring attitudes toward nature ... The Asian record, however, makes clear that, despite the professed values of the literate elite, people have altered or destroyed the Asian environment for longer and on a greater scale than anywhere else in the world, even in the twentieth-century West.[64]

This position appears to resonate with Thornber's, but she reminds readers that Murphey "perhaps overstates the case."[65] Scholars writing about Asia often employ rhetorical strategies such as appeals to balance and acknowledgments of biases to moderate this type of argument, as the following examples show. Murphey, for instance, concludes at the end of his essay that there are voices attuned to Nature and voices against environmental destruction "in both East and West."[66] In "China's Environment: Resilient Myths and Contradictory Realities," published in the same edited volume in 2000, Vaclav Smil questions the accuracy and veracity of PRC government reports on afforestation efforts but attributes their uncritical acceptance among some scholars to "Western biases, misinformation, and wishful thinking."[67] In *The Great Clod: Notes and Memoirs on Nature and History in East Asia*, published in 2016, Gary Snyder reflects on how his early experiences have shaped his reception of Chinese culture:

> I was exposed to East Asian paintings at the Seattle Art Museum and a bit later as a college undergraduate was reading Confucius and Lao-tzu and Chuang-tzu (as well as the Homeric epics and Greek drama and much else). I remember distinctly the time it came to me that perhaps China and Japan were fully developed high humanistic civilizations that had made peace with nature! I didn't let go of that idea for years. ... I thought I had come onto a fully engaged civilization that maintained a respectful and careful regard for the land itself, and the many other beings who already lived there. It turned out that I was wrong, but in a very complex and challenging way.[68]

Such a self-reflective account reveals the complex synthesis of enthusiasm and frustration, and perhaps of allophilia and essentialism, that often underlies academic or creative pursuits. Generalizations about other cultures, whether they praise or criticize (or both), attain the status of enduring truths, as they are reiterated in the form of scholarly arguments. Scholars continue to invest in these ideas, deriving different payoffs in their respective intellectual bargains.

In light of conversations about the politics of knowledge production in academia, scholars have become more cognizant of their positions in relation to the topics they study. Contemporary scholarship on environmentalism in non-Western countries has attempted to address issues such as privilege, appropriation, and representational authority. For example, in *Slow Violence and the Environmentalism of the Poor*, Rob Nixon studies how writer-activists in the Global South speak up against long-term environmental catastrophes that are often ignored or quickly forgotten by the fast-paced media.[69] In the preface of his book, where he recounts his experience studying with Edward Said, Nixon demonstrates an awareness of the privilege with which researchers at elite US institutions carry out their work.[70] As an interpreter of the Global South, Nixon likewise discusses the relative privilege of writer-activists in relation to the communities about which they write.[71] Whereas Nixon views writer-activists as possessing the agency to represent and address environmental concerns through their writing, the ability of East Asian literatures to effect intellectual or material change is more ambiguous in Thornber's assessment. Although Thornber is equally concerned with environmental degradation, especially from a global perspective,[72] she is more skeptical about the role of literature in the pursuit of environmental justice. Because of literature's "regular and often blatant defiance of logic, precision, and unity,"[73] in Thornber's view, it is predisposed toward expressing ambiguity, contradiction, and hypocrisy more often than performing the type of work that makes literary texts strategically vital in confronting environmental problems.

If the writer-activists in Nixon's account are selected for their ability to address environmental problems in their native countries, the authors in Thornber's work tend to highlight the ambiguity, contradiction, and hypocrisy in East Asian peoples' interactions with the environment. In this sense, their collective agency to respond to environmental issues—in ways that are meaningful to them—tends to be superseded by Thornber's broader framework. Moreover, this "ecoambiguity" is not limited to East Asian literatures, as Thornber extends her argument to the category of literature itself:

> Literature rarely offers comprehensive remedies, much less proposes official policies to prevent future or remediate current damage to landscapes; in some cases, creative writing itself might even abet the ecodegradation it deplores.[74]

Indeed, there is nothing inherent in literature that necessitates practical solutions to environmental problems; literature may sometimes serve as an aesthetic veneer for troubling developments in environmental history.

Wallace Stegner, for example, was commissioned by the Arabian American Oil Company (now Saudi Aramco) in the mid-1950s to write a novel about the company's role in discovering oil in Saudi Arabia. Given the environmental impact of petroleum and petroculture in the twentieth and twenty-first centuries, one can reasonably approach Stegner's *Discovery! The Search for Arabian Oil* with the benefit of historical hindsight. As Nixon notes, Stegner "blithely reads Aramco's history in Saudi Arabia as a mostly benign extension of America's own mostly benign frontier development … at pains to distance his paymasters from any intimation of imperial malpractice."[75] Thornber's framework can lend new vocabulary to such a reading by highlighting the "ecoambiguity" of Stegner's novel. For Thornber, ecoambiguity as an interpretation of East Asian literatures, or literature in general, checks naïve claims of literature's positive impact on human experiences and the environment.

Together, Thornber's *Ecoambiguity* and Nixon's *Slow Violence and the Environmentalism of the Poor* outline the spectrum of existing approaches to non-Western literatures about the environment. As Nixon shows, not all forms of writing are implicated in Thornber's phenomenon of "loving nature to death." Nixon articulates literature's vital role in the pursuit of social and environmental justice by illustrating how writer-activists in the Global South confront environmental challenges in their respective countries. In the case of Ken Saro-Wiwa and other activists who criticized Shell's operations in Ogoniland in southern Nigeria, they did so at great personal risk.[76] Their subsequent execution by the military government gives a poignant dimension to Thornber's otherwise ironic dictum of "loving nature to death." Here, Nixon spotlights the "second scramble for Africa," in which American, European, and Chinese corporations engage in the neocolonial practices of resource extraction on the continent. Nevertheless, Nixon offers little analysis of China and virtually none of Japan, South Korea, and Taiwan. This is where Thornber's work complements Nixon's in bringing East Asian literatures to the fore. Thornber assembles a list of East Asian literary texts that convey a sense of impasse and helplessness, reflecting the realities that writers face as their nations pursue economic growth at the cost of the environment. As Thornber explains, her framework "explores the multiple ways fiction and poetry highlight the absence of simple answers and the paucity of facile solutions to environmental problems."[77] Since the publication of *Ecoambiguity*, Thornber's position has shifted, as she focuses more on how Chinese and other Asian writers address environmental crises in proactive ways.[78] For Thornber, environmental humanities cannot afford to ignore the histories, cultures, and realities of more than half of the world's population:

> Asia and Asian studies are important to understanding the Anthropocene—not only because of Asian peoples' long histories of transforming environments, not only because of Asian peoples' long histories of environmental philosophies advocating harmonious human-nonhuman relationships, but also because of Asian peoples' long histories of cultural products exposing and decrying environmental abuse and the vivid global responses to Asia's environmental crises.[79]

Long histories also mean that no single narrative is likely to capture the "essence" of a civilization, people, or culture. For Elvin, his account of China's "three thousand years of unsustainable growth" ultimately results in a paradox. Chinese attitudes and behaviors toward the environment—as incorporated into a grand thesis about Chinese "reality"—inevitably appear multifaceted, complex, or even contradictory. Perhaps, the constant in such a thesis is inconsistency, which then perpetuates itself in other forms as paradox, ambiguity, contradiction, equivocation, and hypocrisy. Yet, claims of cultural specificity confine the resulting discrepancies to Eastern and ethnically marked cultures, while the thesis' broader implications (for Western cultures or humanity as a whole, for example) are left unexamined or undertheorized. If, as Elvin warns, there indeed lurks a "trap" in the study of China and the environment, would it not be wise to explore alternative approaches?

11.3 The Future of China's Sustainability

China's population had fluctuated between 30 and 200 million since the Han dynasty. It was not until the Qing dynasty, however, that China's population began to grow from 72 million in 1657 to 426 million in 1848.[80] By this time, the respective economies of China and Europe had diverged, with China relying largely on ecosystems within its borders for the production of food, fiber, fuel, and construction materials. As historians observe, China's population growth in the later Qing period pushed the limits of labor-intensive agriculture and accelerated environmental degradation at home and abroad through trade. Nonetheless, China's population growth continued in the twentieth century: the 1953–1954 census reported a total of 583 million people, while estimates put the figure at more than 800 million by 1969.[81] In the late 1970s, the PRC government began to enforce its notorious one-child policy among most families, drastically altering the country's demographics over the next decades. During this period, the population continued to grow, albeit at a slower rate, with the ratio of newborns skewing heavily toward males. By the mid-2010s, factors such as a slowing economy, an aging population, and risks of social instability ushered in a policy reversal, ending the one-child policy. Whether or not China's population would continue to grow as it did before, it remains one of the largest in the world in the new millennium, comparable only to India's.

China's population growth and environmental challenges have been closely linked, especially in the twentieth century. During the Republican period, China suffered from a series of environmental disasters and was known as the "Land of Famine" in the West, especially among relief workers.[82] Nonetheless, the Chinese population continued to expand after the founding of the PRC. Given the Maoist position that a large population was a resource, rather than a burden, for the Communist state, it was politically incorrect to advocate population control.[83] Meanwhile, a growing population demanded more farmland and higher yields. Guo Moruo's poem "Wage War against the Earth" ("Xiang diqiu kaizhan") captures the spirit of campaigns in the 1950s, in which veterans and their families settled in the Great Northern Wilderness (*Beidahuang*) of northeastern China to reclaim land for agriculture:

> Chairman Mao once said: We must first breach the surface of the Earth
> And enter the space of its outer layer.
> This boundless battlefield
> Has opened up before our six hundred million people.
> …
> We must reclaim farmland on a massive scale,
> South to Hainan Island, north to Heilong River's banks,
> West to the Pamir Mountains, the roof of the world.
> Ocean floors and mountaintops can also be cultivated;
> We must think of ways to turn the Great Gobi Desert into fertile fields.[84]

The surface of the Earth became a battlefield on which the Chinese people, led by Mao, fought against Nature's constraints. Slogans such as "the foolish old man removes mountains" (*yugong yishan*) and "encircle the lakes and create farmland" (*weihu zaotian*) reflected the disregard for local terrains and ecological conditions in the pursuit of humanity's dominance over Nature. While it is debatable whether Maoist campaigns increased the total amount of farmland at the time, the disastrous consequences of such efforts are well known.[85] Improving yield was another goal of these campaigns. Farmers had relied on a combination of organic and synthetic fertilizers, the latter of which were either imported or produced locally on a small scale. Richard Nixon's visit in 1972 would alter the course of China's agricultural development. In one of the first commercial deals since the normalization of relations with the US, China purchased a series of large-scale synthetic ammonia and urea plants from the M. W. Kellogg Company and its affiliates.[86] The construction of these facilities not only reduced China's traditional reliance on organic fertilizers but also helped China to break through what Vaclav Smil calls the "nitrogen barrier."[87] China began to construct its own synthetic fertilizer plants in the 1980s, achieved near self-sufficiency in the 1990s, and became an exporter by the 2000s. With its agricultural capabilities significantly improved at home, China is taking the "war against the Earth" beyond its borders through the acquisition of farmland and corporations abroad. The aspirational goals of a bygone era now seem plausible, if not within reach.

The surface of the Earth is exploited for not only agriculture but also energy production. Historically, China depended on biomass such as firewood, fallen foliage, and agricultural byproducts, as well as coal, for fuel. In the early twentieth century, China shifted away from a somatic energy regime toward the use of modern fossil fuels in power generation. As China approached the turn of the millennium, coal-fired power plants and hydropower plants were joined by nuclear power plants such as those in Qinshan in Zhejiang province. While official publications tout a series of terms such as "green energy" (*lüse nengyuan*), "clean energy" (*qingjie nengyuan*), "sustainable energy" (*kechixu nengyuan*), "renewable energy" (*kezaisheng nengyuan*), and "low-carbon energy" (*ditan nengyuan*) as environmentally responsible solutions to China's energy needs, they arguably represent the next phase of the PRC's national agenda. Renewable energy schemes such as wind farms in Inner Mongolia, Gansu, and Xinjiang and hydropower projects in Yunnan can be interpreted as instances of China's continuing reach into historically "non-Han" territories, drawing on their climates and terrains to serve the core regions.[88] Meanwhile, the Belt and Road Initiative (*Yidai yilu*) further expands China's energy and economic interests in the South China Sea, Asia, the Middle East, Africa, and Europe as it seeks to overcome the ecological constraints of its own lands and to gain access to new resources and markets. If "ecological civilization" functions as a policy framework with which to discuss China's future, what meanings has the term accrued in light of these developments? Whereas the paradox of China's sustainability sheds light on the history of Chinese interactions with the environment, the future of China's sustainability remains a story to be told.

Notes

1. The Commission was convened by the United Nations in 1983, following the UN Conference on the Human Environment in Stockholm in 1972. The United Nations Conference on Environment and Development (UNCED), also known as the Earth Summit, was first held in Rio de Janeiro in June 1992.
2. WCED, *Our Common Future*, 8–9.
3. See Ehrlich, *The Population Bomb* and Meadows et al., *The Limits to Growth*. These works have also been criticized for being alarmist and unreliable. For the authors' subsequent reflections on their respective works, see Meadows et al., *Beyond the Limits*, 105 and Ehrlich and Ehrlich, "The Population Bomb Revisited."
4. Luke, "Sustainable Development as a Power/Knowledge System," 21–22. See also Basiago, "Methods of Defining Sustainability" and Dobson, *Justice and the Environment*, 33–61.
5. Jackson, *Prosperity without Growth*, 17.
6. Barry, *The Politics of Actually Existing Unsustainability*, 5–8.
7. Dobson, *Green Political Thought*, 62. See also Dobson, *Justice and the Environment*, 33–61.
8. Tilt, *The Struggle for Sustainability in Rural China*, 138–140.
9. "Ecological civilization" was introduced at the 17th Party Congress in 2007. See Parr and Henry, "A New Starting Point"; Wang-Kaeding, "What Does Xi Jinping's New Phrase 'Ecological Civilization' Mean?"; and Geall and Ely, "Narratives and Pathways towards an Ecological Civilization in Contemporary China." The four other guiding principles are "material civilization" (*wuzhi wenming*), "spiritual civilization" (*jingshen wenming*), "political civilization" (*zhengzhi wenming*), and "social civilization" (*shehui wenming*). See Dynon, "'Four Civilizations' and the Evolution of Post-Mao Chinese Socialist Ideology."
10. Chai Jing, a former investigative reporter and television host for China Central Television, describes the effects of China's air pollution on the wellbeing of its citizens and criticizes the inefficacy of government regulations. The film garnered hundreds of millions of views in the first days of its online release and was promptly censored by the Chinese Communist Party. See Wong, "China Blocks Web Access to 'Under the Dome' Documentary on Pollution."

11. China and the US together accounted for approximately 40 percent of energy-related carbon dioxide emissions in 2012. EIA, *International Energy Outlook 2016*, 140. See also EIA, *International Energy Outlook 2017*, 137–140. The 2018 reports limit their analysis to India, China, and Africa only.
12. EIA, *International Energy Outlook 2016*, 92, 3. See also EIA, *International Energy Outlook 2017*, 83–84, 141–142.
13. EIA, *Energy Implications of China's Transition Toward Consumption-led Growth*, 2–3.
14. Office of the Press Secretary, "U.S.-China Joint Announcement on Climate Change and Clean Energy Cooperation."
15. Economy, *The River Runs Black*, 63, 106–117. Tilt, *The Struggle for Sustainability in Rural China*, 69–70, 109–120. See also Ma and Ortolano, *Environmental Regulation in China*.
16. See Kirby, "When Did China Become China? Thoughts on the Twentieth Century." In this book chapter, I use the term "China" to refer to the People's Republic and its predecessors, broadly defined. My usage follows existing arguments that assume a certain historical continuity between ancient, imperial, and modern China. Some may find this nomenclature acceptable, while others less than ideal.
17. Elvin, *The Retreat of the Elephants*, 321. However, "untouched" Nature is itself a construct rooted in culture. As William Cronon reminds us, wilderness is "the creation of very particular human cultures at very particular moments in human history" ("The Trouble with Wilderness; or, Getting Back to the Wrong Nature," 69). Ramachandra Guha observes in his postcolonial critique of Western environmentalism that it is typical to "equate environmental protection with the protection of wilderness" ("Radical American Environmentalism and Wilderness Preservation," 79). Elvin seems to be making a similar assumption here that lands are protected from exploitation when they are "untouched."
18. Elvin, *The Retreat of the Elephants*, 17. For a similar narrative about tigers in southern China, see Marks, *Tigers, Rice, Silk, and Silt*, 324–326.
19. Elvin, "Three Thousand Years of Unsustainable Growth," 32. Compare to Murphey, "Man and Nature in China," 315.
20. For a classic discussion of "Chineseness" and the politics of knowledge production, see Chow, "Introduction: On Chineseness as a Theoretical Problem." See also Shi et al., *Sinophone Studies*.
21. Duara, *Rescuing History from the Nation*, 33–43, 49. Perdue, "The Chinese."
22. Mullaney, *Critical Han Studies*, 2. According to Mark Elliot, the term "Han" originated in the Han period as a political identifier: it signified the subjects of the Han emperor and did not refer to culture, descent, language, or ethnicity. The term began to emerge as an ethnonym in the

Northern Wei dynasty; however, it had various uses from the Tang to the Yuan and did not approach its modern meaning of ethnicity until the Ming dynasty. See Elliot, "*Hushuo.*"

23. Officially called the Office of Chinese Language Council International (*Guojia Hanyu guoji tuiguang lingdao xiaozu bangongshi*), Hanban is affiliated with the PRC's Ministry of Education. In addition to overseeing the Confucius Institute, Hanban is known for administering the *Hanyu shuiping kaoshi* (HSK), translated as the Chinese Proficiency Test.
24. Mullaney, *Critical Han Studies*, 2–3.
25. Ibid., 5. See also Mullaney, *Coming to Terms with the Nation.*
26. Elvin considers the Han encroachment on non-Han peoples in a case study on Guizhou in *The Retreat of Elephant*, 216–232. See also Lattimore, *Studies in Frontier History*, and Di Cosmo, *Ancient China and Its Enemies.*
27. As early as the Zhou dynasty, people of the Middle Kingdom differentiated themselves from foreign tribes not only by their hair, skin, clothing, and shelter but also by whether they cooked their food and ate grains. For example, the people of the Middle Kingdom were called "grain-eaters" (*lishi zhi min*). See Legge, *Sacred Books of the East*, 27:22 [*Liji* 3.III.14], and Watson, *Mozi*, 85 [*Mozi* 7A:5]. See also Dikötter, *The Discourse of Race in Modern China*, 7.
28. Marks, *China*, 401.
29. Scott, *Against the Grain*, 23.
30. Ibid., 135. Scott further suggests that the power of early Chinese states was limited to the arable drainage basins of the Yellow and Yangzi Rivers because nonagricultural subsistence activities beyond these areas were difficult to tax (*Against the Grain*, 134–5).
31. Chang, *The Rise of the Chinese Empire*, 1:211–213. Marks, *China*, 90.
32. Pomeranz, *Great Divergence*, 215.
33. Ibid., 11.
34. Ibid., 12, 306.
35. Marks, *China*, 161, 184, 288. By Pomeranz's calculation, China's forest cover circa 1700 could be as high as 37.2 percent, if the figure excluded the large, sparsely populated regions of Tibet, Xinjiang, Qinghai, and Outer Mongolia, which were far away from the empire's core regions. See Pomeranz, *Great Divergence*, 228. Based on figures in Ling, "Wo guo senlin ziyuan de bianqian," 1983.
36. Pomeranz, *Great Divergence*, 222, 312. As quoted in Williams, "Forests," 181. The figure for France's forest cover in 1789 is nine million hectares. See Cooper, "In Search of Agrarian Capitalism," 21n3.
37. Pomeranz, *Great Divergence*, 227.
38. Ibid., 23.
39. Ibid., 283.
40. Ibid., 239, 285.

41. Ibid., 7.
42. Pomeranz, *Great Divergence*, 287. Marks, *China*, 398. Elvin, "Three Thousand Years of Unsustainable Growth," 32. Totman, *Pre-Industrial Korea and Japan in Environmental Perspective*, 1–3, 126–128. In the case of Japan, fish was an additional source of sustenance and fertilizer.
43. Elvin uses the Gwoyeu Romatzyh (*guoyu luomazi*) Chinese romanization system, in which the additional "r" indicates the second tone. Elvin, "Three Thousand Years of Unsustainable Growth," 32–33. Italics added.
44. Elvin, "Three Thousand Years of Unsustainable Growth," 32.
45. Using the terminology of the American philosopher W. V. Quine, I classify Elvin's thesis as a "veridical paradox." In "Ways of the Paradox," Quine identifies three types of paradoxes: *veridical*, *falsidical*, and *the antinomy*. In a veridical paradox, a proposition that appears to be absurd at first is found to be true. When one discovers the reasoning that makes the proposition sound, the sense of absurdity quickly dissipates. The example that Quine gives is Frederick in *The Pirates of Penzance* who, despite being 21 years old, has experienced only five birthdays. When one realizes that Frederick, who was born on February 29, technically has a birthday once every leap year, the paradoxicality of the proposition is resolved. As Quine notes, veridical paradoxes tend to lose their paradoxical quality over time, as people are increasingly attuned to the underlying reasoning.
46. Marks, *China*, 1, 7, 398.
47. As quoted in Marks, *China*, 159.
48. Marks, *China*, 131. The introduction of quick-ripening and drought-resistant Champa rice varieties, widely planted by the twelfth century, was a major breakthrough in Chinese agronomy. See Bray, *Science and Civilisation in China*, 6.II.493.
49. Marks, *China*, 398.
50. Ibid., 316, 380n26, 327.
51. Callicott and Ames, *Nature in Asian Traditions of Thought*. Swearer, "An Assessment of Buddhist Eco-Philosophy." See also *Environmental Ethics* 8.4 (Winter 1986), a special issue on "Asian Traditions as a Conceptual Resource for Environmental Ethics."
52. Three volumes in this series are especially relevant to Chinese environmental ethics: Tucker and John Berthrong's *Confucianism and Ecology*, N. J. Girardot et al's *Daoism and Ecology*, and Tucker and Duncan Ryuken Williams' *Buddhism and Ecology*.
53. There is an abundance of Chinese-language scholarship on this topic as well. Ren Junhua and Liu Xiaohua's *A Cultural Explanation of Environmental Ethics* (*Huanjing lunli de wenhua chanshi*) and Lu Shuyuan's *The Space for Ecological Criticism* (*Shengtai piping de kongjian*) are particularly helpful in contextualizing Chinese sources in the broader study of environmental ethics.

54. See Tu, "The Ecological Turn in New Confucian Humanism" and Tu, "The Continuity of Being." Also worth mentioning is Chen Yexin's *Confucian Ecological Consciousness and Environmental Protection in Ancient China* (*Rujia shengtai yishi yu Zhongguo gudai huanjing baohu yanjiu*), which situates the emergence of Confucian concepts and practices of environmental stewardship in the pre-Qin and Qin-Han periods. See also Miller, *China's Green Religion*.
55. Essays representing the spectrum of views include: Goldin, "Why Daoism is not Environmentalism"; Cooper, "Is Daoism Green?"; Harris, "An American Appropriation of Buddhism"; and Sponsel and Natadecha-Sponsel, "Buddhist Views of Nature and the Environment."
56. Elvin, *The Retreat of the Elephants*, 323. Elvin credits Roetz, *Mensch und Natur im alten China*, 85. Others such as Rhoads Murphey, Vaclav Smil, Gary Snyder, and Karen Thornber have made similar arguments, as I will discuss later.
57. Elvin, *The Retreat of the Elephants*, 323. Original italics modified.
58. Thornber, *Ecoambiguity*, 1.
59. Ibid., vii. Aldo Leopold addresses an earlier instance of this problem in "Outdoor Recreation, Latest Model," which discusses the building of roads in nature parks for easy access by automobiles. See Leopold, *A Sand County Almanac*, 165–166, 176.
60. Thornber, *Ecoambiguity*, 9, 381, 417.
61. Thornber, *Ecoambiguity*, 19. See also Elvin, *The Retreat of the Elephants*, 323–324.
62. Thornber, *Ecoambiguity*, 3–4.
63. See White, "The Historical Roots of Our Ecological Crisis" and Passmore, *Man's Responsibility for Nature*. For a comparative study of Europe and China, including a response to White, see Tuan, "Discrepancies between Environmental Attitude and Behavior."
64. Murphey, "Asian Perspectives of and Behavior toward the Natural Environment," 35–36.
65. Thornber, *Ecoambiguity*, 19.
66. Murphey, "Asian Perspectives of and Behavior toward the Natural Environment," 55–56.
67. Smil, "China's Environment: Resilient Myths and Contradictory Realities," 174, 180.
68. Snyder, *The Great Clod*, xv–xviii.
69. Nixon, *Slow Violence*, 6–10.
70. Ibid., ix–x.
71. Ibid., 26.
72. Thornber, *Ecoambiguity*, 7, 26.
73. Ibid., 5.

74. Ibid., 15–16.
75. Nixon, *Slow Violence*, 100. Stegner's novel was first serialized in the company magazine *Aramco World*, published as a book in Beirut in 1971, and made available in the US by Selwa Press in 2007. On the controversy surrounding its publication, see Thompson, "Was This 'Discovery!' Meant to Be Found?"
76. Nixon, *Slow Violence*, 104, 107. See also Amnesty International, *A Criminal Enterprise*.
77. Thornber, *Ecoambiguity*, 21.
78. See Thornber, "Literature, Asia, and the Anthropocene" and Thornber, "*Wolf Totem* and Nature Writing," 892–894.
79. Thornber, "Literature, Asia, and the Anthropocene," 989.
80. As quoted in Marks, *China*, 159. See Chao, *Man and Land in Chinese History*.
81. Qu and Li, *Population and the Environment in China*, 174–175.
82. See Mallory, *China: Land of Famine* and Li, *Fighting Famine in North China*.
83. In 1957, the economist Ma Yinchu presented a study on controlling China's population, which was eventually labeled "Malthusian" by the Chinese Communist Party. Ma was forced to resign his official posts and withdrew from public life until his rehabilitation in 1979. See Shapiro, *Mao's War against Nature*, 39–45.
84. Guo, *Complete Works*, 3:251. See also Sun, "War against the Earth."
85. Marks, *China*, 312, 378–379n11. See Smil, *The Bad Earth*; Shapiro, *Mao's War against Nature*; and Dikötter, *Mao's Great Famine*.
86. Wilcke, "Kellogg to Build 8 Plants in China."
87. Smil, *China's Past, China's Future*, 115–116.
88. On renewable energy in China, see Lewis, *Green Innovations in China* and Tilt, *Dams and Development in China*.

Bibliography

Ames, Roger T. 1986. "Taoism and the Nature of Nature." *Environmental Ethics* 8 (4): 317–350.

Amnesty International. 2017. *A Criminal Enterprise? Shell's Involvement in Human Rights Violations in Nigeria in the 1990s*, 28 November. http://www.amnesty.org/download/Documents/AFR4473932017ENGLISH.PDF. Accessed 1 Oct 2018.

Barry, John. 2012. *The Politics of Actually Existing Unsustainability: Human Flourishing in a Climate-Changed, Carbon-Constrained World*. Oxford: Oxford University Press.

Basiago, Andrew D. 1995. "Methods of Defining Sustainability." *Sustainable Development* 3 (3): 109–119.

Bray, Francesca. 1986. *Science and Civilisation in China. Vol. 6, Biology and Biological Technology. Part II: Agriculture*, ed. Joseph Needham. Cambridge: Cambridge University Press.

Callicott, J. Baird. 1986. "The Metaphysical Implications of Ecology." *Environmental Ethics* 8 (4): 301–316.

Callicott, J. Baird, and Roger T. Ames, eds. 1989. *Nature in Asian Traditions of Thought: Essays in Environmental Philosophy*. Albany: State University of New York Press.

Chang, Chun-shu. 2007. *The Rise of the Chinese Empire: Nation, State, and Imperialism in Early China, Ca. 1600 BC–AD 811*. Ann Arbor: University of Michigan Press.

Chao, Kang. 1986. *Man and Land in Chinese History: An Economic Analysis*. Stanford: Stanford University Press.

Chen, Yexin. 2012. *Confucian Ecological Consciousness and Environmental Protection in Ancient China* 儒家生态意识与中国古代环境保护研究 (*Rujia shengtai yishi yu Zhongguo gudai huanjing baohu yanjiu*). Shanghai: Shanghai Jiaotong University Press.

Cheng, Chung-Ying. 1986. "On the Environmental Ethics of the Tao and the Chi." *Environmental Ethics* 8 (4): 351–370.

Chow, Rey. 2000. "Introduction: On Chineseness as a Theoretical Problem." In *Modern Chinese Literary and Cultural Studies in the Age of Theory: Reimagining a Field*, ed. Rey Chow, 1–25. Durham: Duke University Press.

Coggins, Chris. 2004. *The Tiger and the Pangolin: Nature, Culture and Conservation in China*. Honolulu: University of Hawai'i Press.

Cooper, J.P. 1978. "In Search of Agrarian Capitalism." *Past and Present* 80: 20–65.

Cooper, David E. 1996. "Is Daoism Green?" In *Morals and Society in Asian Philosophy*, ed. Brian Carr, 82–91. Richmond: Curzon Press.

Cronon, William. 1995. "The Trouble with Wilderness; or, Getting Back to the Wrong Nature." In *Uncommon Ground: Rethinking the Human Place in Nature*, ed. William Cronon, 69–90. New York: W. W. Norton.

Day, Kristen A., ed. 2016. *China's Environment and the Challenge of Sustainable Development*. New York: Routledge.

Di Cosmo, Nicola. 2002. *Ancient China and Its Enemies: The Rise of Nomadic Power in East Asian History*. Cambridge: Cambridge University Press.

Dikötter, Frank. 2010. *Mao's Great Famine: The History of China's Most Devastating Catastrophe*, 1958–1962. New York: Walker & Co.

———. 2015. *The Discourse of Race in Modern China*. Oxford: Oxford University Press.

Dobson, Andrew. 1998. *Justice and the Environment: Conceptions of Environmental Sustainability and Theories of Distributive Justice*. Oxford: Oxford University Press.

———. 2000. *Green Political Thought*. 3rd ed. London: Routledge.

Duara, Prasenjit. 1995. *Rescuing History from the Nation: Questioning Narratives of Modern China*. Chicago/London: University of Chicago Press.

———. 2014. *The Crisis of Global Modernity: Asian Traditions and a Sustainable Future*. Cambridge: Cambridge University Press.

Dynon, Nicholas. 2008. "'Four Civilizations' and the Evolution of Post-Mao Chinese Socialist Ideology." *China Journal* 60: 83–109.

Economy, Elizabeth. 2010. *The River Runs Black: The Environmental Challenge to China's Future*. 2nd ed. Ithaca: Cornell University Press.

Edmonds, Richard Louis. 2000. *Managing the Chinese Environment*. Oxford: Oxford University Press.

Ehrlich, Paul R. 1968. *The Population Bomb*. New York: Ballantine Books.

Ehrlich, Paul R., and Anne H. Ehrlich. 2009. "The Population Bomb Revisited." *The Electronic Journal of Sustainable Development* 1 (3): 63–71.

Elliot, Mark. 2012. "*Hushuo*: The Northern Other and the Naming of the Han Chinese." In *Critical Han Studies: The History, Representation, and Identity of China's Majority*, ed. Thomas S. Mullaney, James Leibold, Stéphane Gros, and Eric Vanden Bussche, 173–190. Berkeley: University of California Press.

Ellis, Erle C., and Navin Ramankutty. 2008. "Putting People in the Map: Anthropogenic Biomes of the World." *Frontiers in Ecology and the Environment* 6 (8): 439–447.

Elvin, Mark. 1970. "The Last Thousand Years of Chinese History: Changing Patterns in Land Tenure." *Modern Asian Studies* 4 (2): 97–114.

———. 1973. *The Pattern of the Chinese Past*. Stanford: Stanford University Press.

———. 1998. "The Environmental Legacy of Imperial China." *The China Quarterly* 156: 733–756.

———. 2001. "Three Thousand Years of Unsustainable Growth: China's Environment from Archaic Times to the Present." In *Environmental History in the Pacific World*. 1993, ed. J.R. McNeil, 29–68. Aldershot: Ashgate.

———. 2004. *The Retreat of the Elephants: An Environmental History of China*. New Haven: Yale University Press.

Elvin, Mark, and Ts'ui-jung Liu, eds. 1998. *Sediments of Time: Environment and Society in Chinese History*. Cambridge: Cambridge University Press.

Geall, Sam, and Adrian Ely. 2018. "Narratives and Pathways Towards an Ecological Civilization in Contemporary China." *The China Quarterly* 236: 1175–1196.

Girardot, N.J., James Miller, and Xiaogan Liu, eds. 2001. *Daoism and Ecology: Ways Within a Cosmic Landscape*. Cambridge: Harvard University Center for the Study of World Religions.

Goldin, Paul R. 2005. "Why Daoism Is Not Environmentalism." *Journal of Chinese Philosophy* 32 (1): 75–87.

Guha, Ramachanda. 1989. "Radical American Environmentalism: A Third World Critique." *Environmental Ethics* 11 (1): 71–83.

Guo, Moruo. 1982. *Complete Works of Guo Moruo* 郭沫若全集 (Guo *Moruo quan ji*). Beijing: Renmin wenxue chubanshe.

Harris, Ian. 1996. "An American Appropriation of Buddhism." *Buddhist Forum* 4: 125–139.

Hathaway, Michael J. 2013. *Environmental Winds: Making the Global in Southwest China*. Berkeley: University of California Press.

Jackson, Tim. 2009. *Prosperity Without Growth: Economics for a Finite Planet*. London: Earthscan.

Kirby, William C. 2005. "When Did China Become China? Thoughts on the Twentieth Century." In *Teleology of the Modern Nation-State*, ed. Joshua A. Fogel, 105–114. Philadelphia: University of Pennsylvania Press.

Lattimore, Owen. 1962. *Studies in Frontier History: Collected Papers, 1928–1958*. New York: Oxford University Press.

Legge, James. 1885. *The Sacred Books of the East. The Texts of Confucianism*, ed. F. Max Müller, vol. 27. Oxford: The Clarendon Press.

Leopold, Aldo. 1949. *A Sand County Almanac: And Sketches Here and There*. Oxford: Oxford University Press.

Lewis, Joanna I. 2013. *Green Innovation in China: China's Wind Power Industry and the Global Transition to a Low-Carbon Economy*. New York: Columbia University Press.

Li, Lillian M. 2007. *Fighting Famine in North China: State, Market, and Environmental Decline, 1690s–1990s*. Stanford: Stanford University Press.

Ling, Daxie. 1983. "Changes in Our Country's Forest Resources" 我国森林资源的变迁 ("Wo guo senlin ziyuan de bianqian"). *Chinese Agricultural History* 中国农史 (*Zhongguo nongshi*) 2: 26–36.

Lu, Shuyuan. 2006. *The Space for Ecological Criticism* 生态批评的空间 (*Shengtai piping de kongjian*). Shanghai: East China Normal University Press.

Luke, Timothy. 1995. "Sustainable Development as a Power/Knowledge System: The Problem of 'Governmentality'." In *Greening Environmental Policy: The Politics of a Sustainable Future*, ed. Frank Fischer and Michael Black, 21–32. London: Paul Chapman Publishing.

Ma, Xiaoying, and Leonard Ortolano. 2000. *Environmental Regulation in China: Institutions, Enforcement, and Compliance*. Lanham: Rowman & Littlefield.

Mallory, Walter H. 1926. *China: Land of Famine*. New York: American Geographical Society.

Marks, Robert. 1997. *Tigers, Rice, Silk, and Silt: Environment and Economy in Late Imperial South China*. New York: Cambridge University Press.

———. 2017. *China: An Environmental History*. Lanham: Rowman & Littlefield.

Meadows, Donella H., Dennis L. Meadows, Jørgen Randers, and William W. Behrens III. 1972. *The Limits to Growth: A Report for the Club of Rome's Project on the Predicament of Mankind*. New York: Universe Books.

Meadows, Donella H., Dennis L. Meadows, and Jørgen Randers. 1992. *Beyond the Limits: Confronting Global Collapse, Envisioning a Sustainable Future*. Post Mills: Chelsea Green Publishing.

Menzies, Nicholas K. 1992. "Strategic Space: Exclusion and Inclusion in Wildland Policies in Late Imperial China." *Modern Asian Studies* 26 (4): 719–733.

Miller, James. 2017. *China's Green Religion: Daoism and the Quest for a Sustainable Future*. New York: Columbia University Press.

Mullaney, Thomas S. 2012. *Coming to Terms with the Nation: Ethnic Classification in Modern China*. Berkeley: University of California Press.

Mullaney, Thomas S., James Leibold, Stéphane Gros, and Eric Vanden Bussche, eds. 2012. *Critical Han Studies: The History, Representation, and Identity of China's Majority*. Berkeley: University of California Press.

Murphey, Rhoads. 1967. "Man and Nature in China." *Modern Asian Studies* 1 (4): 313–333.

———. 2000. "Asian Perceptions of and Behavior Toward the Natural Environment." In *Landscapes and Communities on the Pacific Rim: From Asia to the Pacific Northwest*, ed. Karen K. Gaul and Jackie Hiltz, 35–57. Armonk: M. E. Sharpe.

Nixon, Rob. 2005. "Environmentalism and Postcolonialism." In *Postcolonial Studies and Beyond*, ed. Ania Loomba, 233–251. Durham: Duke University Press.

———. 2011a. "Slow Violence." *The Chronicle Review*, June 26. http://www.chronicle.com/article/Slow-Violence/127968. Accessed 20 Jan 2017.

———. 2011b. *Slow Violence and the Environmentalism of the Poor*. Cambridge: Harvard University Press.

Parr, Ben, and Don Henry. 2016. "A New Starting Point: China's Eco-Civilisation and Climate Action Post-Paris." Melbourne Sustainability Society Institute Briefing Paper 6, June. http://sustainable.unimelb.edu.au/__data/assets/pdf_file/0010/2756809/MSSI-Briefing-Paper6_China_2016_0.pdf. Accessed 4 May 2019.

Passmore, John Arthur. 1974. *Man's Responsibility for Nature: Ecological Problems and Western Traditions*. London: Duckworth.

Perdue, Peter C. 1987. *Exhausting the Earth: State and Peasant in Hunan, 1500–1850*. Cambridge: Council on East Asian Studies, Harvard University.

———. 2013. "The Chinese." In *Demystifying China: New Understandings of Chinese History*, ed. Naomi Standen, 15–24. Lanham: Rowman & Littlefield.

Pomeranz, Kenneth. 1993. *The Making of a Hinterland: State, Society, and Economy in Inland North China, 1853–1937*. Berkeley: University of California Press.

———. 2001. *The Great Divergence: China, Europe, and the Making of the Modern World Economy*. Princeton: Princeton University Press.

Qu, Geping, and Jinchang Li. 1994. *Population and Environment in China*. Trans. Baozhong Jiang and Ran Gu. Ed. Robert B. Boardman. Boulder: Lynne Rienner Publishers.

Quine, W.V. 1976. "The Ways of Paradox." In *The Ways of Paradox, and Other Essays*, 1–18. Cambridge: Harvard University Press.

Ren, Junhua, and Xiaohua Liu. 2004. *A Cultural Explanation of Environmental Ethics* 环境伦理的文化阐释 (*Huanjing lunli de wenhua chanshi*). Changsha: Hunan Normal University Press.

Roetz, Heiner. 1984. *Mensch und Natur im alten China*. Frankfurt am Main: Peter Lang.

Scott, James C. 2017. *Against the Grain: A Deep History of the Earliest States*. New Haven: Yale University Press.

Shapiro, Judith. 2001. *Mao's War Against Nature: Politics and the Environment in Revolutionary China*. Cambridge: Cambridge University Press.

Shi, Shumei, Chien-hsin Tsai, and Brian Bernards, eds. 2013. *Sinophone Studies: A Critical Reader*. New York: Columbia University Press.

Smil, Vaclav. 1984. *The Bad Earth: Environmental Degradation in China*. Armonk: M. E. Sharpe.

———. 2000. "China's Environment: Resilient Myths and Contradictory Realities." In *Landscapes and Communities on the Pacific Rim: From Asia to the Pacific Northwest*, ed. Karen K. Gaul and Jackie Hiltz, 167–181. Armonk: M. E. Sharpe.

———. 2004. *China's Past, China's Future: Energy, Food, Environment*. New York: Routledge.

Snyder, Gary. 2016. *The Great Clod: Notes and Memories on the Natural History in East Asia*. Berkeley: Counterpoint.

Sponsel, Leslie E., and Poranee Natadecha-Sponsel. 2003. "Buddhist Views of Nature and the Environment." In *Nature Across Cultures: Views of Nature and the Environment in Non-Western Cultures*, ed. Helaine Selin, 351–371. The Hague/London: Kluwer Academic Publishers.

Stegner, Wallace. 2007. *Discovery! The Search for Arabian Oil*. Vista: Selwa Press.

Sun, Xiaoping. 2017. "'War Against the Earth': Military Farming in Communist Manchuria, 1949–1975." In *Empire and the Environment in the Making of Manchuria*, ed. Norman Smith. Vancouver: UBC Press.

Sze, Julie, ed. 2018. *Sustainability: Approaches to Environmental Justice and Social Power*. New York: New York University Press.

Thompson, Bob. 2007. "Was This 'Discovery!' Meant to Be Found?" *The Washington Post*, November 29. http://www.washingtonpost.com/wp-dyn/content/article/2007/11/28/AR2007112802404.html. Accessed 1 Oct 2018.

Thornber, Karen Laura. 2012. *Ecoambiguity: Environmental Crises and East Asian Literatures*. Ann Arbor: University of Michigan Press.

———. 2014. "Literature, Asia, and the Anthropocene: Possibilities for Asian Studies and the Environmental Humanities." *Journal of Asian Studies* 73 (4): 989–1000.

———. 2017. "*Wolf Totem* and Nature Writing." In *A New Literary History of Modern China*, ed. David Der-wei Wang, 889–894. Cambridge: The Belknap Press of Harvard University Press.

Tilt, Bryan. 2010. *The Struggle for Sustainability in Rural China Environmental Values and Civil Society*. New York: Columbia University Press.

———. 2014. *Dams and Development in China: The Moral Economy of Water and Power*. New York: Columbia University Press.

Totman, Conrad D. 2004. *Pre-Industrial Korea and Japan in Environmental Perspective*. Leiden: Brill.

Tu, Wei-ming. 1984. "The Continuity of Being: Chinese Visions of Nature." In *On Nature*, ed. Leroy S. Rouner, 113–129. Notre Dame: University of Notre Dame Press.

———. 2001. "The Ecological Turn in New Confucian Humanism: Implications for China and the World." *Daedalus* 130 (4, Fall): 243–264.

Tuan, Yi-Fu. 1968. "Discrepancies Between Environmental Attitude and Behavior: Examples from Europe and China." *The Canadian Geographer* 12: 176–190.

Tucker, Mary Evelyn, and John Berthrong, eds. 1998. *Confucianism and Ecology: The Interrelation of Heaven, Earth, and Humans*. Cambridge: Harvard University Center for the Study of World Religions.

Tucker, Mary Evelyn, and Duncan Ryuken Williams, eds. 1997. *Buddhism and Ecology: The Interconnection of Dharma and Deeds*. Cambridge: Harvard University Center for the Study of World Religions.

Under the Dome 穹顶之下 (*Qiongding zhi xia*). 2015. Directed by Chai Jing. http://www.youtube.com/watch?v=T6X2uwlQGQM, 1 March. Accessed 20 Jan 2017.

United Nations, General Assembly. 2015. *Transforming Our World: The 2030 Agenda for Sustainable Development*, October 21, A/RES/70/1. http://sustainabledevelopment.un.org/post2015/transformingourworld/publication. Accessed 1 Oct 2018.

United Nations, World Commission on Environment and Development. 2009. *Our Common Future*. Oxford: Oxford University Press.

United States, Department of Energy, Energy Information Administration (EIA). 2016. *International Energy Outlook 2016*, May. http://www.eia.gov/forecasts/ieo/pdf/0484(2016).pdf. Accessed 20 Jan 2017.

———. 2017. *International Energy Outlook 2017*, September 14. http://www.eia.gov/outlooks/ieo/pdf/0484(2017).pdf. Accessed 30 Nov 2017.

———. 2018. *Energy Implications of China's Transition Toward Consumption-led Growth*, July 24. http://www.eia.gov/outlooks/ieo/china/pdf/china_detailed.pdf. Accessed 1 Oct 2018.

United States, Office of the Press Secretary. 2014. "U.S.-China Joint Announcement on Climate Change and Clean Energy Cooperation." *The White House*, November 11. http://obamawhitehouse.archives.gov/the-press-office/2014/11/11/fact-sheet-us-china-joint-announcement-climate-change-and-clean-energy-c. Accessed 20 Jan 2017.

Wang-Kaeding, Heidi. 2018. "What Does Xi Jinping's New Phrase 'Ecological Civilization' Mean?" *The Diplomat*, March 6. http://thediplomat.com/2018/03/what-does-xi-jinpings-new-phrase-ecological-civilization-mean/. Accessed 4 May 2019.

Watson, Burton. 2003. *Mozi: Basic Writings*. New York: Columbia University Press.

White, Lynn, Jr. 1967. "The Historical Roots of Our Ecological Crisis." *Science* 155: 203–207.

Wilcke, Gerd. 1973. "Kellogg to Build 8 Plants in China." *The New York Times*, November 28. http://www.nytimes.com/1973/11/28/archives/kellogg-to-build-8-plants-in-china-company-gets-new-orders-for-five.html. Accessed 1 Oct 2018.

Williams, Michael. 1990. "Forests." In *The Earth as Transformed by Human Action: Global and Regional Changes in the Biosphere over the Past 300 Years*, ed. B.L. Turner II, 179–201. Cambridge: Cambridge University Press.

Wong, Edward. 2015. "China Blocks Web Access to 'Under the Dome' Documentary on Pollution." *The New York Times*, March 6. http://www.nytimes.com/2015/03/07/world/asia/china-blocks-web-access-to-documentary-on-nations-air-pollution.html. Accessed 20 Jan 2017.

CHAPTER 12

Contemplating Land: An Ecocritique of Hong Kong

Winnie L. M. Yee

It is difficult to talk about ecocriticism in the context of Hong Kong, not because Hong Kong is immune to global climate change or disasters resulting from human activity.[1] On the contrary, more and more people are acutely aware of the pollutants entering the city from industries in southern China, more people are "going green," and discussions of animal rights have become much more prominent. Still, the lack of imminent danger seems to have lulled Hong Kong citizens into a comfortable sense that they are, at least temporarily, free from global climate change. This lack of ecocritical concern is the result not only of Hong Kong's escape from major disasters (unlike Japan and Taiwan, which are constantly under the threat of earthquakes) but also of the city's self-imposed cultural imagination, which has been premised on its economic success since the new prosperity of the 1960s. The economic "miracle" of Hong Kong suggests the successful integration of Hong Kong culture into a capitalist narrative where Euro-American values are globally dominant.

W. L. M. Yee (✉)
Department of Comparative Literature, University of Hong Kong,
Hong Kong, China

C. Chang (ed.), *Chinese Environmental Humanities*,
Chinese Literature and Culture in the World,
https://doi.org/10.1007/978-3-030-18634-0_12

Even though capitalist modernization was initiated from the outside, the people of Hong Kong embraced it. In his discussion of the globalization of capitalism, Harry Harootunian observes,

> capitalism was "born of colonization and the world market" and has subsequently "universalized" history, inasmuch as it has established systematic relations of social interdependence on a global scale that have eventually encompassed noncapitalist societies. In this regard, capitalism has managed to fix a standard of measurement—world time—produced by a "single global space of co-existence," within which action and events are subject to a single, quantifiable chronology. But because different social practices remain outside this abstract measure, capitalism has not "unified" history. (2000, 49)

Hong Kong is under the sway of capitalist discourse and its adherence to the model of economic success has become, according to theorist Ackbar Abbas, compensation for its lack of political power (Abbas 1992, 45). However, there is also the persistence of established social practices that Harootunian cites, a persistence that is captured in the emergence of organic communities that espouse a renewed relationship with nature that transcends and challenges Harrotunian's "single, quantifiable chronology." From the 1960s onwards, Hong Kong has experienced extraordinary economic development and rapid urbanization under British colonial authority and has emerged as a major metropolis in Southeast Asia (Lo 2010, 166). This economic success continues to shape and dominate the self-image of Hong Kong: the success of the city has become taken for granted. Its adherence to capitalism has continued beyond the official end of the colonial period and been reinforced by its position as an instrumental city in post-socialist China. This single chronology was not even challenged in 1997 when the sovereignty of the city was returned to the motherland; instead, it was further entrenched to ensure that Hong Kong would remain economically competitive and to support the consumer society.

Since the 1960s, the discourse of British colonizers and Hong Kong post-war boomers has focused on the rise of the "city" and has pointed to the rapid changes in urban Hong Kong as a mark of its superiority to the motherland. At the time that the Sino-British Joint Declaration was signed in 1984, the destiny of the city was decided without the participation of its inhabitants, who were left with an urgent need to redefine their home and their identity. Their concerns persist: Hong Kongers are still asking, what alternative model could be generated to replace or "improve" Hong Kong's modern, urban subjectivity? Is it possible to invent a different imaginary Hong Kong that would liberate itself from the self-indulgence of materialism? The insistence on the search for an alternative is in keeping with the subject of this volume—"the discourse of environing at the margin." While Hong Kong's land and cultural imaginary have been "encircled" and bounded by its dominant economic model, the practices of organic communities, which are the focus of this chapter, call into question the legitimacy of such practices and reimagine the land as a site of local creativity and eco-cosmopolitan imagination—a site where different ideologies can be contested and practiced. Mirroring the political struggle between local citizens and the Hong Kong Special Administrative Region government that has characterized the post-2008 period is the conflict between a commitment to a green sustainable lifestyle and the capitalist logic of urban development. The adherence to green living is widely evident, but we will focus on one manifestation: the formation of organic communities.

The emergence of organic communities, such as the PEACE (Partnership for Eco-Agriculture and the Conservation of Earth—活耕建養地協會) community farm, the Mapopo Community Farm (馬寶寶社區農場), and *Sangwoodgoon* (literally "House of Living" 生活館) heralds an integration of art appreciation (scholarly and aesthetic experience) and active engagement in farming activities (physical and material experience). *Sangwoodgoon*,[2] the focus of this chapter, has been the subject of several recent documentaries. It belongs to the category of "alternative communities," so called because they are not built on a foundation of familial lineage or cultural history. They began as an alternative to the urban lifestyle that is characterized by capitalist logic, developmental discourse, and a neglect of environmental awareness. They are also a direct result of the anti-government movement that began in 2006. Flexible in operation and small in scale, these communities challenge the massive production promulgated by capitalism and resist dependence on the importation of food.

The emergence of alternative communities in Hong Kong was precipitated by political measures that used patriotism and economic growth as justifications for dispossessing people of their homes and farmland. The sacrifice of individual benefits and natural resources was demanded in the name of governmental efficiency. North American ecocritics are familiar with this situation. In *Going Away to Think*, Scott Slovic discusses the importance of "ecocritical responsibility," by which he means "various forms of engagement and retreat, in all pursuits of 'responsibility,' in quest of meaningful response to the world as I experience it and gather information about it" (2008, 3). Ecocritical responsibility must characterize the relationship between humans and nature. Treating nature as an instrument under the capitalistic model has led to dire consequences: climate change, pollution, and the extinction of species, among others. Failure to learn from these consequences and a blind acceptance of the prolongation of Hong Kong's "economic miracle" will lead to a double marginalization. Hong Kong will be marginalized by Mainland China if it loses the advantage of providing a successful economic model for the motherland. Moreover, without a sustainable means of development, Hong Kong will become marginalized globally by its ignorance of, or indifference to, the shared efforts of the global community to create a more sustainable environment. Re-envisioning its relationship with land and promoting a sustainable lifestyle represent new ways of developing Hong Kong's identity and reimagining its future. This ecocritical turn embraces sustainable living, activism, knowledge dissemination, and art: it upholds the necessity of shouldering the "ecocritical responsibility" that will determine the future of Hong Kong.

Sangwoodgoon is the case study we use to present a vision of an alternative future for Hong Kong, one that encourages organic communities and green awareness. "Going green" conveys the intention to respect nature and to reconnect with the non-human world. *Sangwoodgoon* combines farming activities with films, art, and performance, creating a platform that connects the aesthetic portrayal of the land with the physical working of the land. Land is perceived as both a physical site and a rich terrain of imagination, one free of capitalist discourse and linear chronology. The combination of art appreciation and farm-related activities encourages members to reflect on life and environment, nature and development, the local and the global. In his discussion of ecocritical practice, Patrick D. Murphy draws attention to the Bakhtinian "dialogical concepts of answerability and anotherness," which address "the referential versus textual problematic." He maintains that "the dialogical concepts of answerability and anotherness provide a way of talking about how various movements within nature-oriented literatures ground their action and ground their readers in ethically referential situations aware of difference and responsibility" (2009, 33). To Murphy, the importance of nature-oriented literature should not be confined to its representation of nature; rather, it is the situation in which nature is depicted in the text that provides the condition for readers to come to terms with their relationship with nature in everyday life.

Members of the organic community of *Sangwoodgoon* exemplify—through their manner of living, dwelling, and production in the organic community—a "textual reference" for observers and the general public, who are encouraged to realize the significance of interconnection and to invest in their community. The relational politics of connectivity—to the community, to nature, and to the world as a whole—holds that "the local" is defined by, and involves an extension into, the sphere of "the other": it is both distant and proximal, abstract and concrete, rural and urban. This extension to the other sphere expands our vision of the organic community: to branch out and make connections becomes a mode of living. The emphasis on connectedness challenges the institutionalization of knowledge within academia: the proliferation of knowledge should occur in everyday life among the general public. The activities and practices adopted by the members of *Sangwoodgoon* are excellent examples.

To transcend the colonial model of developmentalism and to differentiate its mode of production from that of Mainland China, Hong Kong must expand its frame of reference. This chapter will focus on three aspects of *Sangwoodgoon* that encourage this expansion. The first aspect is practical: the permaculture farming adopted and practiced by farmers in *Sangwoodgoon* can be seen as an ecological and cultural attempt to forge a sustainable and harmonious relationship with the "other." The "other" here refers not only to the non-human or natural world but also to the motherland, with which Hong Kong is struggling to find a sustainable relationship. The second aspect is cross-cultural: the two Food and Farm Filming Festivals curated by the founders of *Sangwoodgoon* and Ying E Chi (a non-profit arts organization[3]) highlight cross-cultural fertilization through the promotion of documentary films around the world. The viewing experiences (including the venue, post-screening discussions, and other events) encourage cross-cultural dialogue. These cross-cultural interactions reinforce the importance of marginal movements by underscoring the shared concerns and similar problems encountered by the community-based activists around the world. The documentary has established its place as the medium where excluded subjects or marginalized groups can be seen and heard. The third aspect is scholarly and pedagogical: the community magazine, *Planting Hong Kong*, provides a forum for intellectual discussion, oral histories, and personal reflections. The magazine was founded by Yuan Yi-tin, an important figure in the recent rejuvenation of the farming industry in Hong Kong. He has been teaching people about permaculture for years through workshops offered at Mapopo Community. One of the founders of *Sangwoodgoon,* Chow Sze-chung learned farming skills from him. *Planting Hong Kong* deals with the histories of farmers and farming activities in Hong Kong and has acquired a wide readership.

The emergence of organic communities stems from an awareness of the stagnant urban discourse that has prevailed in Hong Kong since the economic upturn in the 1970s and that has been reinforced by both colonizers and the motherland. These communities promote an alternative lifestyle, stress cross-cultural connections, and seek to connect Hong Kong people through shared experiences. The movement to revive interest in ecology, farming, and the land is intended to provide Hong Kong people with a renewed sense of home and to create the conditions to perceive Hong Kong's future differently.

12.1 Farming as Living and the Tactics of Everyday Life

As the ecofeminist Greta Gaard points out, "An intersectional ecological-feminist approach frames … issues in such a way that people can recognize common cause across the boundaries of race, class, gender, sexuality, species, age, ability, nation—and affords a basis for engaged theory, education, and activism" (2011, 44). In Hong Kong, the links between ecocriticism and activism have become evident in recent years. Since the return of sovereignty to the motherland, activism in Hong Kong has been ignited by many issues, the most controversial being the construction of the Guangzhou–Hong Kong high-speed rail line, which was the source of civil resistance in 2009 to 2010. This express railway runs from a terminus in West Kowloon and heads north to the Shenzhen/Hong Kong border where it connects with the mainland. The Express Rail Link will eventually connect Hong Kong with the 16,000-kilometer People's Republic of China (PRC) National High-Speed Rail Network, the largest such network in the world.

In November 2008, Choi Yuen villagers received notice that they were required to vacate their village by November 2010. There were no meetings, discussions, or negotiations between the villagers and the government beforehand. The project affected approximately 150 households, representing a population of around 500 and three million square feet of agricultural land. Opposition to the rail line was initiated by media activists, who had been involved in similar campaigns, such as the preservation of the Star Ferry Pier in Edinburgh Place in 2006 and the Queen's Pier in 2007. At the outset, the campaign against the Guangzhou–Hong Kong line focused on saving Choi Yuen village, which was painstakingly built by its residents over four decades. The villagers' demands to stay in their homes went unheeded by the government. This was widely regarded as an example of the sacrifice of the interests of the common people to those of a small coterie of wealthy property developers (*"Xianggang huigui ershi nian," Lianhe yingyin*).

Sangwoodgoon was founded during the years of the movements protesting the Guangzhou–Hong Kong high-speed railway line and defending Choi Yuen village. The founders stated that democracy depends on whether citizens have the freedom to choose and create their own ways of living (*"Yi kuai tian gengzhuochu weilai," The Standnews.com*). The alternative lifestyle it offered, based on organic farming and permaculture, was a direct response to the government's lack of social and ecological responsibility.

According to David Holmgren, "Permaculture is a design system for sustainable living and land use. Its conceptual origins in Australia in the 1970s coincided with the first wave of modern environmentalism stimulated by the oil crises of 1973 and 1979" (2011, 92). It refers to "[c]onsciously designed landscapes which mimic the patterns and relationships found in nature, while yielding an abundance of food, fibre and energy for provision of local needs" (2002, xix). Ka-ming Wu's analysis of the Ma Shi Po Community Farm (also known as Mapopo Community) in Hong Kong demonstrates the importance of the permacultural methods that have been widely adopted in various farming communities in Hong Kong. Wu observes, "The concept of permaculture stresses the philosophy of working with the natural landscape and existing ecosystem while providing food for humans. It advocates the preservation of biodiversity, picking instead of killing bugs, practicing natural composting, and adopting alternate farming… in order to enrich the nutrients in productive land and utilize the least amount of resources for crop production…as a practice that connects urban communities to the rural while also reducing municipal waste" (177). In addition to its ecological and technological benefits, permaculture can be seen as an emblem of the benefits of co-existence.

Permaculture stresses the importance of a balanced ecology. Jenny Lee, one of the founders of *Sangwoodgoon*, observes that permaculture is more than a farming method; it is a life principle. It is about re-establishing a relationship with nature, with others, and with life (*"Dakai chengshi shenghuodi quekou," inmediahk.net*). Adhering to the belief that everything, even what is usually considered waste, is useful, the members of *Sangwoodgoon* use the dregs from the nearby soya bean manufacturing factories to make compost. The community has established a network of recycling throughout the neighborhood and with other food manufacturers, creating strong bonds between consumers and producers and among producers. Organic farming also allows and encourages consumers to question the means of growing and manufacturing food. Jenny Lee notes that this sense of accountability ensures that only healthy products are sold to the community ("*Yu tudide goutong*," *Student Paper of CUHK*).

In her discussion of the Permablitz Melbourne community, Tania Lewis refers to the work on relationality by geographers Samir Amin and Doreen Massey. Lewis points out that both geographers "have focused on the role of relationality… For both scholars, the relational politics of connectivity assumes that 'the local' necessarily is defined by and involves an extension into 'other' spaces, both distant and proximal, abstract and concrete" (Lewis, 351). In her analysis of ecological activism, Lewis argues that the term "lifestyle politics" should be used "to emphasize the role of style and aesthetics and the thorough imbrication of green living practices within a broader culture of lifestyle and consumption practices" (350). This emphasis on green living practices and the close relationship between theory and practice are also clearly reflected in the principles of the *Sangwoodgoon* farmers. The broad and nuanced connotations of "lifestyle" distinguish it from the narrow notions of consumerism that are prevalent. The emphasis on the lifestyle offered by organic communities in Hong Kong highlights the fact that these communities are not only sites of food production but also spaces in which experiences, attitudes toward the community, and new visions of Hong Kong's future are shared and passed on. When Chow describes the importance of achieving a balance of nutrients in the soil in order to grow a range of seasonal vegetables ("*Yu tudide goutong*," *Student Paper of CUHK*), he is also referring to ways to embrace a more inclusive way of life, one that considers all possibilities and allows for harmonious co-existence.

According to Chow Sze-chung, farming is a whole way of life. It involves a wide range of skills and techniques, and its influence extends into all aspects of living. It affects one's lifestyle, consumption patterns, and human relationships ("*Cong nongye gaizao Xianggang: Ji zhongzhi Xianggang fabuhui*," *Paratext*). As such, it offers a new way to remake or reimagine Hong Kong. The notion of sustainability through respect and diversity will help Hong Kong people realize that a healthy and enduring future for Hong Kong must be achieved by striking a balance. This balance must take into consideration all interested stakeholders and all aspects of their lifestyle, affections, and bonds with the community and the motherland. The practices of these farming communities represent "a shift in the nature and *form* of what we recognize as civic politics, from more formal, organized civic movements often tied to local and national structures of governance to embodied, everyday lifestyle practices" (Lewis, 352; emphasis in original). The attitudes toward planting and politics shared by the farmers are embedded in the temporal rhythms, habits, and routines of everyday domestic practices, not in the single chronology of modern capitalism.

12.2 The Film Festival as a Cross-Cultural Connection

Sangwoodgoon is devoted to food production, and, building on this foundation, its members initiated a Food and Farming Film Festival (FFFF) to raise awareness of food- and agriculture-related topics, both locally and globally, and to explore the most basic components of sustainable living. In their manifesto, the members maintain that,

> in the modernized, industrialized age, big corporations and governments are intertwined, taking away from each and every one of us... the chance to grow our own food and distancing us further away from the producers. We end up consuming what Alice Water sees as fast, cheap and easy food. Farmers who originally have the power to live sustainably and independently, have become the ones being exploited. (web[4])

Their aims are to promote sustainable farming as an alternative way of life and, through active engagement with farming, to encourage awareness of local environmental concerns and similar issues in Asia and around the world. By highlighting the shared responsibilities and exploitation of farmers everywhere, the film festival has become a space where people can reflect on their role in Hong Kong and the world.

Between 2014 and 2016, two *Sangwoodgoon* FFFFs have been held. In the first (from December 2014 to January 2015), six independent films from Japan, Taiwan, Thailand, the U.S., and Hong Kong were screened in a range of venues in Hong Kong (see Appendix I). The first festival screened seven documentaries on the farming industry and food movement. Covering a range of farmers from Taiwan, India, Japan, and Hong Kong, these films show the experiences of rural people from different cultural backgrounds facing various hardships. Some wage battles against the predations of big corporations; others devote themselves to the immensely difficult task of sustaining the farming industry in today's global world. Many lament the indifference of the general public to the exploitation of agricultural workers. Among the films shown in the first FFFF were *Bitter Seeds* (2011), a documentary on suicidal Indian farmers, *Struggles for Land: New Territories Developing Stories* (2014), which depicts the confrontations between farmers and real estate developers in northeastern Hong Kong, and *The Organic Dream* (2013), which explores the difficulties experienced in traditional Taiwan farming villages when farmers tried to switch to organic farming. All the documentaries explore communal and democratic solutions to looming problems. The broad range of subject matter highlights the curators' aim to connect the experiences of Hong Kong with those of similar communities and to stress the urgency of collective action.

For the second FFFF (from January 2 to January 30, 2016), "soil" was chosen as the theme. Closely following the International Year of the Soil (2015), the film festival aimed to stress the fundamental importance of soil as the origin of all sorts of life. ("*Nitu bingbu laofeng: shenghuoguan dierjie youzhong dianyingje*," *CINEZEN*) Five long and two short documentary films from Japan, Taiwan, the U.S., and Hong Kong were presented (see Appendix II). The films featured at the festival explore issues ranging from agriculture and genetically modified food to community struggles. A quick look at the titles shows that the aim of the organizers was to increase what Ursula Heise calls "eco-cosmopolitanism." They included *GMO OMG* (2013), which focuses on genetically modified food; *The Garden* (2008), which deals with the struggles of urban farms; *Magino Village: A Tale* (1987), which takes a popular science approach; and *The Symphony of the Soil* (2012), which portrays technologies and skills that are related to farming and planetology.

In her seminal work *Sense of Place and Sense of Planet*, Heise stresses the importance of realizing "how different cultural frameworks … condition quite divergent perceptions of what the local ecology consists of, what it requires from humans, or what an appropriate way of responding to it may be" (2008, 44, quoted in Oppermann 2010, 768). We need to understand local concerns when we think about ecology, and Heise's notion of eco-cosmopolitanism, which argues for "a more nuanced understanding of how both local cultural and ecological systems are imbricated in global ones" (2008, 59), is particularly instrumental. Heise's eco-cosmopolitanism should be viewed not as a call to minimize local considerations but as a means of seeing these considerations as elements in a Latourian network that includes culture, country, language, class, and ecology, among other categories. This positioning of Hong Kong's problems as part of a global phenomenon is crucial, as it encourages its citizens to realize that environmental activism is a reflection of structures, processes, and products that we actively engage with and consume in our day-to-day lives. Alternative communities see themselves as part of a planetary community.

The organizers of the FFFF maintain, "Documentary is not only a way for agriculture and farmers to express themselves; it is even a form of movement, of engagement and of organization for agriculture, farmers and food, for it is simultaneously analytical, intellectual, tasteful, sensual, passionate and revolutionary" (foreword, 1st *Sangwoodgoon* Food & Farming Film Festival brochure). In addition to airing the documentaries, the organizers invite groups to participate in a variety of related activities such as discussion panels, food and recipe sharing, farmers' markets, and guided tours. Bringing to bear these different perspectives on food suggests a dynamic process of material expression, evident in exchanges, bodies, and phenomena. The FFFF provides a space where material ecocriticism successfully integrates with its theoretical dimension—ecocritical activism.

12.3 Planting Hong Kong's Future and Green Activism

As well as curating the two film festivals, some members of *Sangwoodgoon* are also involved in the publication of a magazine, ***Planting Hong Kong*** (2016–), which is funded entirely by the editors. Its aim is to engage Hong Kongers in academic and social debate to determine the future of the city. Most of the editors and contributors are organizers of *Sangwoodgoon* or farmers at Mapopo Community and key figures in the re-centralization of farming in Hong Kong. Focusing on farmers and agricultural activities and history in Hong Kong, the magazine reflects the initial vision of the organic communities—to raise the awareness of local citizens so that they will engage with nature and become environmentally responsible. The aim of the magazine is stressed in the preface to the first issue: founder Yuan Yi-tin outlines the grand narrative of Hong Kong and critiques excessive development at the expense of farmland. Using both carrot and stick policies, the Hong Kong government encouraged farmers to give up their land, which led to the decline of the farming industry. For Yuan, giving up farming and land is equivalent to giving away Hong Kong's unique identity ("*Zhongzhi Xiangangde Fangfa*," 3). He urges the government and all those concerned about the sustainable future of Hong Kong to redirect their attention to their relationship with farming and land (Ibid.).

The magazine features academic discussions of sustainable development in local farm villages, historical studies of the post-war Hong Kong agricultural industry, and profiles of farmers and college students belonging to the community. The first issue focuses on government policy and the future of the farming industry as a whole. There is a detailed historical record of Hong Kong's farming industry in the post-World War II period ("*Nongyu bushi qiong qinqi: huigu Xianggang zhanhou nongyue xilie*" (Farming Industry is Not Our Poor Relative: A Review of the Farming Industry in Post-World War II Hong Kong), 74–79) and an account of the fate of Chan Kwan Kei's seed shop ("*Chen Kunji caizhongxing yige dian: zongguan Xianggang nongye bianjie*" (The Nodal Point of Chan Kwan Kei Seed Shop: An Overview of the Boundary of Hong Kong's Farming Industry), 80–91). Both articles impart a sense of urgency in the face of the decline of farming—an urgency to review, research, and reflect on our disappearing history. "*Xianggnag huayuan chengshi chutan*" (A Preliminary Research on "Hong Kong's Garden City," 8–25) deals with a research project led by Edward Yiu to determine the feasibility of a sustainable lifestyle in the New Territories. Scanning the past and the future, the first issue aims to encourage readers to reflect on Hong Kong's fate beyond the ivory tower.

The second issue is devoted to the narratives of aging farmers. Interviews were conducted with ten farmers from various farming regions. Their stories provide readers with a broader picture of Hong Kong's rural past, which is often neglected or marginalized in the writing of Hong Kong history. The farmers recount the suffering, hardship, and sacrifices that they endured in order to contribute to the success of today's Hong Kong. The magazine provides a cultural and political platform for the principles of ecological citizenship, which incorporate environmental concerns for sustainability and commitment into everyday cultural practices and political work. It is not a leisure magazine promoting organic food: it is more clearly focused on raising collective awareness about the need to participate in designing Hong Kong's future through understanding the past. Like the art produced at *Sangwoodgoon*, which serves as a medium of communication within and between communities, the magazine, which is the result of the collective efforts of community farmers and activists, is a means of spreading the ecocritical message.

12.4 Conclusion

Ursula Heise insists that "The challenge for environmentalist thinking, then, is to shift the core of its cultural imagination from a sense of place to a less territorial and more systemic sense of planet" (2008, 55). It is too early to determine whether the alternative communities in Hong Kong will help shift its cultural image from a place-based economic metropolis to a less territorial component of the planet. Still, the conscious ecocritical turn advocated by these communities provides a platform to imagine Hong Kong differently. This platform is a malleable and flexible space where artistic production is imbued with physical and material experience; its aim is to effect a change in dominant ideological formations and social awareness. As Patrick D. Murphy notes, "Ecocriticism is a form of aesthetic praxis that is in itself not a form of activism in the narrow sense of direct, immediate, and local actions, but by being propagandistic and agitational contributes to the potential success of activism through its effect on social consciousness." (2015, 16) I would argue that there is a strong connection between ecocritical discussion and efforts to effect change: the flexible community of *Sangwoodgoon* represents such a convergence of ecocriticism and social activism. As the founder of *Sangwoodgoon* Chow stresses, "Farming is an autonomous lifestyle, it's personal, local, but also national, international, transnational and global. How to re-position Hong Kong in this whole new wave of neo-liberalism is an urgent task to handle" ("*Zhou Sizhong: houzhiminde di zizhuxing*," *multitude.asia*).

By adopting permaculture, the farmers of *Sangwoodgoon* stress the importance of a sustainable and harmonious relationship both among individuals and between people and nature. Film festivals highlight the connectivity among farmers across cultures, temporalities, and borders. In the process of establishing the community magazines, members of various farming groups have signaled a shift in the nature and form of social movements, from an organized and often hierarchical structure to a focus on everyday activities, folk knowledge, and lifestyles. This chapter argues that the emergence of organic communities during a period of continued appropriation of land by property developers and the adherence to the developmental model are extremely important developments. Through their everyday practices and lifestyles, their appropriation of art forms and film festivals, and their dissemination of knowledge by means of community magazines, these organic communities demonstrate the intricate relationship between environmental responsibility and civic movements.

Rather than emulating organized social movements that were justified by reasons and ruled by hierarchies, organic communities rely on everyday practices, which are more sustainable points of connectivity. It is hoped that the experience of Hong Kong will be discussed within a broader cultural context that includes its Asian neighbors and even the global community so that a more comprehensive environmental awareness will emerge (Tables 12.1 and 12.2).

Table 12.1 The first food and farming film festival 第一屆有種電影節

Film	*Region (Language)*	*Year*	*Duration*	*Director*	*Screening venue*
巨輪下的新界: 東北 Struggles for Land: New Territories Developing Stories	Hong Kong (Cantonese)	2014	106 mins.	原銘翹 Kelvin Yuen 麥嘉熙 Marco Mak	艺鵠 (灣仔軒尼詩道 365–367 號富德樓 14 樓) ACO, 14/F, Foo Tak Building, 365–367 Hennessy Road, Wanchai
行健村的有機夢 The Organic Dream	Taiwan (Hakka, Mandarin Chinese)	2013	74 mins.	許文烽 Col Boom-Fong	油街實現(北角油街 12 號) Oi! (12 Oil Street, North Point, Hong Kong)
小小的家 Abong/Small Home	Philippines (Ilocano, Kankanaey, Tagalog and English)	2003	111 mins.	今泉光司 Koji Imaizumi	艺鵠 (灣仔軒尼詩道 365–367 號富德樓14樓) ACO, 14/F, Foo Tak Building, 365–367 Hennessy Road, Wanchai
只因我們生活在地球:福岡正信印度之旅 For Living on the Planet Earth: Fukuoka Masanobu Goes to India	India (Indian, Japanese)	1998	61 mins.	今泉光司 Koji Imaizumi	馬寶寶社區農場 (粉嶺馬適路) Mapopo Community Farm, Ma Shi Po Village (Ma Sik Road, Fanling)
稻米之歌 The Songs of Rice	Thailand (Thai)	2014	75 mins.	Uruphong Raksasad	香港兆基創意書院 HKICC LSK School of Creativity
苦澀的種子 Bitter Seeds	India (English)	2011	88 mins.	Micha Peled	香港兆基創意書院 HKICC LSK School of Creativity
歲月河山—鄉曲 The Song of a Village	Hong Kong (Cantonese)	1980	45 mins.	黃敬強 Wong King-keung	生活館 (元朗八鄉錦上路謝屋村) Sangwoodgoon, Tse Uk Tsuen, Kam Sheung Road, Yuen Long

Table 12.2 The second food and farming film festival 第二屆有種電影節

Film	*Region (Language)*	*Year*	*Duration*	*Director*	*Screening venue*
Symphony of the Soil	U.S.A. (English)	2012	103 mins.	Deborah Koons Garcia	艺鵠 (灣仔軒尼詩道 365–367 號富德樓14樓) ACO, 14/F, Foo Tak Building, 365–367 Hennessy Road, Wanchai
The Garden	U.S.A. (English)	2008	80 mins.	Scott Hamilton Kennedy	土家(土瓜灣鴻福街 16 號地下)
有米真好 Pure Land	Taiwan (Mandarin Chinese)	2015	27 mins.	陳祥豪 Xiang-Hao Chen 林婉婷 Wan-ting Lin 徐冠軒 Kuan-Xuan Shiu	ToHome, 16 Hung Fook Street, To Kwa Wan, Kowloon
GMO OMG	USA (English)	2013	90 mins.	Jeremy Seifert	風車草劇團排練室(香港九龍觀塘大業街 11 號華凱發展大廈 8 樓) Windmill Grass Theatre (8/F, Kevin Wong Development Building, 11, Tai Yip Street, Kwun Tong, Kwoloon, Hong Kong)
飲食法西斯 Food Fascist	Hong Kong (Cantonese)	2013	30 mins.	葉文希Yip Man-Hay	生活書院 (新界大埔錦山 178 號生活書院—前佛教大光中學) School of Everyday Life (178 Kam Shan, Tai Po, New Territories)
牧野村千年物語 Magino Village: A Tale	Japan (Japanese)	1987	222 mins.	小川紳介 Shinsuke Ogawa	上水鄉土學社 (梧桐河畔華山村旁) SoIL (Society for Indigenous Learning), Sheung Shui (Beside Ng Tung River, near Wa Shan Village)
活在三里塚 The Wages of Resistance: Narita Stories	Japan (Japanese)	2014	140 mins.	代島治彥 Haruhiko Daishima 大津幸四郎 Koshiro Otsu	元朗八鄉錦上路 菜園新村 New Choi Yuen Village, Kam Sheung Road, Pat Heung, Yuen Long

Notes

1. Traditional characters are used in this chapter. As this chapter articulates the cultural phenomena of Hong Kong, the author finds it appropriate to keep the traditional Chinese characters for terms that are used and circulated in the city.
2. For a preliminary discussion of organic communities in Hong Kong, see "Green Activism in the City: Hong Kong Independent Ecofilms in the Post-2008 Era," *Situations* 11:1 (2018): 39–57.
3. Ying E Chi is a non-profit arts organization that was established by a group of Hong Kong independent filmmakers in 1997. As outlined in their website, the institution's mission is to "unite local independent filmmakers, as well as to distribute and promote independent films." Details can be found on their webpage: http://yingechi.org/about/
4. Further information can be found on their website: https://sangwoodgoon.wordpress.com/filmfestival/

Bibliography

Abbas, Ackbar. 1992. "The Last Emporium: Verse and Cultural Space." In *City at the End of Time*, ed. Ping-Kwan Leung, 43–59. Hong Kong: Twilight Books in Association with Department of Comparative Literature, University of Hong Kong.

Gaard, Greta. 2011. "Ecofeminism Revisited: Rejecting Essentialism and Re-Placing Species in a Material Feminist Environmentalism." *Feminist Formations* 23 (2, Summer): 26–53.

Harootunian, Harry. 2000. *History's Disquiet: Modernity, Cultural Practice, and the Question of Everyday Life.* New York: Columbia University Press.

Heise, Ursula K. 2008. *Sense of Place and Sense of Planet: The Environmental Imagination of the Global.* Oxford: Oxford University Press.

Holmgren, David. 2002. *Permaculture: Principles and Pathways Beyond Sustainability*, 2014. Victoria: Holmgren Design Services.

———. 2011. "Weeds or Wild Nature: A Permaculture Perspective." *Plant Protection Quarterly* 26 (3): 92–97.

Horton, Dave. 2006. "Demonstrating Environmental Citizenship? A Study of Everyday Life Among Green Activists." In *Environmental Citizenship*, ed. Andrew Dobson and Derek Bell, 127–150. Cambridge, MA/London: The MIT Press.

Lewis, Tania. 2015. "One City Block at a Time: Researching and Cultivating Green Transformations." *International Journal of Cultural Studies* 18 (3): 347–363.

Lo, Kwai-Cheung. 2010. "Liu Yichang and the Temporalities of Capitalist Modernity." *Journal of Modern Literature in Chinese* 10 (1): 162–176.

Murphy, Patrick D. 2009. *Ecocritical Explorations in Literary and Cultural Studies*. New York: Lexington Books.

———. 2015. *Persuasive Aesthetic Ecocritical Praxis*. London: Lexington Books.

Nichols, Bill. 2010. *Introduction to Documentary*. 2nd ed. Bloomington/Indianapolis: Indiana UP.

Oppermann, Serpil. 2010. "Ecocriticism's Phobic Relations with Theory." *Interdisciplinary Studies in Literature and Environment* 17 (4): 768–770.

Slovic, Scott. 2008. *Going Away to Think: Engagement, Retreat, and Ecocritical Responsibility*. Reno/Las Vegas: University of Nevada Press.

Wu, Ka-ming. 2017. "Farming Against Real Estate Dominance." In *Green Asia: Ecocultures, Sustainable Lifestyles, and Ethical Consumption*, ed. Tania Lewis, 169–183. London/New York: Routledge.

Yiu, Edward. 2016. "Xianggnag huayuan chengshi chutan." 香港「花園城市」初探 ("A Preliminary Research on 'Hong Kong's Garden City'"). *Planting Hong Kong: The Beginning of Autumn*: 8–25.

Yuan, Yi-tian. 2016. "Zhongzhi Xianggangde fangfa" 種植香港的方法 ("Some Ways to Planting Hong Kong"). *Planting Hong Kong: The Beginning of Autumn*: 2–3.

Zhou, Sizhong. 2016. "Nongyu bushi qiong qinqi: huigu Xianggang zhanhou nongyue xilie" 農業不是窮親戚: 回顧香港戰後農業系列 ("Farming Industry Is Not Our Poor Relative: A Review of the Farming Industry in Post-World War II Hong Kong"). *Planting Hong Kong: The Beginning of Autumn*: 74–9.

Zhou, Sizhong, et al. 2016. "Chen Kunji caizhongxing yige dian: zongguan Xianggang nongye bianjie" 陳坤記菜種行一個點; 縱觀香港農業邊界 ("The Nodal Point of Chan Kwan Kei Seed Shop: An Overview of the Boundary of Hong Kong's Farming Industry"). *Planting Hong Kong: The Beginning of Autumn*: 80–91.

Online Materials

"Dakai chengshi shenghuodi quekou" 打開城市生活的缺口 ("Open the Gap in Our Urban Life"). 2010. *inmediahk.net*, March 27. www.inmediahk.net/node/1006553. Accessed 21 June 2017.

"Yi kuai tian gengzhuochu weilai" 一塊田耕作出未來 ("Planting the Future Through the Land"). 2015. *The Standnews*, January 16. https://thestandnews.com/nature/一塊田-耕作出未來/. Accessed 28 Aug 2017.

"Yu tudide goutong" 與土地的溝通 ("Communicating with the Land"). 2015. *Student Paper of CUHK*, January 14. www.inmediahk.net/node/1030521. Accessed 28 Aug 2017.

"Zhou Sizhong: Houzhiminde di zizhuxin" 周思中: 後殖民地的自主性 ("Zhou Sizhong: the autonomy of the decolonized"). 2014. *multitude.asia*, July 31. http://multitude.asia/archives/156. Accessed 28 Aug 2017.

He, Dingzhao. 2017. "Xianggang huigui ershi nian" 香港回歸二十年 ("Twenty Years After the Reunion with Mainland China"). *Lianhe yingyin*, June 27. https://video.udn.com/news/711454. Accessed 28 Aug 2017.

Li, Zhuoqian. 2016. "Cong nongye gaizao Xianggang: Ji zhongzhi Xianggang fabuhui" 從農 業改造香港: 記《種植香港》發佈會 ("Transforming Hong Kong Through Farming: The Launch of Planting Hong Kong"). *Paratext*, December 19. http://paratext.hk/?p=300. Accessed 28 Aug 2017.

Zhou, Sizhong. 2015. "Nitu bingbu laofeng: shenghuoguan dierjie youzhong dianyingje" 泥土 並不老奉: 生活館第二屆有種電影節 ("We Cannot Take Soil for Granted: The 2nd Food and Farming Film Festival"). *CINEZEN*, December 31. www.cinezen.hk/?p=5128. Accessed 15 Sept 2016.

CHAPTER 13

The Intersection of Sentient Beings and Species, Traditional and Modern, in the Practices and Doctrine of Dharma Drum Mountain

Jeffrey Nicolaisen

The seven fig trees (moraceae ficus, deciduous arbor, also known as the white flesh fig and Philippine fig) in front of the Library & Information Center (LIC) are native to Dharma Drum Mountain. Master Sheng Yen called them the "Seven Tathagatas." Having taken root here over 100 years ago, all seven have been deemed rare and old, and placed on the official protection list by the New Taipei City Government.

Chinese characters are included where it may assist a reader familiar with Chinese to identify pronouns or specialized terms. Traditional Chinese characters are used to preserve the format of the original material. Pinyin romanization is used, but where an established proper name uses the Wade-Giles romanization system, the established form is used. In these cases, the pinyin romanization is included in parentheses.

J. Nicolaisen (✉)
Graduate Program in Religion, Duke University, Durham, NC, USA

C. Chang (ed.), *Chinese Environmental Humanities*,
Chinese Literature and Culture in the World,
https://doi.org/10.1007/978-3-030-18634-0_13

> *To protect these fig trees the LIC's original site was moved back and its orientation altered, creating a picturesque contrast between man-made and natural elements, as well as the stirring scenery that has become a Dharma Drum Mountain Complex landmark.*
> *The "Seven Tathagatas" have been infected by Phellinus noxius, a pathogen that causes brown root rot, which is common in tropical and semi-tropical forests. After trees are infected, they are usually cut down and burned to avoid further spread of the disease. But because of the special causes and conditions that surround these seven trees and Dharma Drum Mountain, in April of 2011 the organization opted to employ tree surgery instead. During the surgical operations, steel ribs were set up to support the trees and prevent the danger of collapse. The infected areas were then removed and the infected surrounding soil replaced (the diseased soil had to be fumigated). After the operation the trees need three to five years to recuperate, and it is hoped that the "Seven Tathagatas" will regain their original vitality, with a new lease on life.*
> —*Treating Trees in Distress*, signboard in front of trees at Dharma Drum Mountain

The language of this signboard is carefully crafted to reflect the deep concern for the preservation of the natural environment at Dharma Drum Mountain (*Fagushan*; DDM). The vocabulary of the sign reflects several different environmental concerns. On the one hand, the trees are "rare and old." They are also "natural elements" that contrast with "man-made elements." They contribute to the "stirring scenery" at DDM. The purposes described in the sign are not exhaustive of the diverse functions of the trees. Rather, they point to a multi-layered relationship of humans with the trees. This relationship with trees is also not in isolation. The sign does not only mention the *moraceae ficus* fig tree species, it also mentions *Phellinus noxius*, a fungus. This fungus is described as a "pathogen" that infects the trees. This terminology of disease contrasts with the Buddhist epithet of "Seven Tathagatas." *Tathāgata*, the Thus Come One, is a word that refers to a buddha. It is not clear that Sheng Yen's epithet actually suggests that he thinks the trees are buddhas, but it does denote reverence for the fig tree, as opposed to aversion for the "pathogenic" fungus. Neither the fungus nor the trees are animals, as understood by modern science, but there are a series of further relationships also suggested by this sign. The diseased soil was fumigated. Although the fumigation agent is not specifically identified, it seems safe to assume that the soil insects, nematodes, fungi, and other categories of species in the soil were killed to preserve these trees.

Although fungi and trees are not animals, nematodes and insects are. These animals are generally understood to be sentient beings in Buddhist discourse. Thus, in order to preserve seven non-sentient beings, many sentient beings were killed. Through this process, humans select certain species to preserve and certain species to destroy, which blurs the "contrast between man-made and natural elements." The "natural" elements are human-selected ones that appear in a human-engineered landscape. This problematizes words such as *natural* and *man-made* and reveals a complex series of relationships that does not fit into a single Buddhist or biological framework. The definition of terms such as *animal* and *sentient being* overlap but also require further examination.

Sheng Yen (1930–2009) is the founder of DDM located in New Taipei City, Taiwan. In his spiritual environmentalism, he described natural environmentalism in terms of the protection of both individual bodies of sentient beings, as well as species. This approach extends beyond a practice that is merely concerned with salvation of sentient beings to include a concern for the diversity of life on earth and the preservation of species as a discrete category. This approach, however, does not reject earlier notions that include all six realms of existence. In other words, two ways of environing collide at DDM. Sheng Yen must negotiate pre-modern Buddhism with the modern notion of environmental protection (*huanjing baohu*) in order to forge his spiritual environmentalism. In the practices and doctrine of DDM, modern notions of ecology and species thus do not replace but operate in a complex relationship with traditional notions that include ambiguous boundaries between animals, humans, and the sentient beings in other realms of existence. In this chapter, I show how the result of these complex relationships is an alliance between pre-modern Buddhism and modern science that redefines boundaries and margins to form a mutualistic alliance of cosmologies.

13.1 Animals

> Animal, (kingdom Animalia), any of a group of multicellular eukaryotic organisms (i.e., as distinct from bacteria, their deoxyribonucleic acid, or DNA, is contained in a membrane-bound nucleus). They are thought to have evolved independently from the unicellular eukaryotes. Animals differ from members of the two other kingdoms of multicellular eukaryotes, the plants (Plantae) and the fungi (Mycota), in fundamental variations in morphology and physiology. This is largely because animals have developed muscles and hence mobility, a characteristic that has stimulated the further development of tissues and organ systems.[1]

This definition from the *Encyclopedia Britannica* presents a biological definition of the word *animal.* In Chinese, this definition corresponds most closely with the word *dongwu*. This definition contrasts sharply with the Buddhist definition of animal. In Buddhism, the traditional definition of animal relates to the six realms of existence. These realms consist of hell-beings, hungry ghosts, animals, asuras, humans, and devas. In Chinese, the word *chusheng* refers to those beings that inhabit the "animal" realm. The word *chusheng* better correlates to the word *beast* in English because *chusheng* live a life of suffering due to karmic consequences from a previous life.[2] Another significant distinction is that, biologically speaking, humans are subsumed within the category of *dongwu*, whereas in Buddhist cosmology, *chusheng* and humans abide in distinct realms because of karmic conditions. Humans have earned their birth in the human realm through positive karma from previous lives and thus experience less suffering than animals. Finally, in Buddhist cosmology, all distinctions are much more fluid. Creatures in all six realms of existence are sentient and are subject to the cycle of rebirth. Distinctions are only temporary, and each sentient being may be born in any one of the six realms in its next life, depending on her karma—just as she has been reborn many times in the past.

From an ethical standpoint, the five precepts[3] and the *vinaya*—the code of conduct for monks—prohibit killing. However, in some versions of the vows, the severity of the offense is different for animals and humans. For example, in the Pali Theravadin Vinaya, there is a distinction between killing animals and humans. Killing a human is one of the four grave offenses called *pārājika* (*siboluoyi*) offenses. If a monk or nun commits one of these offenses, he is expelled from the sangha. However, killing an animal is an offense that requires expiation. This type of offense is called a *prāyaścittika* (*boyiti*) offense and requires confession to three other monks or nuns. Therefore, killing a human is a much more serious offense then killing an animal, but nonetheless, both forms of killing are offenses.[4] On the other hand, Mahayana vows varied. For example, the *Brahma Net Sutra* prohibited the killing of all sentient beings equally.[5]

At DDM, someone who accepts the full set of lay precepts, including the five precepts known as the Bodhisattva Precepts, also strives to awaken all sentient beings.[6] Thus, DDM initiates vow to engage in a special relationship with all sentient beings, including animals. They must assist animals in the process of awakening.

The biological definition of animals thus differs from the Buddhist definition in several key ways. An animal is generally defined based on its morphology and physiology and perhaps is most distinguished by muscular features that allow it to move. This type of animal is defined by its material characteristics. The definition suggests nothing about whether an animal is sentient, and it implies nothing about its history of deeds, whether it can be reborn, or whether it possesses the capacity to achieve realization. On the other hand, Buddhist texts often emphasize sentience and rebirth over detailed morphological and physiological traits. For example, the twentieth vow of the *Brahma Net Sutra* admonishes disciples of the Buddha to reflect, "throughout the eons of time, all male sentient beings have been my father, all female sentient beings my mother. I was born of them. If I now slaughter them, I would be slaughtering my parents as well as eating flesh that was once my own."[7] The physiological and morphological details of a sentient being are temporary karmic conditions, and presume a moral relationship with other sentient beings. Animals also possess the ability to achieve buddhahood, although that may require being reborn as a human first. The biological definition is better suited for identification, whereas the Buddhist definition is related more to identifying a karmic status and moral obligation. However, the lack of emphasis on how to identify an animal in Buddhism does not necessarily imply that the biological definition, which focuses more on identification, is suitable in identifying sentient beings for Buddhists.

13.2 Are Trees Sentient?

In May and June of 2014, I spent five weeks at DDM to conduct research. I was a visiting graduate student and lived in the student dormitories at Dharma Drum Buddhist College. Dharma Drum Buddhist College was a secular university that offered undergraduate, masters, and doctoral degrees in Buddhist Studies. It has since been renamed Dharma Drum Institute of Liberal Arts, as it is opening a new campus that will offer majors in a liberal arts curriculum in addition to Buddhist Studies. Dharma Drum Buddhist College is located at the DDM World Center for Buddhist Education. The complex also includes a four-year monastic training university called Dharma Drum Sangha University and a practice center for lay people. While at DDM, I interviewed staff, professors, and students and observed the practices there. In interviews, I asked about environmental practices and philosophy, as well as the global activities of the organization.

When I asked monks, nuns, students, and professors at DDM about the environmental practices there, many of them discussed the great effort which has been committed to preserving trees. One nun provided a very detailed account of the tree transplanting process and justification for preserving the trees. In regard to the seven fig trees—the Seven Tathagatas—she explained that the library was originally planned to be built in the location of the seven fig trees in front of the current library, but the plan for the building was set back so as not to damage the trees. She further explained that when trees grow large enough, they can provide a dwelling place for formless spirits. The trees in front of the library are the home to 70 billion spirits. Sheng Yen protected the trees to protect the spirits. Now those spirits may serve as dharma protectors for Sheng Yen and DDM. Many monastics have had dreams about the trees prior to becoming monastics. Without being familiar with Sheng Yen or DDM, these men and women have found their way to DDM, drawn by the trees.

The nun explained that Sheng Yen compared the location of DDM to a Ming chair. The location of the campus has one tall mountain behind it and a mountain on each side. There are two streams that flank both sides of the campus. Originally, these were too narrow for good *fengshui* (the proper flow of energy, or *qi*, through the landscape), but in the late 1990s, an earthquake and typhoon caused the two streams to become wider. The nun explained that it was the dharma protectors that caused these streams to widen and improve the *fengshui*.

In another case, the nun explained DDM was beginning construction for the new campus, but there was an old tree there. The tree was to be moved from its location at the site of the planned campus to a spot in front of the administrative building. Sheng Yen knew the tree did not want to move, so he spoke to it the day before it was to be moved. Then just before it was moved, the monks and nuns gathered around it to chant. When it was moved, it was loaded on a 14-ton truck, but the truck got stuck. Sheng Yen spoke to the tree again. A 50-ton truck was brought in to move it the next day. When it got to the hole where it is was to be replanted, it again would not come off the truck. Sheng Yen spoke to it, and it eventually came off the truck. It would not go in the center of the hole, but chose a spot where it was more comfortable. Later, the nun who told me this story met a Jewish woman who studied Native American religion and had come to visit DDM. The woman took a look at the tree and said it was the most powerful tree she had seen. It was sick, but she hoped it would become well again.

These stories are based on one nun's account, and represent one of many perspectives. When I asked another nun about the story, she seemed rather taken aback by the story. While the first nun said Taiwanese regulations prohibit altering rivers—suggesting that the rivers must have been widened by the earthquake and typhoon—the second nun noted that stones had been placed by the river for river improvement. Indeed, the rivers are heavily engineered. Both nuns referred me to a video that captured the tree transplanting process.

In the video, Liao Chiu-Cheng (Liao Qiucheng), Associate Professor of Forestry at National Chiayi University, narrates the transplanting process. The process begins with the cutting of roots on October 12, 2003, followed by the actual transplanting on June 25, 2004. The tree that is transplanted is a *luhua tree* (*luhua shu*, *Scolopia oldhamii* Hance) that is nearly 100 years old. The video focuses on the technical aspect of the move, describing and showing the technology in detail. It shows the tree being lifted onto a flatbed of a tractor trailer by crane; the tree is pulled to the new location; and the tree is dropped into its new place by a crane. It does not show the monks or nuns chanting or Sheng Yen talking to the tree. The video begins and ends with a statement by the President of Dharma Drum Buddhist College, Zeng Jiqun. President Zeng explains that DDM is preserving the primary forest on its campus and is explicit that the trees are sentient:

> According to the spirit of our [form of] Buddhism, living things are sentient. Even if they are trees, they still have the same sentience.

Therefore, although the video focuses on the technical side of the transplant, it is clear that trees are viewed as sentient. At the end of the video, President Zeng states:

> We at Dharma Drum Mountain emphasize environmental protection—the meaning of environmental protection. We think that each and every element of the natural world has meaning in its survival. Thus, in this transplant process, we want to provide a very good example for the people of the world. That is, living things are sentient. Toward all living things, we all wish to say [we] can respect them, cherish them, and relate with them. That is a big and important message for society to be drawn from this transplanting of the primary forest. [That] is what Sheng Yen often emphasizes.[8]

When President Zeng discusses the natural world, he talks about it collectively. When he mentions the survival of each and every element, he talks about the meaning of their survival, rather than individual spirits. Even when he mentions the sentience of the forest, it is the collective of trees together that possess sentience, which might imply a different type of sentience than the sentience of individual beings.

The nun who described the 70 billion spirits spoke in a very different way. To her, the spirits were discrete, active agents. The *luhua tree* itself had the power to actively resist movement, and the spirits that lived in the Seven Tathagatas were able to actively protect DDM. They were able to widen the rivers to improve the *fengshui* at DDM. Both the view of a sentient collective and the view of plants as individual active agents have precedence in the Buddhist literature. I discuss briefly the active role of plants in this section, and discuss the collective view in a later section.

The nun's explanation suggests that some trees possess sentience or they provide an abode for a number of formless spirits. This sort of story is not unprecedented in Buddhist literature. For example, there are several stories in the *Jātaka* tales—stories of the Buddha's previous lives—in which the Buddha appears as trees or even a tuft of *kusa* grass.

In the *Kusanjāli Jātaka* tale, the Buddha is a tuft of *kusa* grass that dwells at the base of a wishing tree. King Brahmadatta of Banaras had only one pillar supporting the roof of his palace, and the pillar had become shaky. He sent some carpenters out to prepare a new pillar. They searched all over for a suitable tree, but the only suitable tree was the wishing tree. Despite the carpenter's hesitation to cut down the wishing tree, the King ordered them to cut it down. They begged the tree for forgiveness, and told her they would come back the next day to cut her down. The wishing tree burst into tears. The following day, the Buddha-grass transformed into a chameleon and moved up the side of the tree to make it look like it had holes in it. When the carpenters saw the tree was full of holes, the carpenters decided not to cut it down because it was rotten. After telling this story, the Buddha reveals that he was the *kusa* grass sprite, and Ānanda was the tree sprite.[9]

In the *Rukkhadhamma Jātaka*, the Buddha is the spirit of a sal tree in the Himalayas. He admonishes the trees, shrubs, bushes, and plants of his family to take up abode in the forest. Some vegetative spirits, however, defied his advice and inhabited giant trees in open spaces. One day, a storm snapped the branches and upturned the roots of the trees in open spaces, while the trees in the forests were left undamaged. The spirits who had lost their homes, weeping, took their children in their arms, and set out for the Himalayas.[10]

In both tales, the sprites and spirits are active agents, but in the first tale, the future Buddha and the future Ānanda function as if their bodies are the kusa grass and the tree. This is similar to how the nun at DDM described the *luhua tree* resisting those that were moving it. In the second story, the spirits are mobile. They take the trees as homes and are able to leave when their homes are destroyed. This is similar to how the nun described the spirits reside in the Seven Tathagatas.

From the multiple interviews I have conducted at DDM, this nun's account of the tree spirits and the dharma protectors is the only description of its kind—although I did not ask all my informants specifically about the trees. This could be because it represents only one nun's understanding, but considering that there was an organized chant and Sheng Yen spoke to the tree, this view may be more widespread. The video did not include the narrative the nun described, so this view at least does not appear to be the understanding that DDM is promoting in their official media. However, the *Jātaka* tales demonstrate that the nun's account is not an unprecedented one, and it does represent at least one of the many perspectives within the DDM organization.

It is not completely clear how formless spirits and sentient trees fit into the six realms of existence. There is not a vegetal realm. They could belong to a number of realms, devas perhaps or hungry ghosts, but they do seem to be sentient. From a Buddhist perspective, the quality of sentience is key to determining proper conduct. Since they are formless, they cannot be directly killed, but on the other hand, judging from the nun's account, preserving their home does help these beings and bring good fortune to DDM. These relationships also reflect a set of priorities in relation to animals. It is possible that maintaining good relationships with these spirits is a tradeoff for the killing involved in fumigating the roots of the Seven Tathagatas.

While the case of the Seven Tathagatas demonstrates the broad and ambiguous range of sentient beings, it does not directly address the realm of the beasts. One area in which there exists a significant literature directly related to animals is the recent discourse on the animal release (*fangsheng*) ceremony. In the next section, I briefly discuss the *fangsheng* discourse to identify and develop the varied views on animals.

13.3 Animal Release

In 1999, Lucia Liu Severinghaus and Li Chi published an article in the journal *Biological Conservation* that demonstrated that the practice of *fangsheng* was common in Taiwan. They brought the attention of the scientific community to a mode of introduction of invasive species that had previously not drawn much attention from scientists. Animals, including non-native turtles and fish, were being bought from pet stores or local markets to be released. In a global market, pet stores and markets were carrying more species non-native to the region. Later articles confirmed this practice was not confined to Taiwan. For example, in a 2012 article, Liu, McGarrity, and Li demonstrated that *fangsheng* has been a significant mode by which breeding populations of non-native bullfrogs have been established in Yunnan Province, China.[11]

However, by the time of the publication of the Severinghaus and Li article, leaders of the Taiwanese Buddhist community were already imploring followers not to participate in the ritual in the way that it had frequently been practiced. For example, the largest Buddhist organization in Taiwan is Buddha's Light Mountain (*Foguang Shan*), and the founder of Buddha's Light Mountain Hsing Yun (*Xingyun*) criticized the way *fangsheng* was being conducted.

In 1998, Hsing Yun wrote, "Freshwater fish are set free in the ocean, and saltwater fish are released in ponds or streams. People have even released poisonous snakes where they can harm people... To pay such intensive attention to releasing life is not to release life. It may go by the pretty name 'releasing life,' but, in truth, it is not moral. One ought to promote the release of life according to conditions, not according to prearranged design."[12] In his book *Chan de Shijie* published in 1999, Sheng Yen also implores his followers not to participate in animal release. He cites ecological reasons similar to Hsing Yun's about how animals are released into unsuitable environments, but he also argues against the suffering of individual animals as well. For example, he provides an example of how people will buy a turtle to release, but the turtle endures the suffering of having the releasers' initials carved on her shell. He continues that these turtles are sometimes captured again, and rereleased, whereas if they were never caught and rereleased at all they would survive just fine. As an alternative, Sheng Yen proposed that money that would otherwise be spent on *fangsheng* be donated to the rehabilitation of endangered species, so they can be nursed back to health and returned to the wild.[13]

The ceremony had also been critiqued in the scholarly religious studies literature, especially related to Japan. There have been a variety of views expressed in both popular and scholarly literature that have associated Buddhism and Japanese culture with an affinity for nature as opposed to the Western dominion over nature. In response, some scholars have used the *fangsheng* ceremony to demonstrate that the Japanese affinity for nature is a modern construct that does not consider the considerable countervailing evidence. For example, Duncan Williams examines the most well-known medieval site for the ritual, Iwashima Hachiman Shrine, and demonstrates that the ritual was used by the Shogunate—Japan's military authorities—to pay obeisance to a powerful potential adversary in order to maintain the status quo. The result was that many fish were caught to be rereleased with as many as two-thirds dying in the process.[14] Barbara Ambros argues that *fangsheng* and other Buddhist rituals historically tended to relate to the propitiation and salvation of animals, rather than care for them in the present life. It was more important that animals attained a good rebirth than that their welfare was improved in their current life.[15]

Neither of these historical arguments fits Sheng Yen's position. Sheng Yen's descriptions are not only related to the welfare of individual animals but also to the preservation of the ecosystem. There is an overlap between these two purposes, but they are both simultaneously present. It is clear that the reasons that Sheng Yen offers to critique the modern practice of *fangsheng* relate to the welfare of animals in their current life. He is concerned that an animal is introduced to an environment where it can thrive—for example, a cold-water fish is not introduced to a warm water stream. He also rejects the notion that an animal should be caught and rereleased just for the purpose of performing the *fangsheng* ritual. The alternative he proposes of nursing and rearing endangered species also provides benefits to the individuals of the endangered species. Yet, at the same time, his concerns also entail consideration for the health of the ecosystem. Releasing animals into the ecologically appropriate ecosystem also presumably enhances the viability of the ecosystem. This is even clearer in the fact that he does not suggest nursing any unhealthy animal back to health but specifically endangered animals. This requires thinking about animals as more than individuals—not only in terms of the health of another sentient being but also in terms of the robustness of the species. Thus, Sheng Yen promotes both individual animal welfare and species-level conservation practice.

This view, particularly the species-level conservation component, develops in response to modernity. Sheng Yen himself adopts the language of modern ecology. Thus, his critique of the *fangsheng* ritual is stated in modern terms in response to modern problems, but it is rooted in Buddhist doctrine. Instead of focusing on salvation as Ambros suggests has been the historical focus of *fangsheng*, Sheng Yen focuses on ecological conservation and animal welfare. Sheng Yen's interest in ecological conservation is demonstrated by his concern for the fitness of an animal to a particular environment and the preservation of endangered species. This interest fits neatly into a conservation biology paradigm as represented by scientists such as Severinghaus and Li who have criticized the *fangsheng* ritual as a mode of entry of invasive species. At the same time, Sheng Yen is also concerned with the welfare of individual animals. Both of these views find precedence in Buddhism. The concern for the welfare of sentient beings was discussed above, and the concern for the integrity of the ecosystem as part of the Dharmabody is discussed below. These two concerns, however, also often clash.

To illustrate this point, the Seven Tathagata trees provide multiple functions to the ecosystem. If they die, birds may lose their place to nest, caterpillars may be deprived of their food source, and the atmosphere would lose another source of oxygen. In short, the ecosystem would be altered. Moreover, if the entire primary forest that DDM occupies was eliminated, the new ecosystem would be entirely different and likely support a less diverse group of species. Therefore, fumigating the roots of the tree may kill some individual insects, nematodes, fungi, and other organisms, yet by saving the tree, not only will the trees be able to continue to support the species it has been supporting, but nematodes and insects that live in the soil will be replaced in the future by individuals of the same species. If the surgery is successful, only the species *Phellinus noxius* described as the "pathogen" on the signpost in front of the Seven Tathagata will be eliminated.

13.4 Sheng Yen's Ecological Views

In order to understand Sheng Yen's broader ecological views, it is also important to understand a broader movement of which he is a part, Humanistic Buddhism. This form of Buddhism is generally attributed to the great Chinese reformer Tai Xu (1890–1947) and his student Yin Shun (1906–2005), who coined the term *Humanistic Buddhism* (*renjian fojiao*) and brought the movement to Taiwan.[16] Three of the four major Buddhist organizations in Taiwan—Buddha's Light Mountain, Tzu Chi (*Ciji*) Compassion Relief, and DDM—are all associated with the movement. Even some monastics in the fourth organization, Chung Tai Chan Szu (*Zhongtai Chan si*), also claim their organization is a form of Humanistic Buddhism.[17] The Humanistic Buddhism movement puts much more energy into education and charitable service and focuses more on this-worldly activities than the afterlife. One component of this is the Pure Land on Earth[18] (*renjian jingtu*) doctrine. Unlike Amitābha's Pure Land Sukhāvatī, the Pure Land on Earth is on this earth, in this present life. Thus, a practitioner does not need to wait to be reborn to another Pure Land, she only needs to purify her mind in order to see that this world already is a Pure Land. One of the doctrinal foundations for this type of Pure Land is the Vimalakīrti Sutra, from which Sheng Yen often quoted the phrase "When the mind becomes pure, the Buddha land also becomes pure."[19]

According to Sheng Yen, although the best way to purify the mind is traditional Buddhist practice, those who are not ready for a more traditional practice can practice the Four Environmentalisms.[20] These Four Environmentalisms—spiritual environmentalism, social environmentalism, living environmentalism, and natural environmentalism—translate Buddhist concepts into non-sectarian language. Underlying these Four Environmentalisms is the idea that each part of the universe is intimately connected. In his 2000 speech at the Millennium World Peace Summit of Religious and Spiritual Leaders, Sheng Yen described this environmentalist message:

> The Buddha told us in the sutras and precepts that we should take loving care of animals, and that we should not harm the grass and trees, but regard them as the home where sentient beings lead their lives. In the stories of the Buddha's past lives, when he was following the Bodhisattva path, he was once reborn as a bird. During a forest fire, he tried fearlessly to put out the fire, disregarding his own safety by bringing water with his feathers. In the Avatamsaka Sutra it is said that mountains, waters, grass, and trees are all the manifestation of the great bodhisattvas. So, Buddhists believe that both sentient beings and non-sentient things are all the Dharmabody of the buddhas. Not only do the yellow flowers and green bamboo preach Buddhist teachings, but rocks can also understand Buddhist doctrines. Therefore, Buddhists regard our living environment as their own bodies. The Buddhists' life of spiritual practice is by all means very simple, frugal, and pure.[21]

Through this passage, Sheng Yen portrays a kind of sentience of the entire universe, including non-sentient beings such as rocks and waters. The trees are not sentient beings, but they are home to sentient beings. At the same time, as part of the Dharmabody, even rocks can understand the dharma. The Dharmabody to which Sheng Yen refers has several different doctrinal explanations. One explanation is its existence as one of the three bodies (*trikāya*, *sanshin*), which consists of the dharma body (*dharma-kāya*, *fashen*), reward body (*saṃbhoga-kāya*, *baoshen*), and transformation body (*nirmāṇakāya*, *huashen*). The dharma body is the body of the buddha that encompasses the entire universe due to transcendence of self. It can be equated with emptiness or buddha nature. The reward body is the joy body derived from the merits of the bodhisattvas. The transformation body is the body that enables a buddha to appear in the world and teach other sentient beings. The three bodies are not separate but unified. According to this understanding, the mountains, waters, grass, and trees appear as part of the transformation body in order to teach sentient beings, but they are simultaneously part of the Dharmabody, which encompasses the whole universe. What Sheng Yen is establishing here is that all beings are part of the same body, so that hurting the environment is hurting oneself. Individual sentience beings possess an individual sentience, but they also are part of the Buddhahood of the entire universe.

This is possibly what underlies statements of President Zeng in his remarks about moving the *luhua tree* quoted earlier. He says that even trees are sentient. This sentience is perhaps through the universal buddhahood of the universe. Indeed, one of the phrases that introduce the video of the transplanting of the tree is "Build a Pure Land on Earth."[22] The project of moving the trees is not viewed in isolation. It is viewed as creating a Pure Land on Earth. President Zeng speaks of trees in the plural, almost as the entire primary forest itself, which suggests he is thinking more of a collective sentience.

While on the one hand Sheng Yen and President Zeng discussed Buddhist conceptions of sentience, Sheng Yen also adopted the modern language of species. To Sheng Yen, these species are like organs of an even greater Dharmabody. The interests of individual animals, species, and sentient beings of other realms do not always appear to be the same. It is clear that in the case of the Seven Tathagatas, the surgery and fumigation killed fungi, nematodes, insects, and possibly other organisms in order to save a tree that may be an abode not only to animals, but also possibly billions of spirits. However, the outcome of balancing the multiple interests does not always play out in the way it did with the Seven Tathagatas and the organisms dwelling in the ground. Sheng Yen provides an example of a case involving an apparent conflict of interest between insects and trees:

> Ecological resources are very strange. At Dharma Drum Mountain, there is a banyan tree. Two weeks ago, it was still full of leaves, When I went back to see it again last week, the leaves had already been gnawed bare. This tree was eaten by very pretty caterpillars. After eating, the caterpillars became butterflies. I think this is worth it. The tree leaves, after having been eaten by the caterpillars, can support many butterflies. Even though most of the caterpillars died after eating and did not become butterflies, this is the self-regulation of the ecology of the natural world. Even though the leaves are eaten bare, next year it will definitely again grow a full tree of green leaves.
>
> From a different perspective, if we spray these caterpillars with pesticides, whether or not the tree will be protected is one thing, but the ecological cycles will be damaged.
>
> Therefore, at Dharma Drum Mountain, we do not intentionally attend to these caterpillars. They themselves look for food to eat. Life and death is a natural cycle. Humans know not to do things that jeopardize their own survival, but must take one step further to respect the right of other organisms to survive.[23]

Although in the case of the Seven Tathagatas, the insects were sacrificed for the trees, in this case, the tree is sacrificed for the insects. There are prohibitions against killing at DDM, but there also seems to be a great concern in protecting species and ecosystems and in building a Pure Land on Earth.

Although President Zeng begins and ends the video of the transplanting of the trees with remarks on the sentience of trees, the majority of the video is devoted to showing and describing the technical process of moving the tree. While the religious rituals, such as the monastics chanting, are left out, Liao Chiu-Cheng, the professor of forestry, describes the technical aspects of moving the tree in detail. These editing decisions not only suggest that DDM seeks to identify itself with science, but that it is actively using scientific and technological methods to achieve its goals. On the one hand, Sheng Yen used the language of spiritual environmentalism as a way to communicate with non-Buddhists, but on the other hand, he also employs it and, in some cases, modifies practices based on scientific or ecological knowledge.

13.5 Conclusion

There are multiple layers of understanding animals in the DDM organization. There are individual animals, species, and the entire Dharmabody. Unlike in the biological understanding of animals, animals fall under a greater umbrella of sentient beings, which is a more relevant category to Buddhist morality than animals. Otherwise, to be consistent with the biological definition of animals, animals would need to be split into at least two Buddhist categories, humans and beasts. Yet, these two categories do not exhaust the categories of the six realms of existence, which also include devas, asuras, hungry ghosts, and hell-beings. In at least one nun's view, these formless spirits also are worthy of protection, and take an active role in the ecosystem.

These multiple relationships demonstrate that environmental decisions are not represented well by a continuum of environmentally destructive versus environmentally friendly. Both fungi and trees are living things, and DDM made the decision to protect one over the other, even at the cost of killing creatures dwelling in the soil. Environmental decisions seem to require a more dynamic approach of weighing multiple interests. There were likely multiple views and perspectives within DDM about how to manage the Seven Tathagatas. The categories that I have identified are based on a comparison of diverse and disparate materials. Sheng Yen, President Zeng, and the nuns with whom I spoke, each presented a personal understanding of DDM's form of Buddhism. At the same time, the nuns and President Zeng refer to Sheng Yen as the authority.

There are also a number of non-human actors. According to one nun's account, some tree sprites that lived in the Seven Tathagatas altered the two rivers flanking the DDM campus. Sheng Yen notes the mountains, waters, grass and trees are all manifestations of the bodhisattvas. The environmental decisions are not only made in reference to multiple interests, but these environmental actors are also active in participating in the decision-making process. This suggests that the "picturesque contrast between man-made and natural elements" indicated on the signboard in front of the Seven Tathagatas is not as deep and static as it may first appear. Just as the library was set back to accommodate the Seven Tathagatas, the surgical intervention of humans is what allowed the trees to survive. Humans are engaged in dynamic relationships with a diverse group of non-human agents—as organs of the same Dharmabody, as a species in an ecological web, and as individuals that deserve humane treatment.

Modern science and technology play an unmistakable role in DDM's ecology. Not only does DDM adopt technical procedures, such as tree surgery and tree transplanting, Sheng Yen also promotes new forms of Buddhist practice based on the principles of conservation biology. He supports the targeted rehabilitation and release of endangered species rather than a more traditional *fangsheng* ritual focused on the mutual salvation of the animal and the releaser. Science and technology are used as methods to communicate with and relate to people. The video of the tree transplanting emphasizes the advanced technology that DDM uses, while not addressing chanting and other Buddhist rituals involved in moving the tree. This strategy fits into Sheng Yen's overall framework of spiritual environmentalism, in which he uses a secular vocabulary to communicate Buddhist principles. Sheng Yen does rank traditional Chan practice above the practice of the Four Environmentalisms, but both forms of practice may be used side by side. Indeed, the monks and nuns may perform chants for the success of the tree transplant, while scientific technologies are used to perform the transplant.

Not only decision-making but also problem-solving involve a balancing of multiple systems. Vows against killing, concerns for animal welfare, and species-level conservation each represent values that on occasion may conflict, but these different value systems are applied flexibly. Moreover, modern scientific and traditional Buddhist techniques are also integrated into approaches to ecological problems at DDM. This creates a flexible and dynamic process for protecting the environment and a multi-dimensional approach to the understanding of animals. The modern biological and traditional Buddhist concepts of animals differ, but, in practice, they can operate simultaneously. Whereas the physicalist biological definition of animals provides a basis for ecological discourse, the sentience-based Buddhist model focuses on moral behavior. Sheng Yen took seriously the physical ecological demands of organisms, while, at the same time, integrating the understanding of their function within traditional Buddhist discourse. In this way, he also aligned DDM with the ethics of conservation biology and used the appeal of science, technology, and environmental discourse to communicate with non-Buddhists. In this ecology, professors of forestry, dharma protectors, and caterpillars sculpted the landscape together.

Rather than using one way of environing to exclude the other, Sheng Yen and DDM found a way to integrate modern science and pre-modern Buddhism. The global environmental movement became a way of reconfiguring identity and regenerating tradition in a way that spoke to an audience of a new generation. An environing of inclusion pulls a classical Buddhist cosmology of dharma protectors and tree sprites from the margins and harmonizes it with a hegemonic scientific cosmology and global environmental movement. For the Chinese environmental humanities (CEH), DDM's spiritual environmentalism (*xinling huanbao*) provides a positive example of how careful alliances can elevate the profile of communities on the margins, and give voice to local people, local trees, local dharma protectors, and their shared cosmologies.

Notes

1. "Animal," *Encyclopedia Britannica*.
2. For further discussion on beast/*chusheng*, see Ambros, *Bones of Contention*, 35–8.
3. DDM discusses the five precepts on its website: DDM, "The Five Precepts."
4. An English translation of the Pali Theravadin Vinaya is available online: Thanissaro Bhikkhu (Geoffrey DeGraff), *The Buddhist Monastic Code I*, https://www.accesstoinsight.org/lib/authors/thanissaro/bmc1.pdf. Accessed 3 Jan 2015.
5. For an English translation of the Brahma Net Sutra, see Sutra Translation Committee of the United States and Canada, *Brahma Net Sutra*.
6. DDM, "Bodhisattva Precepts."
7. See the Sutra Translation Committee of the United States and Canada, *Brahma Net Sutra*.
8. "Fagushan dashu banjia—Luhuashu banjia shiji quanjilu."
9. Chapple, "Animals and Environment in the Buddhist Birth Stories," 140–142.
10. Ibid., 140–142.
11. Liu, McGarrity, and Li, "The Influence of Traditional Wildlife Release on Biological Invasions," *Conservation Letters* 5 (2012): 107–114.
12. Hsing Yun, ed., *Foguangxue*, 327, quoted in Stuart Chandler, *Establishing a Pure Land on Earth*, 149.
13. Sheng Yen, *Chan de shijie*, 314–315.
14. Williams, "Animal Liberation, Death, and the State," 149–162.
15. Ambros, "Animals in Japanese Buddhism."
16. Shi Zhiru. "Buddhist Responses to State Control of Religion in China at the Century's Turn," 125–157.

17. Schak and Hsiao, "Taiwan's Socially Engaged Buddhist Groups."
18. This is the standard English term that DDM uses. However, the Chinese equivalent 人間淨土 is more literally translated as Pure Land of the Human Realm. This could suggest an anthropocosmic view. For a more detailed description of this concept, see DDM, "A Pure Land on Earth," http://old.shengyen.org/e_content/content/about/about_02_1_1.aspx. Accessed 30 Nov 2017.
19. See Shi Sheng Yen. *Renjian jingtu sixiang*. Find excerpts translated to English online at DDM, "A Pure Land on Earth," http://old.shengyen.org/e_content/content/about/about_02_1_1.aspx. Accessed 30 Nov 2017.
20. Sheng Yen, "Cong dongya sixiang tan xiandairen de xinling huanbao," 41–61.
21. "Speech Presented in Environmental Protection Workshop," 7–8.
22. "Fagushan dashu banjia—Luhuashu banjia shiji quanjilu."
23. Sheng Yen. *Chan de shijie*, 313–314.

Bibliography

Ambros, Barbara. 2012. *Bones of Contention: Animals and Religion in Contemporary Japan*. Honolulu: University of Hawaii Press.

———. 2014. "Animals in Japanese Buddhism: The Third Path of Existence." *Religion Compass* 8 (8): 251–263.

Bhikkhu, Thanissaro (Geoffrey DeGraff). 2013. *The Buddhist Monastic Code I*. Valley Center: Metta Forest Monastery.

Chandler, Stuart. 2004. *Establishing a Pure Land on Earth: The Foguang Buddhist Perspective on Modernization and Globalization*. Honolulu: University of Hawaii Press.

Chappel, Christopher Key. 1997. "Animals and Environment in the Buddhist Birth Stories." In *Buddhism and Ecology: The Interconnection of Dharma and Deeds*, ed. Mary Evelyn Tucker and Duncan Ryūken Williams, 131–148. Cambridge: Harvard University Center for the Study of World Religions Publications.

Dharma Drum Mountain. "Bodhisattva Precepts." http://www.dharmadrum.org. Accessed 9 Dec 2014.

———. "The Five Precepts." http://www.dharmadrum.org/content/chan_garden/chan_garden3.aspx?sn=39. Accessed 3 Jan 2015.

Encyclopedia Britannica. "Animal." http://www.britannica.com. Accessed 7 Dec 2014.

Fagushan dashu banjia—Luhuashu banjia shiji quanjilu 法鼓山大樹搬家-魯花樹搬家事紀全紀錄 ("A Big Tree at Dharma Drum Mountain Relocates to a New Home—The Complete Recording of the Luhua Tree Home Relocation Event"). https://www.youtube.com/watch?v=uj2VWYnBW-I; https://www.youtube.com/watch?v=eWMWMRftckc; https://www.youtube.com/watch?v=eGzZ1ei9i_M. Accessed 7 Dec 2014.

Liu, Xuan, Monica E. McGarrity, and Yiming Li. 2012. "The Influence of Traditional Buddhist Wildlife Release on Biological Invasions." *Conservation Letters* 5 (2): 107–114.

Schak, D., and H.-H.M. Hsiao. 2005. "Taiwan's Socially Engaged Buddhist Groups." *China Perspectives* 59: 43–55.

Shi, Hsing Yun 釋星雲. 1998. *Foguangxue* 佛光學 (*Buddha Light Studies*). Kaohsiung: Foguang.

Shi, Sheng Yen 釋聖嚴. 1999. *Chan de shijie* 禪的世界 (*The World of Chan*). Taipei: Fagu wenhua.

———. 2003a. "Cong dongya sixiang tan xiandairen de xinling huanbao" 從東亞思想談現代人的心靈環保 ("Discussing the Spiritual Environmental Protection of Modern People in Relation to East Asian Thought"). *Xueshu lunkao* II: 41–61.

———. 2003b. *Renjian jingtu sixiang* 「人間淨土」思想 (*An Explanation of the Pureland in the Human Realm*). Taipei: Fagu wenhua.

———. 2011. "Speech Presented in Environmental Protection Workshop." In *Living in the 21st Century: A Buddhist View*, ed. Dharma Drum Mountain Culture Center, 7–14. Taipei: Sheng Yen Education Foundation. http://ebooks.dila.edu.tw/books/n/DDM_en_9-16. Accessed 30 Nov 2017.

Shi, Zhiru. 2010. "Buddhist Responses to State Control of Religion in China at the Century's Turn." In *Buddhist Approaches to Human Rights: Dissonances and Resonances*, ed. C. Meinert and H.B. Zöllner, 125–157. Bielefeld: Transcript Verlag.

Sutra Translation Committee of the United States and Canada. 2000. *Brahma Net Sutra: Moral Code of the Bodhisattvas.* New York: Buddha Education Foundation.

Williams, Duncan Ryūken. 1997. "Animal Liberation, Death, and the State: Rites to Release Animals in Medieval Japan." In *Buddhism and Ecology: The Interconnection of Dharma and Deeds*, ed. Mary Evelyn Tucker and Duncan Ryūken Williams, 149–162. Cambridge: Harvard University Center for the Study of World Religions Publications.

CHAPTER 14

An Exposition of the Buddhist Philosophy of Protecting Life and Animal Protection

Chao-hwei Shih

14.1 Translator's Introduction

Taiwan is a state on the margins. China conditions formal relations with other states on the recognition that there is only one China, which includes the province of Taiwan. In reality, Taiwan functions as a sovereign state with a democratic government that operates independently of China but without formal diplomatic relations with most of the countries of the world. Following the fall of martial law in Taiwan in 1987, Taiwan's continued existence depends on maintaining allegiances with the United States and other free democratic societies, so it distinguishes itself from China as an open democratic society. Buddhist organizations flourished in the new democratic system. Some scholars argue that Taiwan's global

Translated by Jeffrey Nicolaisen

The author is Professor of Religion and Dean of the College of Social Science at Hsuan Chuang University.

C. Shih (✉)
Department of Religion and Culture, Hsuan Chuang University,
Hsinchu City, Taiwan

C. Chang (ed.), *Chinese Environmental Humanities*,
Chinese Literature and Culture in the World,
https://doi.org/10.1007/978-3-030-18634-0_14

Buddhist organizations filled the role of goodwill diplomacy that the government could not formally fill.[a] In parallel with the expansion and globalization of Taiwan's Buddhist organizations, the democratization of Taiwan's erstwhile military state unleashed the latent energy behind the global environmental movement. As Taiwan reinvented itself as the only open democracy in the Chinese cultural sphere, Buddhist organizations embraced the environmental movement as part of their modern global platform. Among the boldest of the Buddhist environmental activists was Chao-hwei Shih, who embraced democratic egalitarian values while arguing that human equality was incomplete and insufficient. For Chao-hwei, unlike the North Atlantic concept in which all that was nonhuman *environed*—constituted the environment for—the human, for Chao-hwei all that was nonsentient *environed* the sentient. In her academic work, Chao-hwei has endeavored to distinguish the Buddhist position from North Atlantic hegemonic theology and philosophy, while in her activist work, she has allied herself with global animal liberation and animal rights movements. Taiwan lies at the margins of China as a bare state where its sovereignty lacks the recognition of the global political order, just as for Chao-hwei, animals lie at the margins of equality as bare life without recognition in the same global political order. This marginality juxtaposes the premodern and modern definition of the Chinese word *huanjing*—the area outside a political boundary and the natural environment—as two contested spaces. As Taiwan's economic development preceded China's, the post-development resurgence of traditional teachings, such as Buddhism, in Taiwan foreshadowed the nascent revival of traditional teachings including Buddhism in China. While the diverse traditions of China represent a smorgasbord of options from which to construct China's post-development state, Taiwan's experience provides a possibility for the future development of China. In the following text, Chao-hwei explains one of these possibilities, the Buddhist concepts of "equality of life" and "protecting life." While Chao-hwei originally used a technical Buddhist language in her essay, I have aimed to provide a translation readable in English while taking minimal artistic liberties in order to best preserve Chao-hwei's original meaning.[b]

14.2 Introduction

To this day, the standard Western criteria for bioscience research protocols (including medical science) have made animal experiments a virtually essential component. Otherwise, the accuracy of the research faces intense skepticism, and the results are frequently dismissed. For example, if policy stipulates that the release of every new drug must undergo a certain period and a certain quantity of animal testing and then enter the human trial phase before it can finally reach the market, then no pharmaceutical developer or manufacturing plant would dare not fully comply.

The field of Eastern pharmacology and medical science originally emphasized the mysterious experience of empathic identification with the other (the cultivation of the skills of visual examination, auditory and olfactory examination, interviewing, and touch examination), and making one's own body the subject of experiment (as in the good tradition of "The God-Farmer Tastes One Hundred Types of Grass"). Now traditional Chinese medicine (TCM) universities and hospitals in mainland China have actually imposed standard protocols for animal experiments, which have even become a point of pride.[1]

This type of phenomenon illustrates the cruel fact of ideological colonization. Following the lead of a scientifically and technologically powerful Western civilization, what the East imports from the West is not at all limited only to a massive amount of science and technology, rather it also includes a deep logic of scientific and technological thought. People see animals as moving machines and prop up the superiority of human rationality. It is the West's anthropocentrism that makes things like this happen. Therefore, even in the laboratories of the Eastern scientific community, the Eastern philosophy that sees animals and protects them as innately moral has no place to establish a foothold because it has already accepted the ideological training of complete Westernization and the rigorous demands of standard operating procedures.

On the other hand, in the same way, capitalism originated in the West and engulfed the entire world, leaving nowhere untouched. Under the ideology of the dominance of profit, the azure skies, green prairies, and natural diet that the animals traditionally enjoyed before being slaughtered have already become an unreasonable request. From birth to death, commercial livestock suffer incalculable torments and abuses that would make your hair raise. Even more, the administration of large quantities of illegal drugs, such as antibiotics, hormones, and leanness-enhancing agents, severely impacts the health of consumers of meat products. Therefore, the meticulous scrutiny and comprehensive academic treatment of animal ethics are indispensable.

According to Buddhist philosophy, just like common knowledge and experience tells us, the Buddha teaches the equality of sentient beings.

Therefore, Buddhists, naturally, will protect animals. So how should Buddhism respond to the contemporary ethical controversy over animal protection and produce a comprehensive analysis at the levels of concept and practice? This is the focus of the present chapter.

This chapter, first, returns to the fundamental Buddhist principle of *dependent arising*: all phenomena form, exist, change, and fade away according to the support and restriction of causes and conditions. Only then, according to the three great principles, does this chapter justify the concept of protection of life (at the most passive, not injuring or killing animals). Finally, in regard to practice and implementation, I raise the concept of the Middle Way as the basis by which to judge whether to accept or reject my arguments.

14.3 Dependent Arising, Emptiness, Protecting Life, and the Middle Way: The Definition and Connotation of Core Concepts

Because the basis of this chapter's discussion is built on the following core concepts—dependent arising (*yuanqi*; Skt. *pratītya-samutpāda*), emptiness (*xingkong*; Skt. *śūnya*), protecting life (*husheng*), and the Middle Way or middle path (*zhongdao*; Skt. *madhya-mārga*,), I will first examine the definition and connotation of each term one by one.

1. *Dependent arising* indicates the law that the causes and conditions of phenomena in this world arise and cease. Originally, the characteristics of all people, events, and objects inevitably emerge as *this* or as *that* according to the harmonies and limitations of causes and conditions. There is absolutely no exception. This is the *empirical* law that the Buddha discovered through penetrating insight and explained to the people after his enlightenment.
2. *Protecting life*, simply put, is nothing more and nothing less than protecting sentient beings. Originally, according to the law of dependent arising, phenomena that are produced by causes and conditions—with the qualities of impermanence, nonseparation, and nonreality—include sentient beings (animals) and nonsentient beings (plants and inanimate objects). As such, do the animals, plants, and inanimate objects that all fall into this same category require identical ethical treatment? The answer is, clearly, that they

do not. Those for whom the Buddha was deeply concerned were sentient beings. The purpose of teaching disciples was that he wanted to make sentient beings "escape suffering and attain happiness, and transform that of the mundane into that of the sage."

3. The Middle Way, to be precise, is not just the path "in the middle," but rather, it indicates the impartial and correct way. In early Buddhist scriptures, the Middle Way is a paired concept that appeared simultaneously with dependent arising, with which it has a logical relationship at the empirical level: Relying on the penetrating insight of the law of dependent arising, one could possess impartial insight into reality, and in regard to the prejudices that sentient beings universally are apt to commit, the Middle Way provides proper guidance in regard to concepts and behaviors.[2]

To put it simply, *dependent arising* articulates an empirical rational law, *protecting life* elucidates a normative ethical assessment, and the Middle Way suggests a normative, pragmatic guiding principle. This chapter relies on these three major core concepts in order to construct a systematic theory of Buddhist ethics and offer an exposition of Buddhist philosophy for protecting life and animal protection.

14.4 The Difference in Treatment Between Sentient Beings and Nonsentient Beings

According to the law of dependent arising, the animals that roam the world not only have the capacity of awareness, but they also are naturally self-centered. In terms of emotions, they have fierce self-love; their intellect is biased; and they have a firm view of selfhood. With great determination, they constantly value themselves to an extreme degree and exude self-pride. As a result, they generate an intense instinct to pursue life and fear death and to pursue happiness and avoid suffering. Due to this instinct to adapt to life and pursue happiness, people must protect life. Since the term *life* in *protecting life* indicates sentient beings that have the capacity of awareness, even if animals other than humans are flying, squirming, or wriggling sentient bugs, each and every one fits into the scope of ethical concern, and it is not appropriate to favor one at the expense of another. Therefore, there will be the exquisite sentiment to "always leave food for a rat and leave a lamp off for pity of a moth."[3]

Therefore, it can be said that intolerance for the suffering of sentient beings is a kind of naturally expressed ethical intuition. Based on the empirical conditions encapsulated within these ethical feelings, ***protecting life*** self-consciously increases, deepens, and expands, thereby producing the normative will to act. The ethical behavior produced can also be divided into two types: passive and active. In regard to the passive type of behavior, the self-regulation never to injure or kill sentient beings is the essential meaning of upholding the precepts. In regard to the active type of behavior, spreading compassion everywhere in order to save and protect sentient beings is the spirit of the Buddhist practice of giving.

The word *life* in the term ***protecting life*** still primarily indicates the sentient beings with the capacity of awareness, that is, animals. It is not at all the case that we do not need to protect and cherish plants and inanimate objects, but the capacity to be aware is not as strong as that of animals, and most have the capacity to grow back after being cut. Therefore, although the Buddhist precepts also require the monks and nuns to protect and cherish plants—not allowing them to be arbitrarily cut down—the precepts still primarily make animals with the capacity of awareness the object of ethical concern.

14.5 The Categorization of the Academic Discipline of Animal Ethics[4]

The definition of *life* and the recognition of the status of humans naturally influences the categorization of the academic disciplines. For example, does animal ethics research fundamentally belong to the category of biological ethics or environmental ethics?

Ordinarily, the investigation of bioethics (or the ***Life Education*** currently popular in the Taiwanese academic world) most often takes humans as primary, covertly holding the ideology of anthropocentricism and asserting that the ethical status of humans is higher than other things. Therefore, the topic of life that the field of bioethics discusses is primarily human life. Each type of principlism is still tailor-made to the specific characteristics of humans (for instance, the principle of autonomy or the principle of consent, the principle of justice, and the principle of nonmaleficence). At most, when touching upon the topic of animal experiments, the ethical status of laboratory animals is discussed, but there is still no way to take each principle in the field of bioethics and apply it with all animals according to how it is applied with humans. As for nonlaboratory animals (such as wild animals, commercial animals, and companion animals), they all are relegated to the field of environmental ethics. Clearly, humans are seen as the only species with lives deserving of our concern.

But even bioscience researchers would have a hard time denying animals have signs of life. Supposing animals do belong to the category of *life*, what reason do we humans have to equate animals with plants and inanimate objects and only regard them as one part of the environment humans occupy? There is a double standard for life, and there is clearly no way to fill in the gaps and make this theory whole.

Next, let's look at the field of environmental ethics! Although environmental ethics has offered much concern for the plight of wild animals, in general, it has fixed its sight on the consideration of ecological balance. Although recently it has begun to propose "Save the globe with vegetarianism," the attitude on which it has set its sights is "Save energy, reduce carbon" and "Love the earth" but not "Protect animals." Ecological conservation advocates often can enjoy a steak while chastising people for putting rhinoceros horn in medicine. While they awaken the human conscience by screening a horror scene of a rhinoceros falling dead, spilling blood in the midst of the reverberation of gunshots, they can completely ignore the bloody slaughter that livestock cattle must suffer.

Additionally, based on the consideration of ecological balance, government ecological conservation policies establish standards and controls for protected animal types and quantities. If the numbers are low, it is true that they strengthen conservation. If the numbers are high, they do not eliminate the possibility of slaughter in order to achieve a "reasonable culling." Therefore, most ecological conservation advocates do not oppose hunting. They tend not to utter a word about the brutal cruelty of the hunting process and the suffering of the hunted animals. However, in regard to the quantitative management of wild animals, their efforts very possibly can be reduced to an ecological fairy tale of wishful thinking for the benefit of humanity.

What merits attention is that in the logic of environmentalism's holism, for the benefit of the whole ecosystem, innocent individual lives are forced to be sacrificed, even in the name of ethical legitimacy. This is similar to what Mark Sagoff asserts: "Animal liberation and environmental ethics: Bad marriage, quick divorce!" Thus, environmentalists are not animal liberationists, and animal liberationists are not environmentalists.[5]

Therefore, whether due to anthropomorphism (i.e. human chauvinism) and striving for human sustainable use of the natural environment or due to ecocentrism and making sustainable development of the ecosystem the priority of ethical concern, most people that care about environmental ethics maintain a skeptical attitude toward biocentrism (i.e. biocentric ethics)[6] and exclude commercial animals and companion animals from the scope of ethical concern.

In this way, bioethics and environmental ethics not only have no way of removing the serious threat of humans toward laboratory animals and wild animals, they even join hands to exclude commercial animals and companion animals. Anthropocentricism and environmentalism join hands to suffocate animals. The plight of animals is awkward: they have become the intentionally forgotten orphan occupying the space between bioethics and environmental ethics.[7]

14.6 Animal Ethics and Ethical Principles

Among the four ethical principles of bioethics—the principle of autonomy (giving rise to the principle of consent), the principle of justice, the principle of nonmaleficence, and the principle of beneficence—there is not one that can easily be applied to animals. It is very awkward to superimpose animal ethics on bioethics and make a set of principles with a double standard. However, according to the Buddhist perspective, since sentient beings include nonhuman animals, then the discussion of animal ethics ought to belong to bioethics, and the four ethical principles of bioethics naturally should also be applicable to animals other than humans. Only when research is conducted with this kind of fundamental attitude will the tragic price that animals pay for so much human welfare not be ignored when discussing human welfare (especially the market supply and demand for blood, flesh, skin, and feathers; brutal and heartless animal experiments; exploitative genetic engineering; and the most fantastic varieties of organ transplants).

We can respect the contributions of the global animal protection movement, hold discussions on bioethics, and occasionally touch upon the question of the ethical status of laboratory animals, but there is no way to comprehensively transplant the above ethical principles onto animals regardless of who is assigned to do it. May I ask what experimental procedure respected the laboratory animals' right to autonomy by asking them "Yes" or "No"? Who is there that can inform the subject animals and request their consent before an animal experiment is conducted?

Many animal experiment procedures already make subjects experience conditions so wretched that life is not better than death. What is even harder for people to bear is that after the experiment is over, not only is every one of the animals in the experimental group killed, but even the healthy and untreated animals in the control group are killed without exception and without mercy. Thus, as long as there still is even one day of animal experimentation, is it still possible to boast about the so-called principle of nonmaleficence or the principle of beneficence? Of course, contravening the animals' will to survive for the so-called "welfare of humans" absolutely does not satisfy equitable treatment's principle of justice.

Therefore, the field of bioethics at first only framed the category of bioethical discussion within the human species, only subsequently putting forth the aforementioned four ethical principles. Most animals other than laboratory animals (such as wild animals, commercial animals, and companion animals) are thus relegated to environmental ethics. What is clear is that the field of bioethics washed their hands of the matter by placing animals within the fabric of the environment and relinquished all of the difficult questions of animal treatment to environmental philosophy.

However, it would be difficult for even a bioscience researcher to deny that animals possess the same types of biological characteristics as humans. Especially in the advanced genomic research of the present day, we have come to realize, even more than ever, the very high homogeneity of animal and human DNA sequences. Indeed, what Mencius said came true, "The differences between humans and the birds and beasts are very small" (*Mencius*, *Lilou* Chapter, Second Part).

Supposing that animals are a form of life, what reason do we have to equate these animals with plants and inanimate objects, only treat them as the background or the resources in an environment occupied by humans, and not conduct the principled ethical consideration described above? Supposing the criteria of the use of ethical principles are based on special characteristics (such as rationality) that humans do not share with other animals, then how do we also face the sharp skepticism of the animal liberationists? Can the treatment of people that have an IQ approximately equal to (or less than) an orangutan, dog, or cat—fetuses, infants, the learning disabled, and mentally handicapped—compare with that of animals? This is the objection for which bioethics researchers have no way to evade without providing a detailed justification.

14.7 Justification for the Obligation to Protect Life

This type of indictment, however, is still not one that people can accept. They tend to strongly object: "Since we are humans, why should we protect animals other than humans?" That being said, is protecting animals actually silly talk, meddling, and sticking your nose in other people's business? Or the virtue and good deeds of humanity? Or an ethical obligation that should be applied to its fullest? Below, three principles are used in order to provide justification. *Protecting animals*—at its most passive, not injuring or killing sentient beings—is the ethical obligation that people that act as an ethical agent (and not just Buddhists) should implement.[8]

1. The self-transcending method

Protecting animals comes from identifying with the other, from the capacity to be aware of another's point of view. Buddhism calls this the *self-transcending method* (*zitong zhi fa*). As the *Saṃyuktâgama-sūtra* says:

> What is the self-transcending method? Those called the disciples of the Buddha study it in this way. I think this thought: "If there is someone that wants to kill me, I am not happy. If I am not happy, he is also not happy. What do you have to say about killing another?" Having realized this, accept not killing life, do not take pleasure in killing life. As explained above…[9]

The *self-transcending method* is to use one's own state of mind, to discern the state of mind of other sentient beings, and to highly value the instinct to pursue life and fear death and to pursue happiness and avoid suffering. The self-transcending method is extremely similar to Confucians' *conscience* (*liangzhi* 良知). In ethical studies, it is called the Golden Rule Theorem. This, originally, is the Golden Rule faithfully obeyed and commonly shared in different cultures, but Western mainstream culture, on the contrary, tends to be based on divinity or rationality and still does not take animals as the self-transcending method's object of concern.

Since animals and people similarly have the capacity to experience suffering and the moods of happiness, anger, sorrow, and joy (i.e. the capacity of awareness), as in the feeling of taking no joy in being killed, they clearly see the self-transcending method, and the self has no way to escape the empathy and sympathy toward the plight of animals. Therefore, the capacity of awareness should be the criteria for determining who is covered by the Golden Rule. Our companions whom we should not treat as we do not want to be treated (or the companions with whom we are willing to share what we want ourselves) are not limited to humans but should be expanded to include animals other than humans. We should not just care for lives with whom we have relationships (such as pets that we take care of ourselves), but rather, we should expand the baseline of ethical concern to encompass all sentient beings. Moreover, it is only possible to do all one can do to equally attend to the benefits preferred by each individual, but it is not suitable to injure any one innocent bystander in order to pursue maximum utility.

2. The interconnectedness of the manifest form of the law of dependent arising

The theory of dependent arising is an empirical law. It reminds us that every single sentient being is a body existing in the form of a network connected by infinite causes and conditions. Life cannot independently survive and cannot attain the happiness and comfort of self-fulfillment without the support of the appropriate causes and conditions. Therefore, regardless of whether it is for self-fulfillment or to feel gratitude for the causes and conditions that make the wish of self-fulfillment come true, we all should take care of others and not just see one's own needs.

This explains the essence of the ethical rules (the self-transcending method). According to the theory of dependent arising, consciousness is not the most fundamental principle, and there is no way to prove that it comes from external divine revelation. Rather, it is the product of the reciprocal blending of subject and object. That can be proven on the basis of the more fundamental truth of dependent arising. The manifest form of the law of dependent arising relies on *co-existence* and *the equality of the nature of all things*. These two laws were considered the roots of compassion by Master Yin Shun.[10]

To explain this more deeply, from the perspective of the law of dependent arising, any phenomena that exist (including sentient beings) are not originally separated, isolated, and able to exist by themselves. They must have all causes and conditions coalesce in order to come into existence. Therefore, the dependently arisen phenomena themselves have a codependent and coexistent complex network with other dependently arisen phenomena. Based on this premise, how is it possible that a subtle and unimpeded link between life form and life form—each coming into existence by mutually supporting each other's causes and conditions—does not exist?

There is not a moment that our nine orifices and seven apertures[c] are not in a state of interconnecting with the external environment. Our expressions, language, and behavior exchanges even more elaborate information with others. That is why, although life has separate and distinct bodies, the space between one body and another is not insulation. Thus, based on this, the self-transcending method of empathy and sympathy arises.

3. The equality of the nature of all things

Protecting life is the essence of Buddhist ethics, but if you want to understand the attitude of the Buddhist teaching toward animals, you must take one step further to fully grasp the concept of *equality*. The reason that the Buddhist teaching grants deep and broad ethical concern for life is not only because of the single factor of compassion. Compassion and pity generally occur in a top-to-bottom or superior-to-inferior situation. Simultaneously, the recipient of compassion and pity does not necessarily form an equal type of relationship with the giver of compassion. Even the Christian religion has a nonmainstream theological discourse related to the protection of animals, but here, the animal's position under systematic theology is still the object of the management and command of humans based on the command that humans received from God. In Christian theology, humans and animals still have no way of achieving a verdict of equality. However, the *protecting life* taught in Buddhism clearly delineates the reasons for the equality of sentient beings.

In the second volume of *Buddhist Meta-Ethics*, I mentioned that equality, as taught in Buddhism, has three levels:

1. Life uniformly has the capacity of awareness. We must equally respect the experience of suffering and joy and the strong aspiration to escape suffering and achieve happiness. This is the essential meaning of *equality of sentient beings* (*zhongsheng pingdeng*).
2. All life also arises, ceases, changes, and transitions according to causes and conditions. This is nothing other than the fact that, among distinctive elements, the many things produced by causal conditions have no self-nature and are equal and non-dual. This is

the essential meaning of the *equality of the nature of all things* (*faxing pingdeng*).

3. Not only do the ordinary people inherit suffering and joy and transmigrate through birth and death according to causes and conditions, sages also attain nirvana and complete the path to Buddhahood according to causes and conditions. Under the law of dependent arising, the ordinary person has the possibility of liberation, and also has the possibility of becoming a Buddha. This is the essential meaning of *equality of Buddha-nature* (*foxing pingdeng*).[11]

In principle, the three principles described above—protecting life, or at a minimum, not injuring or killing sentient beings—are the obligations that humans should fulfill. Not protecting life does not satisfy the mind (the self-transcending method). Not protecting life also does not satisfy rationality (the nature of dependent arising, the correct principle of form). Thus, the consequences are that the individual's *conscience* is lost, and one's enlightened nature is hindered or even degenerates, which is something that injures both oneself and others. In regard to the principle of retribution (*yingbao faze*)—even if injuring and killing each other and chewing and eating each other could make people tremble in fear of the karmic effects so that they would not dare to rashly injure living creatures, ordinary people are unable to verify true insight without being endowed with the insight of the heavenly eye (*tianyan tong*) and the knowledge of prior lifetimes arising from deep meditative contemplation. This goes beyond common sense and experience, so we put this aside for now.

14.8 Middle Way: Situational Considerations of the Principle of Equality

Buddhist teachings are not only a set of theories. Normative concepts of protecting animals, as well as the normative function of the Middle Way, can be obtained from the empirical law of dependent arising. The Middle Way is a complete set of practical theories on interdependence and relating to one's environment. It is also a practical guiding principle for promoting public policy and accomplishing individual ideas. On the one hand, the agent must be equipped with a just and selfless attitude resulting from the self-cultivation of loving kindness and wisdom. On the other hand, there must be an as-complete-as-possible situational appraisal of the recipients of action or the things that are employed. Further, the agent must grasp the essential points to make resolute choices. I have synthesized the content of the scriptures and composed the following definition of the term Middle Way:

> Selflessly making the relatively best decisions according to the causes and conditions that can be seen, heard, recognized, or known.

Those who speak of what can be seen, heard, recognized, or known foresee that there are limits to the senses and cognition. The agent must intuit the causes and conditions of what can be seen, heard, recognized, or known but not exhaustively determine all causes and conditions. Then at the time of making a decision, the agent would not dare to boast that all his decisions are absolute trüth but can humbly admit that his is just the relatively best decision at this point in time. This type of humble attitude, even if it cannot guarantee the absolute correctness of his decision, can reduce the chance of obstinately repeating mistakes.

Moreover, the self-transcending method—or even the power of penetrating insight into the dependently arising no-self—that is produced by contemplation on the theory of dependent arising is also helpful to put self-profit aside and make a selfless contribution for the benefit of all sentient beings when we are making a decision.[12]

Originally, in terms of ethics, the various principles to which the spirit of protecting life leads (such as the principle of justice, the principle of nonmaleficence, and the principle of beneficence) have made Buddhist bioethics tend toward principlism. However, in regard to the conflict of interest or the difference in value between people and animals, applying ethics must appropriately make assessments and decisions of accepting or rejecting actual practices according to the pragmatic guiding principle of the Middle Way. This is due to the fact that since the concept of protecting life covers all sentient beings, including the "sentient creepers and crawlers" (e.g. insects and worms), it is simply not possible to encompass them all in actual practice. Sometimes, we still must face major epidemics such as diseases that humans and livestock can transmit to each other. We must first sufficiently possess the biological wisdom of the Middle Way—involving interdependence and relating to one's environment—and then adapt to subjective and objective factors to make a limited situational appraisal.

Why emphasize that it is limited? Because behind the situational appraisal, we still must strictly conduct a self-examination and remove the shadow of self-interest. Otherwise, the situational appraisal will transform into opportunism (searching for a program of action that most benefits oneself).

14.9 The Middle Way Thinking on Reasonable Self-Defense

If we make use of the pragmatic guiding principle of the Middle Way to address the problem of reasonable self-defense in regard to a major epidemic such as a disease that humans and livestock can transmit to each other, we can conduct an assessment according to the difference in the level of spiritual attainment cultivated by the ethical subjects. A sage, having transcended self-views and self-love and attained the stage of no-self, of course, can choose to abandon reasonable self-defense in order to protect others, but the ordinary person is limited by self-views and self-love. It is truly difficult to abandon the opportunity to live by reasonable self-defense. Even though this is true, we still should reflect on the situation according to the three aspects below:

1. Only when there is no other way to solve the problem can one injure another in self-defense. In other words, if there is another way to solve the problem, even if it will take more time, money, and energy, one must try to use methods in which it is possible to achieve the results of self-defense without harming others because the value of life is still more important than the convenience of executing the task or possessing resources.

   For example, clearly, the foot-and-mouth disease that infects pigs is not a disease that animals can transmit to humans. However, it, of course, does not meet the Middle Way law of balance to summarily execute tens of thousands of healthy pigs whether it is to remove Taiwan from the list of infected areas as quickly as possible, to avoid a steep decline in the opportunity to export meat products and the quantity exported, or to avoid reducing the profits of farmers, owners, and staff. This mentality that "I'd rather kill one-hundred by mistake, than accidentally let one go," simply does not meet the Middle Way standard. Is it not a serious numbness to a sense of morality and serious confusion about values to see the money of humans as more important than the lives of animals with foot-and-mouth disease, to take life for money, or to plot murder for money?

Now, let's take as an example the recent case of avian influenza at the mention of which Taiwanese people turn pale-faced in horror. Obviously, humans are fond of eating bird meat. They build large numbers of intensive farms that raise animals, causing these outbreaks to transmit rapidly. The viral strain also continuously mutates, but once an outbreak appears, the Bureau of Animal and Plant Health Inspection and Quarantine collectively culls tens of thousands of innocent birds. The low-level disease prevention staff kill until their hands throb and go limp with exhaustion, and they have frequent nightmares. Actually, if someone takes up a vegetarian diet, then what necessity is there to face the risk of transmission of avian influenza? Even more, what is the necessity for manufacturing the enormous killing industry?

2. Even if at present there is truly no other method, we must frequently embrace a sense of shame and do what we can to continue research and wholeheartedly pursue an alternative plan that not only does not require harming others but that also can protect ourselves. With this aim, it may be possible to depend on human intelligence and wisdom to come up with a win-win solution for us and them: for example, developing various animal vaccines so that vaccination can substitute for culling.
3. What is most scary is taking any behavior that harms others as standard procedure for the reason of reasonable self-defense or human benefit (e.g. claiming that doing cruel animal experiments is for human welfare or for maximizing utility and that culling large quantities of dogs, cats, and ferret-badgers is for preventing rabies). Once legitimacy or rationality for harming others has been bestowed on ethical grounds, it will make people lose the ethical feeling of identifying with the other, and thereby neglect the research and development of alternatives.

The topic of animal experiments is a notable example. People sincerely describe the many kinds of human benefits (or the maximum utility) of animal experiments, but they do not rethink whether or not they can find an alternative. This really is limited by the scope of Western mainstream thought. Specifically, people have already hidden the value system of Western mainstream theology or mainstream philosophy within the research process itself. They maintain that animals do not possess divinity, a soul, or rationality, and, therefore, they can do whatever they want to their bodies.[13]

14.10 Adjusting to Differences in the Level of Development of Enlightened Nature

Finally, I want to throw out a question. Does Buddhism also have a tendency toward anthropomorphism?

Originally, in principle, all-encompassing compassion extends to all sentient beings, but in a situation in which causes and conditions are limited, practical work must have an order of priority. Even at times of necessity, we still must assess values and make yes or no decisions. Therefore, Buddhist disciples can draw up a Middle Way principle starting from the near and small and expanding outward, according to the differences in value between animals and humans and between common people and sages.

The distinctive feature of human chauvinism is for humans to occupy the status of ruler and to treat the global ecological environment and even other nonhuman animals as the object of human control and exploitation. According to this standard, the Buddhist teaching has no view that can verify the legitimacy of the dominion over animals.

From the perspective of the current human situation, humans—in comparison to other animals—possess three forms of strengths: rationality, emotion, and will. Accordingly, humans can expand and purify these strengths, perfect their enlightened nature, and be endowed with the Buddha's great wisdom, great compassion, and great heroism.[14] However, even though this is true, the Buddha always vigorously advocated for the animals and forthrightly opposed the sacrificial killing activities of Brahmanism.

Then, the person who is the ethical agent is situated within human society and must plant his feet firmly within human society. Only then can he begin to march forward in order to do things that benefit others. But this is not the "Gradations of Love" theory exemplified by "Love parents affectionately, and care for the people. Care for the people, and be fond of creatures."[d] It is also not the human transcendence theory. Rather, it is the balancing of the Middle Way under the limitations of causes and conditions.

To explain this more deeply, there are two points that we must recognize regarding the value propositions relating to humans and animals:

1. In regard to the "karma body,"[e] we cannot maintain the myth of being the "chosen ones": we do not take too much pride in the fact that humans have the aforementioned three special strengths and see humans as the chosen ones among the diverse species because the path of humaneness is what liberates all life equally. If reason,

emotion, and will evolve to a certain level, then it is easy to enter into the human realm. Conversely, if reason, emotion, and will regress to a certain level, it is also possible to fall again into the realm of other species of nonhuman sentient beings, even the realms of hungry ghosts or hell.

2. In regard to karma, we cannot only be concerned with humans. When comparing humans with other nonhuman sentient beings, indeed, humans possess greater ethical awareness, but we should not establish limitations according to this fact and recognize only humans as the object of ethical concern. Instead, accordingly, we should recognize that humans do not have the right to purely act on instinct alone and that we must see ourselves as agents who possess the capacity of ethical awareness. In the words of Tom Regan, people are "moral agents" who possess ethical judgment and the capacity to practice virtue, but all sentient beings, including animals, should be taken as "moral patients" who are treated morally based on their ability to perceive suffering.[15]

14.11 Conclusion

In summary, animal ethics based on the Buddhist point of view must be established under the principal axis of thought and theoretical framework comprised of *dependent arising*, *protection of life*, and the Middle Way. According to the ethical feelings of the *self-transcending method*, the ethical introspection of *interdependence and interbeing*, and the ethical concept of the *equality of sentient beings*, a fair appraisal of benefits must be made for animals based on each ethical principle of bioethics (the principle of autonomy, the principle of nonmaleficence, the principle of beneficence, and the principle of justice).

But on the level of practical application, when the moral agent is restricted by time constraints, physical capacity, mental capacity, and financial capacity, she must borrow the wisdom of the Middle Way and selflessly make the relatively best decision under the causes and conditions that can be seen, heard, recognized, or known. The most important point is to start from the near and small.

First, sentient beings at the stage of the human realm display varying degrees of special strengths and capacities (but not eternally unchanging special strengths and intrinsic qualities) related to the attributes of reason, emotion, and will. These special strengths and capacities are sufficient enough to awaken their enlightened nature and transform common people into sages. When humans and animals encounter conflicts of interest that threaten lives and there is no other choice, even those with discernment stemming from the awakening of enlightened nature can do nothing more than reluctantly make the Middle Way decision that results in the least injury to animals.

Next, when objective and subjective measures are limited in the face of active rescue, people still must rely on the discernment of the awakening of enlightened nature in order to set an order of priority in which humans and animals (even common people and sages) receive benefits, but this absolutely does not suggest that people can murder every innocent animal for the benefit of humans.

Therefore, because the gate to the human realm, or even the Buddha realm, is constantly opened by all sentient beings, even if Buddhism discerns human value through the awakening of enlightened nature, it, of course, cannot see the human—this relatively stable individual composed of spirit and matter—as a permanent, independent, and real entity. This *entity* cannot be obtained itself, let alone dominate others. In the ethical evaluation of animal treatment, Buddhism absolutely does not have the leeway to endorse anthropocentrism.

Translator's Notes

a. For example, see Richard Madsen, *Democracy's Dharma: Religious Renaissance and Political Development in Taiwan* (Berkeley: University of California Press, 2007), as well as Stuart Chandler, *Establishing a Pure Land on Earth: The Foguang Buddhist Perspective on Modernization and Globalization* (Honolulu: University of Hawai'i Press, 2004), Chapter 10, especially 280–1.

b. Chao-hwei style of writing borrows from the Buddhist sutras and commentaries, the Confucian classics, and modern ethical philosophy. The translator attempts to translate her writing as directly as possible while making it fit into a readable English diction and grammar. As her essay deals with Buddhist theory and ethical philosophy, a looser translation could lose some of the reasoning and nuance of her arguments. While some of the style may be unfamiliar to readers without some experience with the relevant topics, it is the translator's hope that retaining the original style and content will convey the logic and context by which Chao-hwei presents her arguments.

c. The nine orifices are the two eyes, two nostrils, two ears, mouth, anus, and genitals. The seven apertures are the two eyes, two nostrils, two ears, and mouth.
d. This Mencius quote refers to the Confucian concentric theory.
e. The karma body is the body received according to karmic merits. In the context of this paragraph, it refers to the human body.

Notes

1. As Zeng Congyan's work *Launching Animal Experimental Research and Raising the Standard of Hospital Scientific Research—Recording My Hospital's Chinese Medicine Medical Experiments* proudly expresses: "Animal Experiments are one important component to launching medical research. By employing animal experiments in medical research, it is possible to simplify the many very complex problems of human bodies and conduct detailed investigations for each type of variable, and this is difficult for clinical research to do...The establishment of a TCM laboratory brought our hospital's standard of research to a new level, allowing us to leap to the top tier of Zhongshan City's and even Guangdong Province's medical care system and to lay ground for several important breakthroughs." For further details, see Guangzhou Chinese Medical University's affiliate Zhongshan Hospital's website: http://www.zsszyy.com/html/html/zkzb/yxb/201115/1089.html
2. "'Yinqizhongdao' zhi 'zhongdao' zhi genyuanyi yu yanshengyi," 91–128.
3. Su Shi, "Ciyun Dinghui Qin Zhanglao Jianjii Bashou (bingyin)."
4. "The Division of the Academic Discipline of Animal Ethics" below is excerpted from "Fojiao shengming lunlixue zhi yanjiu fangfalun," 90–92.
5. Sagoff, "Animal Liberation and Environmental Ethics," 297–307.
6. Biocentricism can again be divided into the animal-centered theory that places animals that possess the capacity of awareness within the scope of ethical concern and life-centered theory that asserts that all life (including animals, plants, and microorganisms) has an ethical status.
7. See "Fojiao shengming lunlixue zhi yanjiu fangfalun," 90–92.
8. For details about the three types of principles for protecting animals, see *Buddhist Normative Ethics*, 84–93.
9. *Saṃyuktâgama-sūtra*, vol. 37 (Taishō 2, 273, middle-bottom), *Samyutta—Nikāya* 55; "Yu liu xiang ying" (*Nan chuan juan* 16, bottom, 236).
10. For details, see Yin Shun, "Cibei wei Fazong-ben," 120–123.
11. In relation to the debate on equality, see the following works:

    1. *Buddhist Ethics*, 81–82.
    2. *Buddhist Normative Ethics*, 91–93.
    3. *Buddhist Meta-Ethics*, 60–61.

In relation to the debate on difference (i.e. cultural relativism and ethical relativism), see *Buddhist Normative Ethics*, 107–112.

12. *Buddhist Normative Ethics*, 95–113.
13. In regard to ethical inquiry into animals and human experiments, the details need not be repeated here. See the two Chao-hwei articles below:

    1. "Fojiao shengming lunlixue zhi yanjiu fangfalun," 102–103.
    2. "Diqiu dalong zhong de chaoji bailaoshu," 143–144.

14. Yin Shun, 1952, "Renxing," *Fo zai renjian*, 94–96, *Yinshun fashi foxue zhuzuo ji* 2006 compact disc version.
15. The two points above are based on the previous work below, but abbreviated and augmented. See Chao-hwei, "Fojiao shengming lunlixue zhi yanjiu fangfalun," 90.

Bibliography

Beauchamp, Tom L., and James F. Childress. 1994. *Principles of Biomedical Ethics*. New York: Oxford UP.

Chandler, Stuart. 2004. *Establishing a Pure Land on Earth: The Foguang Buddhist Perspective on Modernization and Globalization*. Honolulu: University of Hawai'i Press.

Madsen, Richard. 2007. *Democracy's Dharma: Religious Renaissance and Political Development in Taiwan*. Berkeley: University of California Press.

Sagoff, Mark. 1984. "Animal Liberation and Environmental Ethics: Bad Marriage, Quick Divorce." *Osgoode Hall Law Journal* 22 (2): 297–307.

Shi, Yin Shun 釋印順. 1949. *Fofa gailun* 佛法概論 (*Introduction to the Buddhist Teachings*). http://www.mahabodhi.org/files/yinshun/21/yinshun21-00.html. Accessed 26 Dec 2018.

———. 1952. *Fozai renjian* 佛在人間 (*Buddha in the Human Realm*). http://www.mahabodhi.org/files/yinshun/16/yinshun16-00.html. Accessed 26 Dec 2018.

———. 1980. "Zhongdao zhi fojiao" 中道之佛教 ("Buddhism of the Middle Way"). *Fofa shi jiushi zhi guang* 佛法是救世之光 (The Buddhist Teachings are the Light that Saves the World). *Yinshun fashi foxue zhuzuo ji* 印順法師佛學著作集, 145–56. http://www.mahabodhi.org/files/yinshun/24/yinshun24-00.html. Accessed 26 Dec 2018.

———. "Cibei wei fofa zongben" 慈悲為佛法宗本 ("Compassion is the Fundamental Tenet of the Buddhist Teachings"). *Xuefo san yao* 学佛三要 (The Three Essentials for Studying Buddhism), 117–140. http://yinshun-edu.org.tw/zh-hant/Master_yinshun/y15. Accessed 26 Dec 2018.

Shih, Chao-hwei. 1995. *Fojiao lunlixue* 佛教倫理學 (*Buddhist Ethics*). Taipei: Fajie chubanshe.

———. 2003. *Fojiao guifan lunlixue* 佛教規範倫理學 (*Buddhist Normative Ethics*). Taipei: Fajie chubanshe.

———. 2006. "Fojiao shengming lunlixue zhi yanjiu fangfalun" 佛教生命倫理學之研究方法論 ("Research Methodology of Buddhist Bioethics"). *Xuanzang foxue yanjiu* 5: 81–106.

———. 2007. "Fojiao 'shengming lunlixue' yanjiu: Yi dongwu baohu yiti wei hexin" 佛教「生命倫理學」研究:以動物保護議題為核心 ("Buddhist Bioethics Research: With a Focus on the Issue of Animal Protection"). *Yingyong lunli yanjiu tongxun* 43: 28–44.

———. 2008. *Fojiao houshe lunlixue* 佛教後設倫理學 (*Buddhist Meta-Ethics*). Taipei: Fajie chubanshe.

———. 2009. "Diqiu dalong zhong de chaoji bailaoshu—Yi fofa guandian lun yizhong jiyin zhuanzhi sheji renti shiyan zhi lunli zhengyi" 地球大籠中的超級白老鼠——依佛法觀點論異種基因轉殖涉及人體實驗之倫理爭議 ("Super-Lab Rats in the Big Cage of Earth: Using the Perspective of Buddhist Teachings to Discuss Trans-species Gene Transplant, Touching on the Ethical Controversy of Human Experiments"). *Yingyong lunli pinglun* 46: 137–146.

———. 2012. "'Yinqizhongdao' zhi 'zhongdao' zhi genyuanyi yu yanshengyi—Yi foxue lunli guandian zuo shenceng tansuo" 「緣起中道」之「中道」之根源義與衍生義——依佛教倫理觀點作深層探索 ("The Original and Derivative Definitions of *the Middle Way* in the *Middle Way of Dependent Arising*—Conducting a Deep Exploration from the Perspective of Buddhist Ethics"). *Xuanzang Foxue Yanjiu* 18: 91–128.

Singer, Peter. 1996. *Dongwu jiefang* 動物解放. Trans. Meng Xiangsen 孟祥森 and Qian Yongxiang 錢永祥. Taipei: Guanhuai shengming xiehui.

Su, Shi 蘇軾. "Ciyun dinghui qin zhanglao jianji bashou (bingyin)" 次韻定慧欽長老見寄八首(並引) ("An Eight-Verse Response to the Eight Verses Ding-hui Sent in Respect to his Elders [with Quotation]"). *Su Shi quanji* 23. http://www.bestory.com/novel/11/101604/15956.html. Accessed 8 Dec 2012.

Za a han jing 雜阿含經 *Saṃyuktâgama-sūtra*. *Taishō* 37(2): 273, middle-bottom.

Zeng, Congyan 曾聰彥. 2011. "Kaizhan dongwu shiyan yanjiu tisheng yiyuan keyan shuizhun — ji woyuan zhongyao yaoli shiyanshi" 開展動物實驗研究提升醫院科研水準——記我院中藥藥理實驗室 ("Developing Animal Experiment Research in Order to Raise the Scientific Research Standards for Hospitals—Documenting the Traditional Chinese Medicine Pharmaceutical Research Laboratory at Our Hospital"). Guangzhou zhongyiyao daxue fushu Zhongshan Yiyuan, January 5. http://www.zsszyy.com/html/html/zkzb/yxb/201115/1089.html. Accessed 20 Oct 2011.

Afterword, *Chinese Environmental Humanities: Practices of Environing at the Margins*

Scott Slovic

When I arrived in China for the first time in June of 2006 to teach a class on ecocriticism and environmental literature at the Guangdong University of Foreign Studies, the slopes of *Baiyunshan* (White Cloud Mountain) near the campus were strangely silent. Guangzhou was in the grips of an avian flu panic at the time, and authorities had set up a vast net across the mountain, aiming to capture all potentially disease-bearing birds. That was a "silent summer" for me, strangely evocative of Rachel Carson's *Silent Spring*, which warned about a dystopian future without birds in North America, due to the unchecked use of harmful pesticides. During my six-week stay in China, I came to realize the magnitude of environmental and public health concerns in that country. The vast Pearl River (*Zhujiang*) flows through the mega-city of Guangzhou, but most citizens refuse to go near the water, let alone swim in it, as they are fearful of the industrial pollution carried in its murky, brown flow—I learned at the time that vital water resources for nine out of ten Chinese cities were dangerously polluted or over-exploited. As is my practice, I left my apartment each morning and went running in the thick, humid air for half an hour on the *Guangwai* campus, but, sometimes, I found myself gasping for breath,

S. Slovic
University of Idaho, Moscow, ID, USA

C. Chang (ed.), *Chinese Environmental Humanities*,
Chinese Literature and Culture in the World,
https://doi.org/10.1007/978-3-030-18634-0

unable to breathe even though the air was not visibly dirty. Later in the summer, I traveled to the Three Gorges Dam on the Yangtze River (*Chang Jiang*) in Hubei Province, a highly controversial structure that had just closed its floodgates during the week of my visit, beginning the process of creating a reservoir some 600 kilometers in length, displacing well over a million people, and threatening numerous plant and animal species.

I came to realize quickly that China was a fascinating and beautiful country, but also one in which many of the world's environmental problems were concentrated—rampant air and water pollution, leading to numerous cancer clusters, the loss of access to water resources, and the opportunity to enjoy outdoor spaces without fearing illness; tremendous, chronic deforestation and habitat loss for animals; pavement everywhere, even the paving of pathways up mountains; small mountains of solid waste collected on city streets each morning by sweepers using bamboo brooms, then carted away to be burned; and the lack of public engagement with these environmental issues. But what I also quickly learned was that China was on the verge of a significant cultural shift. When I first visited China in 2006, many of the environmental humanities scholars I met—such as Wang Nuo in Xiamen, Chen Hong in Wuhan, and Cheng Hong in Beijing—had not yet had the opportunity to work together. There was a small community of scholars—including Wei Qingqi in Nanjing, Song Lili in Beijing, and Liu Bei in Jinan—who were aware of each other's work and had been inspired by the writings and mentorship of Wang Ning and other pathbreaking figures in the field. But there was not yet a network of people doing ecological literary studies, and the ecocritics were not working in tandem with the environmental philosophers such as Zeng Fanren and Cheng Xiangzhan in Jinan. No one I met was familiar with colleagues in environmental history or religion. The idea of collaborating with people in Taiwan and Hong Kong to think about the environmental and social crises facing Chinese society never came up during my first visit to the mainland in 2006.

But this quickly changed. Although China prides itself as an ancient civilization, where ideas from such classical philosophers as Laozi, Zhuangzi, and Kongzi continue to influence present-day thinking about the place of human beings in the world, it is also a civilization, in my experience, that moves quickly to embrace current realities and respond with new technologies and ideas. I believe the 2008 conference called "Beyond

Thoreau," hosted by Tsinghua University in Beijing and organized by Song Lili, Wang Ning, and their colleagues, was a kind of tipping point in Chinese environmental humanities. It brought together many foreign and Chinese researchers, including scholars from Taiwan. Late one evening during the conference, the senior Chinese ecocritic Lu Shuyuan, who had been publishing significant books in the field of ecocriticism since the 1980s and who had a research center for social ecology at Suzhou University, organized a tête-à-tête at a tea house near the Tsinghua campus for a dozen people, including, most importantly, several of the leading ecocritics from mainland China and Taiwan. This was a stunning moment of bridge-building that has led to numerous international conferences and workshops, teaching exchanges, and joint publications in the past decade. *Chinese Environmental Humanities*, too, is a product of this collaborative spirit, combining the best ideas of researchers from the mainland, Taiwan, Hong Kong, and other parts of the world to appreciate both classical insights into human-nature relationships and new ideas that might help Chinese society (and societies elsewhere in the world) respond to the ecological challenges we have created for ourselves and the rest of the planet's species.

I greatly appreciate the multidisciplinary thrust of this book, as the concept of "environmental humanities" has, it seems to me, arrived somewhat belatedly to China. Until the past few years, most of the conferences I attended in Beijing, Wuhan, Jinan, and Shanghai consisted primarily of philosophers speaking with philosophers or literary scholars conversing with their own kind. More recently, though, there is emerging in China, and throughout the world, a "we're all in this together" sensibility, bringing ecologists, filmmakers, historians, and textual critics together in Brazil to discuss the plight of Amazonia and literary scholars, historians, and ethnographers together in Finland to consider the future of the Arctic in the age of climate change. Chia-ju Chang has done a remarkable job in this book of compiling lively and diverse approaches to both familiar and surprising aspects of Chinese culture vis-à-vis the natural and built environments.

The subtitle of this book, "practices of environing at the margins," seems particularly playful and insightful to me. Of course, the Chinese word for "China" is *Zhongguo*, which literally means Middle Kingdom. So the suggestion that environmental humanities scholarship in or about

China might be something happening at the "margins" rather than at the center, or in the ***middle***, of society is fascinating and prescient. Even though the Chinese government sends hundreds of researchers abroad every year to study environmental research methodologies and cutting-edge trends of environmental thought through the China Scholarship Council, and even though China's major universities host conferences and courses and journals and book series exploring the latest ideas in the environmental humanities and similar disciplines, there is still the perception that environmental research is somehow peripheral to mainstream Chinese culture, marginalized by the juggernaut of Chinese economic and industrial power. The subjects explored in the essays collected here—from silkworms to migrant workers, from ethnic minorities to art films, from sustainable development in Hong Kong to animal rights in Taiwan—cover a startling range of cultural issues that are intrinsic to contemporary Chinese culture, yet often ignored in discussions of geopolitical power.

These are essential topics, yet strangely—and perhaps tragically—"marginal" to the power centers of Chinese society. Likewise, for Western readers, Chinese environmental culture might appear to be a marginal subject, something far away across the horizon of the Pacific Ocean, relevant mostly to nearby countries in East Asia. Yet all who understand the implications of a growing environmental movement in a geographically and economically immense country like China, a nation with some 1.4 billion people (more than four times the population of the United States), realize that "environing" (or thinking and behaving with environmental consciousness) in China could never be peripheral to the well-being of our planet. In 2010, Karl Gerth published the book *As China Goes, So Goes the World: How Chinese Consumers Are Transforming Everything*, focusing distressingly on how Chinese citizens are developing "an American-style consumer culture." What is less well known, though, is that Chinese academics and artists (and Chinese specialists in other parts of the world) have also been developing not only a Western-style environmentalism but combining traditional Chinese ideas with innovative approaches to sustainability and social and environmental justice in pursuit of a so-called ecological civilization (***shengtai wenming***).

A decade ago, the phrase "ecological civilization" sounded a lot like cultural greenwashing from the Communist Party of China, and it may still be a kind of greenwashing to a certain extent—but there is also a verifiable and vibrant trend in Chinese intellectual and artistic circles toward powerful ecological thinking. This book is an engaging and heartening snapshot of this important social movement. I hope it will also serve as a foundation for future collaborative, cross-disciplinary, multinational partnerships that may help our species pull back from the brink of an apocalyptic precipice ... or at least understand what's happening as we step over the edge.

Index[1]

[1] Note: Page numbers followed by 'n' refer to notes.

C. Chang (ed.), *Chinese Environmental Humanities*,
Chinese Literature and Culture in the World,
https://doi.org/10.1007/978-3-030-18634-0

The manufacturer's authorised representative in the EU is Springer Nature Customer Service Centre GmbH, Europaplatz 3, 69115 Heidelberg, Germany. If you have any concerns regarding our products, please contact ProductSafety@springernature.com

Printed and bound by CPI Group (UK) Ltd, Croydon, CR0 4YY
30/09/2026
02225216-0016